MW01517897

Advances in Industrial Control

Other titles published in this Series:

Andrzej W. Ordys, Damien Uduehi
and Michael A. Johnson (Eds.)

with Nina Thornhill, Alexander Horch, Dale Seborg, David Laing,
Michael Grimble, Pawel Majecki, Hao Xia, Marco Boll,
Marcus Nohr, Sandro Corsi, Martin Bannauer and Gerta Zimmer

Process Control Performance Assessment

From Theory to Implementation

 Springer

Andrzej W. Ordys, PhD, CEng
Faculty of Engineering
Kingston University
Friars Avenue
Roehampton Vale
London
SW15 3DW
UK

Damien Uduehi, PhD
BG International Ltd.
100 Thames Valley Park Drive
Reading
RG6 1PT
UK

Michael A. Johnson, PhD
Industrial Control Centre
University of Strathclyde
Graham Hills Building
50 George Street
Glasgow
G1 1QE
UK

British Library Cataloguing in Publication Data
Process control performance assessment : from theory to
   implementation. - (Advances in industrial control)
   1. Supervisory control systems - Evaluation 2. Automatic
   data collection systems 3. Process control - Data
   processing 4. Benchmarking (Management)
   I. Ordys, A. W. (Andrzej W.), 1956- II. Uduehi, Damien III.
   Johnson, Michael A., 1948- IV. Thornhill, N. F.
   670.4'275
ISBN-13: 9781846286230
ISBN-10: 1846286239

Library of Congress Control Number: 2006936887

Advances in Industrial Control series ISSN 1430-9491
ISBN 978-1-84628-623-0   e-ISBN 978-1-84628-624-7   Printed on acid-free paper

© Springer-Verlag London Limited 2007

MATLAB® and Simulink® are registered trademarks of The MathWorks, Inc., 3 Apple Hill Drive, Natick, MA 01760-2098, USA. http://www.mathworks.com

9 8 7 6 5 4 3 2 1

Springer Science+Business Media
springer.com

Advances in Industrial Control

Professor Emeritus O.P. Malik
Department of Electrical and Computer Engineering
University of Calgary
2500, University Drive, NW
Calgary
Alberta
T2N 1N4
Canada

Professor K.-F. Man
Electronic Engineering Department
City University of Hong Kong
Tat Chee Avenue
Kowloon
Hong Kong

Professor G. Olsson
Department of Industrial Electrical Engineering and Automation
Lund Institute of Technology
Box 118
S-221 00 Lund
Sweden

Professor A. Ray
Pennsylvania State University
Department of Mechanical Engineering
0329 Reber Building
University Park
PA 16802
USA

Professor D.E. Seborg
Chemical Engineering
3335 Engineering II
University of California Santa Barbara
Santa Barbara
CA 93106
USA

Doctor K.K. Tan
Department of Electrical Engineering
National University of Singapore
4 Engineering Drive 3
Singapore 117576

Professor Ikuo Yamamoto
Kyushu University Graduate School
Marine Technology Research and Development Program
MARITEC, Headquarters, JAMSTEC
2-15 Natsushima Yokosuka
Kanagawa 237-0061
Japan

To my wife Anna and my sons Szymon and Bart
Andrzej Ordys

*To my lovely wife Diane Elizabeth and my son and little x-man
Zachary Xavier, you are my inspiration for everything, I thank God
for both of you*
Damien Uduehi

*To my brother, David and his wife, Manya for their love and support
over the years*
Michael Johnson

Foreword

The industrial imperative for control systems stems from a need to meet quality, safety, environmental and economic goals related to asset utilization. The justification for control projects and subsequent analysis of the effectiveness of commissioned schemes is assessed against these criteria. The installed cost of control systems is considerable, and several studies have reported what many practicing control engineers know - many control systems fail to meet their basic objectives. Indeed, a great many control loops may actually increase, rather than reduce variability.

During the past fifteen years there has been considerable academic and industrial interest in developing methods to analyze and improve the performance of control systems. Typical objectives are to identify poorly performing loops and to diagnose causes for this unacceptable performance. Most of the academic developments have focused, quite appropriately, on developing the underlying theory and the mathematical and statistical tools - descriptive statistics, time series and spectral methods, that are required for this analysis. There is great industrial interest in this topic. Commercial packages are available and a number of industrial perspectives on the topic of performance monitoring and analysis have been reported. Not unexpectedly, considerable effort is required to translate the underlying theory into practice.

Control Performance Assessment: From Theory to Implementation provides both a business and technological perspective on this topic. Throughout the monograph, control assessment, monitoring and diagnosis are viewed as essential approaches for satisfying the business constraints under which control schemes are implemented. Benchmarking is used extensively in many private and public sector institutions. Drawing upon the extensive literature and practice in this area, the authors of this monograph provide compelling arguments for interpreting control performance monitoring and assessment in this light. In doing so, they identify a number of advantages and limitations of existing theory and provide a framework for interpreting and extending performance measures.

The contributors of *Control Performance Assessment: From Theory to Implementation* have considerable industrial and academic experience. After presenting the business imperative for benchmarking, the basic algorithms and a number of performance measure extensions are clearly explained and illustrated using case studies of realistic complexity. A concise compilation of academic research on performance monitoring, analysis and diagnosis is included. This monograph is intended for practicing engineers and advanced degree students. It is part of the series *Advances in Industrial Control* whose aim is to report and encourage technology transfer in control engineering. Practicing engineers will appreciate the business perspective, the applications to difficult problems and the recognition of the challenges in applying the theory to industrial processes. The industrial perspective on this topic will be of considerable interest to graduate students who wish to advance the theory and practice of controller performance monitoring and analysis.

T.J. Harris
Department of Chemical Engineering
Queen's University
Kingston, Ontario
Canada

Preface

In the commercial sector, benchmarks, which are derived from observations across a range of processes and companies, have been applied in different forms to assess the performance of business processes. The results of these benchmarking exercises are used to formulate best practice and other strategies to improve business process performance. This performance assessment paradigm is long standing and from it can be extracted a generic assessment framework: *model*, *specify*, *measure*, *analyse* and *improve* performance. Among the philosophies that embody this framework, those of business process re-engineering and business process benchmarking are probably the best known.

In the manufacturing and production sectors, optimisation and control strategies have been used to obtain improved process performance. Optimisation has been an important tool in generating the designs for advanced control systems and has been used to provide operating guidelines at the higher levels in the process control hierarchy. But, there has been no real overarching process performance assessment and improvement methodology available as there has been in the business process domain. For example, little has been done to try and link control system performance to company financial performance goals and there is an absence of a simple, transparent holistic philosophy for determining what control performance could actually be achieved by an installed control system.

The situation for manufacturing and process industry systems (for example, petrochemical refineries or chemical plants) is of thousands of control loops, simple and complex, all contributing to the overall productivity of the plant and ultimately, the financial success of the company. It has always been known that if these loops could be optimised then each concomitant improvement in performance would lead to energy savings, reduction in raw material usage, improved product quality and tolerances and lower material wastage. The obstacles to achieving these outcomes were (and still are) the sheer logistics of tuning thousands of loops and the lack of technical input to comprehend the structural raison d'etre for the control system installed.

A key technical and practical breakthrough was made by Harris [1989] and colleagues who proposed and implemented methods for control loop benchmarking that were simple, data-driven and quantified the scope for tuning improvement in an existing control loop. The benchmark measure chosen was process output variance and the simple idea was to show how close the control loop tuning was to minimum variance control. This seminal work provided the impetus for devising control loop assessment procedures and also ignited research interest in process performance procedures *per se*. By the end of the 1990s, companies like Honeywell and Matrikon were offering various performance assessment and monitoring packages that are now widely used by process companies.

The focus of this monograph is on the broader context of control system performance assessment. Control loop benchmarking is just one (important) component of the toolbox that is needed to assess the performance of installed control systems and processes. It is thought that by taking a broader view, a better understanding is gained of the successful tools and where there are gaps suitable for future research input and development.

To gain insight into the structure of the monograph, it is useful to recall the generic steps in performance assessment: *model, specify, measure, analyse* and *improve* since these tasks inform and permeate so much of the research reported in the monograph. The opening chapter of the monograph reviews the tools and procedures of the business process performance paradigm and in the second chapter a new holistic assessment method to link company financial goals with control system performance is proposed. These chapters establish a broad context for control system and process assessment and the industrial control engineer should be able to learn some new general methods to assist with a preliminary assessment of performance. The reader might also be surprised to see the techniques of Statistical Process Control being used by the business process community.

A major group of chapters (3, 4, 5 and 8) is devoted to variations and extensions of the Harris methodology and pursue the detail of new control loop performance benchmarking procedures. The work begins from the minimum variance benchmark, which although easy to compute from online data, does not necessarily represent a desirable measure of performance. This is because minimum variance control can exhibit over-vigorous control action and some processes have restrictive stability conditions. From minimum variance control, the methods of generalised minimum-variance and predictive control are investigated. The idea of a "restricted structure optimal control" is introduced to allow a comparison of the performance of an existing controller against an optimal controller of the same structure. Many real processes are multivariate in nature and the extension of minimum-variance, generalised minimum variance and predictive control benchmarks to multivariable systems is pursued. This allows the methods to incorporate the impact of interactions between control loops on the overall performance of the plant. The breadth of the coverage includes data-driven methods (Chapter 3) inspired by the original Harris procedures, new model-based methods (Chapter 4), an illustrative industrial simulation study (Chapter 5) and a look at the technical conditions that may be required for the new procedures (Chapter 8).

A study of the business process assessment literature shows that the more difficult steps in the assessment procedure are those of performance 'analysis' and 'improvement'. A similar situation exists in the control field where less attention has been given to developing methods that purposefully unmask the barrier to control and prescribe what action should be taken to improve the control performance. In this monograph two chapters (6 and 7) are devoted to this type of development. Plant-wide disturbances, oscillations in control loops and identifying the presence of stiction as a root cause are the issues considered in these chapters.

The last chapter of the monograph recaps the progress covered in the monograph and surveys the potential directions for future research.

As far as possible the Editors have tried to achieve a common framework for the chapters presented. It was felt that an important feature should be easy access to the various procedures being used by the different researchers. Thus the contributors were all asked to produce precise algorithms descriptions for their procedures. In this way it is hoped that industrial and academic readers will easily be able to learn and use the new routines in their own work. The Editors also tried to encourage the contributors to introduce academic examples to illustrate technical points and larger industrially inspired example to show the procedures at work in more realistic situations. Chapters 3, 5, 6, and 7 all have good industrial studies that should illustrate the real industrial potential of the new procedures proposed in the monograph. The Editors would like to thank all the contributors for their forbearance in meeting the many requests for clarity and transparency that we hope will make the volume accessible to a wide range of industrial and academic readers.

Finally, we should like to thank Professor T.J. Harris (Queen's University, Canada) for providing such an inspiring Foreword and Oliver Jackson (Editorial Staff) of Springer London for his help in smoothing the path to publication.

Glasgow, Scotland   *Andrzej W. Ordys*
                                                          Damien Uduehi
                                                          Michael A. Johnson

Acknowledgements

This book is primarily the result of four years of academic and industrial research funded through EU, EPSRC and ETSU on new control system benchmarking and diagnostic theory and algorithms. The study looked at enhancing the ability of non-intrusive and intrusive benchmarking techniques for optimising control systems using performance assessment and diagnostic tools and, at the same time, optimizing profits through increased throughputs and product quality.

The book contains the major results from two projects, for research and development of commercial process controller performance benchmarking software. Two other, earlier, research projects stimulated that development. In addition to these projects, the book presents research carried on in ABB Corporate Research Centre in Germany and in University College London. The work contains significant contributions from industrial partners across Europe. These contributions have helped to focus the research and hence the results on the "industrial view" of what a benchmark should be and how it should be applied in industry. The authors would like to acknowledge the support of the following:

Project Funding

EU Framework 5 Project: IST-29239 (PAM) *Performance Assessment and Benchmarking of Controls.*
EPSRC Grant: GR/R65800/01 *Optimising Petrochemical and Process Plant Output Using New Performance Assessment and Benchmarking Tools.*
EU Framework 4 Project: ESPRIT 26880 (IN-CONTROL) *Integrated Supervisory Control for Production Plan.*
ETSU Grant: (Low Carbon Technologies Energy Efficiency Programme) *New Advanced Control Techniques to Reduce Energy/CO_2 Emission in the Process Industries*

Industrial Partners

ABB Energy Automation S.p.A, ABB Industries (UK) Ltd, Babcock Process Automation GmbH, BASF Aktiengesellschaft, BP Chemicals Ltd, British Energy

Generation (UK) Ltd, Centro Elettrotecnico Sperimentale Italiano Giacinto Motta SpA (CESI), ENEL, GlaxoSmithKline, Industrial Systems and Control Ltd, Marathon Oil UK Ltd, Omron UK Ltd, Performance Improvements Ltd, Red Electrica, Scottish and Southern plc, Shell UK Ltd, Siemens AG Power Generation.

Andrzej Ordys would like to acknowledge that most of the work on the book was done whilst he was British Energy Senior Lecturer in Control Systems, Department of Electronic and Electrical Engineering, University of Strathclyde, Glasgow, UK.

Damien Uduehi would like to acknowledge that most of research reported in this book was performed whilst he was with Industrial Control Centre, University of Strathclyde, Glasgow, UK.

Last but not least, we would like to thank Sheila Campbell for the help with typing the text, and Bart Ordys for the final editing of many of the figures.

Contents

1

Benchmarking Concepts: An Introduction

Michael Johnson [1] and Dale Seborg [2]

[1] Emeritus Professor, Industrial Control Centre, Department of Electronic and Electrical Engineering, University of Strathclyde, Glasgow, UK,
[2] Professor and Vice Chair, Department of Chemical Engineering, University of California, Santa Barbara, USA.

1.1 Introduction

Most process and control engineers often marvel at the technical complexity and sheer size of the processes found in the process industries. Indeed it is often the fascinating technological challenges that drew them into a career with the industry in the first place. But, it should never be forgotten that these processes are *business* processes operating in a competitive global marketplace. A process will remain operational within a commercial enterprise so long as its *performance* is economically competitive and is contributing to the financial health of the company.

The performance of business processes has been the focus of a new business culture paradigm, which emerged around the mid-1980s. Techniques like quality circles, business process re-engineering and business process benchmarking have evolved from this movement and generated an extensive published literature. However, it is only in the last decade that interest in performance monitoring, assessment and improvement for *technical* processes and process control has really taken root. It is thought that process and control engineers still have much to learn from the business process literature and so that is the starting point for this chapter: What is the business process viewpoint for performance monitoring, performance assessment and performance improvement?

1.1.1 Setting the Scene for Performance Assessment

Processes whether commercial, industrial, service or institutional can all be modelled by the simple input, process, and output activity sequence. A block diagram representation would look like the lower part of Figure 1.1 which is also the standard representation familiar to process and control engineers. However, this simple process representation is lifted into a whole new realm of utility if an anthropomorphic interpretation is added to the process components. The process can then be interpreted as a *business* process and it is possible to consider variables and factors representing the performance and interests of the suppliers, the process owners and the customers.

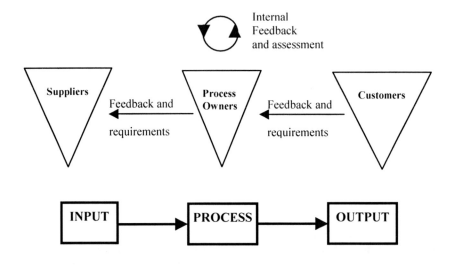

Figure 1.1. Re-interpreting the input-output process model (after Andersen and Pettersen, 1996)

In many ways it is the recognition of this simple interpretational structure for the input-output process model, and the concomitant need for a systems engineering paradigm for all aspects of these so called business processes that has led to the development of Total Quality Management (TQM) ideas. TQM is a group of methods and techniques devised to describe and analyse different aspects of the business process. As will be seen in this chapter, some of the techniques from the TQM toolbox are quite familiar to process and control engineers but others have been devised to deal with the 'peoples' aspect of the business process model, a dimension which is not usually considered in conventional process and control engineering studies.

A key property of universal significance in any branch of systems engineering is performance. Consequently, business performance, service performance, quality performance and process performance are of special interest in Total Quality Management studies. In whatever way performance is measured, enterprises and organizations are always seeking to improve their performance. In the commercial world, if a business does not improve its performance indicators then usually, there are other competitors who will and the result will be lost customers. The tools of TQM are designed to aid this continual drive to make performance advances, i.e., continued improvement. Another feature of performance is its often slow decline as organizations and enterprises become routine and structurally rigid over time. Consequently there is a need to have techniques that reinvigorates the business or organization and enables an assessment and revival of performance levels leading to updated targets and strategies to achieve them. The generic concept of 'customer' applies universally to many sectors including commerce, industry, healthcare, government and education and customer expectations are always rising. Thus performance levels must increase whilst costs must decrease. In this way, customers are retained and customer complaint levels are reduced. In summary, however measured, performance improvement is a ubiquitous objective of business and organisations worldwide.

Total Quality Management as a form of system engineering has a duality of techniques: some for people activities and others for technical tasks. This is another key difference to conventional systems engineering methods. The technical tools are concerned with modelling the process to be investigated, rigorously specifying performance targets, collecting and analyzing performance data, identifying problem causes and devising appropriate solutions. However, most of the TQM methods are structured around a cycle of people activities. These are generally based on the generic cycle: plan, do, check, act; the so-called Deming Wheel of activities [Deming, 1985]. Many authors have adapted this cycle and renamed the activities for use in other TQM techniques. For example, in the method of business process benchmarking, a Benchmarking Wheel [Andersen and Pettersen, 1996] has the stages: Plan, Search, Observe, Analyse, Adapt as a cycle of activities to be undertaken by a team of people.

The second aspect of the 'people activity' in TQM techniques is that the 'Suppliers', 'Process Owners', and 'Customers' interpretation of the input, process and output blocks of the process diagram are often identifiable people. Consequently, in considering performance improvement in such cases it is common to find that a team approach is taken involving representatives from each group. Sometimes the suppliers, process owners, and customers are simply different departments within the same organisation, and then the performance assessment team comprises representatives from the interested departments. Thus, apart from the algorithmic or cyclic aspect of people activities in TQM, there are also techniques concerned with the structure and interaction of the assessment teams in TQM procedures.

1.1.2 Key Techniques in Performance Improvement

The view that performance improvement is part of the systems engineering paradigm is a useful way to categorizing the steps and tools used. This also links the categories to equivalent steps prevalent in conventional process and control engineering methods and highlights the differences too.

The performance improvement exercise within TQM usually begins by rigorously defining the process to be investigated and determining the precise specification of the performance targets to be achieved. The middle stages involve collecting performance data, analyzing the performance reached, and uncovering the problems that are preventing the attainment of the desired performance levels. Once the causes of poor performance have been uncovered, the final stages will instigate a programme of curative actions and possibly install mechanisms to continue to monitor subsequent performance, for example, a quality control committee or a process operations committee. This grouping of the techniques and activities is shown in Figure 1.2.

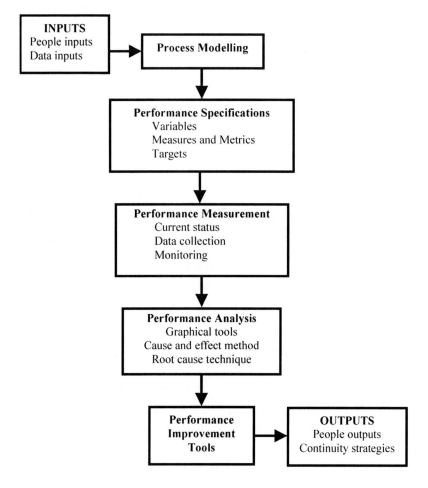

Figure 1.2. Steps and tool groups in performance assessment

It is useful to summarise each of the activity steps and the tools used within these steps as follows.

People Input Tools The extensive range of processes that can be treated for performance assessment and improvements often requires input from a wide range of company/organisation personnel. Special tools and techniques can be used to elicit the necessary information efficiently. This is especially true when the process is an 'action' sequence rather than an industrial process.

Process Modelling   Processes from the business and organization field often look very different to those of the process industries. However, tools are still needed to rigorously define the process under investigation.

Performance Specification   This is the activity of deciding which process variables can be considered to capture the specification of desired performance and then defining the appropriate performance measures and metrics.

Performance Measurement   These are techniques for collecting and displaying measured performance data. The monitoring of performance measures is included in this category.

Performance Problem Analysis   A most difficult task is to identify the actual cause of barriers to ultimate performance achievement. Some tools to assist with this task are given in this step.

Performance Improvement Tools   This class of tools divides into two groups. One group of tools for performance improvement are simple and almost heuristic. A second group are new philosophical approaches to the whole problem of performance improvement; one is Business Process Re-Engineering [Hammer and Champy, 1993] and a second is Business Process Benchmarking [Andersen and Pettersen, 1996].

People Output and Continuity Tools These are people-based mechanisms for implementing the outcomes of the performance assessment exercise, for continuing the monitoring of performance, and for disseminating the outcomes to a wider audience.

1.2 Process Models and Modelling Methods

1.2.1 Business Process Models

The business process model is a simple input-process-output operational sequence augmented with a supplier-process owner-customer interpretational layer, as shown in Figure 1.1. The model without the additional layer is a simple input-output process block diagram that is well known in conventional process and control engineering fields. A definition of a *business* process is [EQI, 1993]: *A chain of logical connected repetitive activities that utilizes the enterprize's resources to refine an object (physical or mental) for the purposes of achieving specified and measurable results or products for internal or external customers.*

The motivation for the interest in business process models lies in the way enterprises, business, organization and government arrange and conduct their activities. Historically, when a business became larger and more complex, it was often organized on a functional basis into departments. But then typically, departmental development and politics became more important than the business itself so that the organization lost sight of the performance of its business objectives. The business process model focuses strictly on input, process transformation and output and on supplier input, process owner performance and customer output, all of which lie at the core of a business or organizational activities. It is a very important point that departments partition an organization into vertical activities whilst business processes travel through an organization horizontally to deliver outputs. This is graphically shown in Figure 1.3, where the figure also intimates why it is often necessary to construct cross-departmental teams in order to have any effect on business process performance.

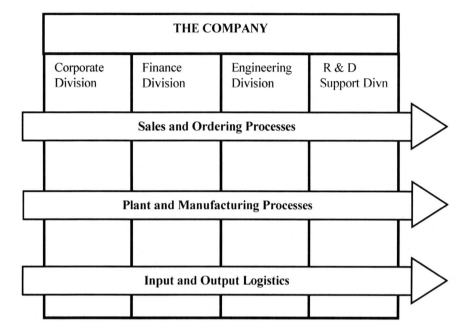

Figure 1.3. Organisational structure and business processes (after Andersen and Pettersen, 1996)

The surprising power of the business process model comes from its wide applicability. Indeed at an operational level, it seems difficult to find processes, which do not fit the business process model structure. Table 1.1 shows the characteristics of some business processes, which were modelled for benchmarking exercises, as described in a volume of such exercises edited by Camp [1998].

1.2.2 Constructing a Business Process Model

The wide range of business process model applications and the cross-departmental structure of business processes usually mean that models have to result from the discussions of a focused cross-departmental group of experts. Thus, organized and targeted discussion sessions may be needed to construct a business process model. For a complicated business process, questionnaires may be formulated to gain an insight to the inherent structure that needs to be identified and documented in a business process model.

Simple tools that can be used to initiate the process of identifying the business process are freely drawn block diagrams, questionnaires and simple attribute tables. A more formal version of the block diagram is known as Relational Mapping.

Table 1.1. Characteristics of some business process models

Company	Business Process	Characteristics	Technique
Sector - MANUFACTURING			
Allied Domecq	Information Delivery	Discrete, Decisions	Block diagram
NIIT, India	Hardware Procurement	Discrete, Decisions	Cross-functional flowchart
Ingwe Coal Corporation	Continuous Mining Operation	Discrete, Decisions	Flowchart
Sector - SERVICES			
Boots the Chemist	Sales Promotion	Attribute set	Tabular Format
IBM Italy	Procurement Process	Discrete, Decision	Block diagram
Sector - NOT FOR PROFIT ORGANISATIONS			
Northern New England Cardiovascular Disease Study Group	Cardiac Surgery	Discrete, Decision	Block diagram
Singapore Productivity and Standards Board	On-the-Job Training	Discrete	Block diagram
Sector - GOVERNMENT			
US Government	Complaint Process	Discrete, Decisions	Tabular Format
Post Office, UK	Supply Chain	Discrete, Decisions	Block diagram
City of Monash	Payroll Production Rate Collection	Discrete, Decision	Block diagram
Sector - EDUCATION			
Babson College, USA	Enrolment	Attributes, Discrete	Tabular format
Oregon State University	Student Advisory	Attributes, Discrete	Tabular format
Queensland University of Technology, Australia	Research Supervision	Discrete	Block diagram

Relational Mapping
This is a discussion-based activity that uses the block diagram to identify those activities and/or enterprises that impact a particular business process. There are no standards for this type of diagram and an example is shown in Figure 1.4. The relationships between the different parties (departments, work units, individuals) in the process are given on the figure by different arrow types. This discussion

activity is usefully performed before a formal block diagram, or flowchart is constructed.

The relational map is a useful tool for initiating a discussion about a business process model but an in-depth analysis will require a more precise tool for describing and documenting the detail of the process. A widely used tool is the flowchart or one of its variants.

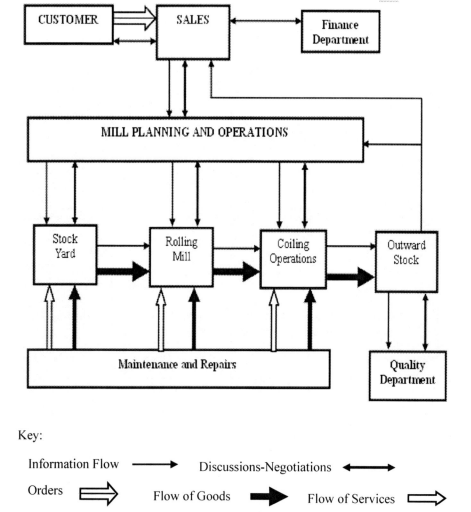

Figure 1.4. Relational map for a rolling mill operation from the steel industry

Flowcharts
The more complex business process generally involves a sequence of discrete operational steps combined with dynamics arising from simple binary decisions.

Typical examples are procurement processes, sales ledger, laboratory procedures, and discrete process operations. Flow charts can capture the discrete nature of these processes quite successfully.

Flowcharts use a simple set of symbols that represent different types of 'activities'; connections between these activities are presented by connecting arrows. Figure 1.5 shows a dictionary of symbols and a simple flowchart example is shown in Figure 1.6.

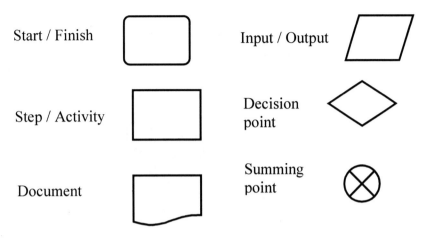

Figure 1.5. Dictionary of simple flowchart symbols

As processes become more complicated and larger in scale, a flowchart can grow to an unwieldy size as more and more detail is documented. A solution to this problem is to use *hierarchical linked flowcharts*. The process operation is simply decomposed into a number of layers of information and each layer has an associated set of flow charts. A two-layer scheme is shown in Figure 1.7.

Simple flowcharts and hierarchical flowcharts describe and document 'what' is happening in a process but they do not indicate 'who' might be performing or be responsible for the decisions, transformations, or actions taking place. A cross-functional flowchart captures both of these aspects of a process.

Cross-function Flowchart
The cross-functional flowchart is a powerful business process modelling tool which details what is happening in a process and relates it to who is active in the stages of the process operation. An example is shown in Figure 1.8.

Using the experience of the modelling group and some preliminary data collection exercises, a flowchart can be augmented with additional quantitative information, for example:

Time spent in an activity
Costs (material, personnel, etc) in an activity
Value added in an activity
Degree of completion of a product item
Cumulative times, cost, etc.

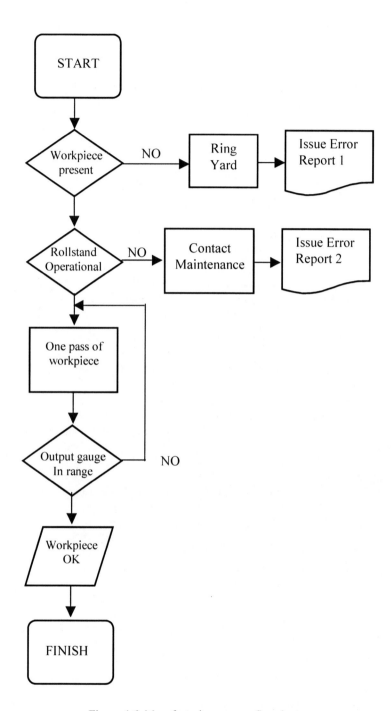

Figure 1.6. Manufacturing process flowchart

PROCESS Top Layer   *First Sub-process layer*

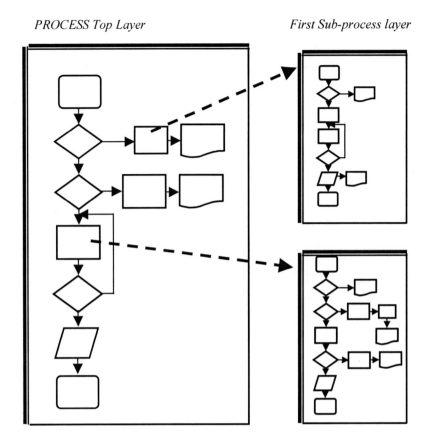

Figure 1.7. Hierarchical flowcharts for process models

1.2.3 Business Processes and Process Control

Process control primarily deals with the operation of industrial systems. However, in the wider context an industrial system is simply part of the holistic business process, and as such is also amenable to the type of performance assessment procedures applied in the wider enterprise. One view is to consider an industrial process as one of the input-output blocks that contribute to the global enterprise wealth creation structure. To operate a process unit also requires information flows, quality objectives, people input, planning input, maintenance support, research support as well as the material and energy flows that are the usual domain of the process control engineer. The business process analyst would be completely familiar with these wider aspects of the process unit. However, there is another level to industrial systems and that concerns the process engineering and the technology that make them work. This more specialized area could be termed the *technical process* and is the concern of process and control engineers. The process and control system models for the technical process levels are far more specialized and usually highly mathematical. Nonetheless, even the technical levels of a process unit can be subject to a performance assessment leading to directions for performance improvement. In the process control field these assessment techniques are particularly well developed for the analysis of control loop performance as will be described in later chapters.

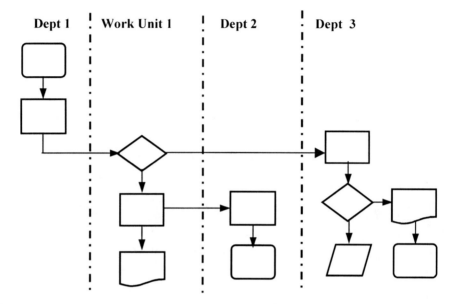

Figure 1.8. Cross-functional flowchart

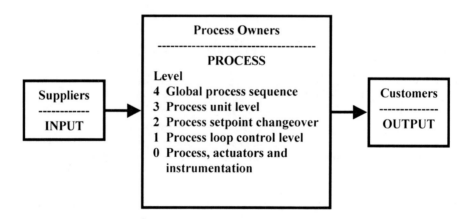

Figure 1.9. Business process and process control hierarchy

Thus, a major thesis of this book is that a holistic assessment of performance encompasses both business process levels and the technical process levels in any enterprise. To embed the technical process aspects into the business process structure, the standard process control hierarchy can be used as shown in Figure 1.9. In this figure the business aspects of suppliers, process owners and customers are retained but for the technical process, performance can be considered according to the process objectives at the different levels of the process control hierarchy.

1.3 Performance Measurement

To understand how well complex processes are being managed, it is necessary to monitor and analyze a representative range of performance metrics. The specific *type* of metrics will be process dependent but to capture the state of a process, it is the careful selection of a range of performance indicators that is important. A common classification is into Financial and Non-Financial performance measures. Typical examples of financial performance measures are profitability, sales, unit costs whilst non-financial indicators might include employee retention rates, customer satisfaction levels, and product defect rates. A second classification is the quantitative-qualitative divide into 'hard' and 'soft' performance measures. Hard performance metrics are those strictly computable quantities based on numerical measurable data; these range from financial measures like unit cost and sales per day to non-financial quantities and technical measures like process plant downtime, product defect rates and product physical properties (temperature, flow, dimensions, etc). In contrast 'soft' performance indices are metrics of more difficult to measure quantities like customer satisfaction variables and are often captured by a set of linguistic variables such as {very poor, poor, satisfactory, good, excellent...}.

1.3.1 Business Process Performance Measures

Determining the range of performance measures to be monitored for any particular process is a more demanding exercise. The objective is to choose a set of performance measures that capture the operational state of the process. This usually means that a wide variety of process issues are available for investigation. The first list in Table 1.2 encompasses generic performance areas that are considered appropriate for business process performance assessment [Bogan and English, 1994]. Whilst the second list in Table 1.2 gives the equivalent generic areas that might be appropriate for the performance assessment of a unit from the process industries. What is interesting from the comparison is the close similarity of underlying principles; all that is different is the terminology and context. This is due to the input-process-output model adopted by a systems engineering approach to process modelling (as seen earlier in Figure 1.1); this common model structure is present in both the business process and the technical process control fields.

Table 1.2. Generic areas for performance assessment

Business Process	Process Industry Unit
Customer service performance	Quality of outputs to the next process unit
Product/Service performance	Quality performance of output product
Core business process performance	Process unit performance
Support processes performance	Process unit maintenance performance
Employee performance	Process operator and support staff performance
Supplier performance	Quality of input materials to process unit
Technology performance	Equipment-machinery-controls performance
New product/service performance	Process unit changeover performance
Cost performance	Energy/material costs performance
Financial performance	Financial performance

The division of performance measures into financial-non-financial and hard-soft categories has been captured in Table 1.3. One implication of the hard-soft divide is in the area of data collection. Hard numerical performance measures may be easier to collect from computer data logging systems whereas soft performance measures may need interviews and questionnaires for collection and may be subject to more uncertainty in quality.

Table 1.3. Performance measure categories

	FINANCIAL	NON-FINANCIAL
HARD (Quantitative)	PM_{hf}	PM_{hnf}
SOFT (Qualitative)	PM_{sf}	PM_{snf}

The very act of modelling a process leads to systematic lists of variables, factors and sub-processes that are involved in generating the outcome of a complete process. Performance measures are usually defined from the underlying domain of process variables to try to quantify the essential qualities important to the process operator and the process customer. In many instances it is possible to capture the essential outcomes of a process in a few qualitative statements and quantitative indices but it is often difficult to determine which of the process variables and sub-processes most impacts the process performance. Identifying the most important process variables for use as a performance measure is a first task in performance assessment and this problem is considered next.

1.3.2 Prioritizing Performance Measurements – Criteria Testing

It is often the case that the desired outcomes of a process can be summarized as four or five qualitative statements or quantitative indices. More difficult is deciding which process variables, factors and sub-processes have a critical impact on the performance measures. The method of Criteria Testing has been devised to assist with this prioritizing task. In the commercial world, the desired outcomes of a process are termed *critical success factors* and these may be defined as [Andersen, 1999]: *The limited number of factors that to large extent impact the organisation's competitiveness and its performance in the marketplace.*

In the examples given for critical success factors, it becomes clear that these are top-level descriptions of what makes an organization and/or its products successful. In the technical world of the process industries and process control, a process critical success factor may be defined by paraphrase as: *The limited number of factors that determine the competitiveness of an industrial process's operational performance and the quality of its process outputs.*

Consequently, once the process critical success factors have been determined, the method of criteria testing can be used to rank the importance of the factors, variables, performance measures, and sub-processes from a given all-encompassing list. The criteria testing is outlined in Procedure 1.1.

Procedure 1.1 Criteria testing

Step 1   Define and describe the critical success factors for the process under investigation. Let nf be the number of critical success factors.

Step 2   List the critical success factors denoted $\{CSF_1, ..., CSF_{nf}\}$ and assign each factor a representative weighting that defines its contribution to the overall process success. Denote the set of weightings as $\{wt_1,...,wt_{nf}\}$

Step 3   List all the possible inputs whether process variables, factors, or performance measures that possibly contribute to the achievement of the total success of the process. Denote these inputs $\{PI_1, PI_2,...,PI_{NI}\}$ where NI is the total number of contributing possible inputs.

Step 4   For each possible input, PI_i, assign a weighting $PI_i(j)$ which expresses its importance or contribution to the jth critical success factor CSF_j. Repeat the procedure for each possible input and each critical success factor.

Denote the results array of weightings, $\{PI_i(j)\}$, i=1,...,NI; j=1...,nf

Step 5   For each possible input construct the score:

$$score(i) = \sum_{j=1}^{nf} PI_i(j) * wt_j \quad ; \quad i = 1,..., NI \tag{1.1}$$

Step 6   Identify the most significant contributing inputs as those having the highest scores.
Procedure end

When the procedure is dependent on a subjective ranking of qualitative variables like 'variable importance' then the numerical scale [0,1] or [0,10] can be used. To obtain a reliable subjective ranking of factors, effort should be devoted to obtaining sufficient expert input into the weighting assessment exercise. An impression of a Criteria Testing tableau can be seen in Table 1.4. From the score column of the table it will be possible to see which inputs to achieving the critical success factors are important. These can then be used as performance measures.

Table 1.4. Criteria testing

Critical Success Factors	CSF_1	CSF_2	. . . .	CSF_{nf}	
CSF Weightings	wt_1	wt_2	. . . .	wt_{nf}	
Possible Inputs					
Variable 1 PI_1	$PI_1(1)$	$PI_1(2)$	. . . .	$PI_1(nf)$	Score (1)
Variable 2 PI_2	$PI_2(1)$	$PI_2(2)$	. . . .	$PI_2(nf)$	Score (2)
Factor 1 PI_3	$PI_3(1)$	$PI_3(2)$	. . . .	$PI_3(nf)$	Score (3)
.	.	.			
Sub-Process 1 PI_l	$PI_l(1)$	$PI_l(2)$	. . . .	$PI_l(nf)$	Score (1)
.	.				
Performance Measure PI_{NI}	$PI_{NI}(1)$	$PI_{NI}(2)$	. . . .	$PI_{NI}(nf)$	Score (NI)

1.3.3 Benchmark Performance Measures

The actual *procedure* of benchmarking as it applies to business and technical processes will be discussed in a later section. At this juncture the emphasis is on defining a performance measure that is a benchmark value. This is quite simply the value of a defined performance measure that is the 'best in a class'. Two contextual interpretations will now be presented, one for generic process assessment and one for specific technical situations.

A generic process benchmarking procedure
In performance assessment exercises, the assessors will determine a set of performance measures that are best thought to capture the holistic state of a process. These performance measures will comprise a mix of indices from the matrix of categories shown in Table 1.3. The indices take on benchmark values when an exercise has been performed to find values that are 'best in a class'. This exercise of establishing a class of processes from which to take the benchmark performance measure values is known as *benchmarking*. Further there are several

different ways in which the 'class' is determined and this gives rise to different benchmarking methods. A simple example is internal benchmarking where a company may have several identical process lines operating at different locations and in different work cultures. These company process lines will form the 'class' over which best performance measures will establish the benchmark performance values for that type of process line.

A technical process benchmark
The 'best in a class' idea is retained for the specific case of a technical process but it is re-interpreted as being captured as the solution of a formal optimization problem. The benchmark value for this performance property is then taken as the optimal value of the optimization problem cost function. In many cases this benchmark optimal value represents an ideal but possibly practically unrealizable value; nonetheless, it represents a benchmark against which actual performance can be measured.

The classical example of a technical process benchmark in the process industries is the capture of controller performance in a minimum variance cost function [Desborough and Harris, 1992]. Subsequent chapters of this book develop technical process benchmarking in considerable depth.

1.4 Assessment of Performance

The main focus of this section is on the task of identifying that there is a limitation to performance achievement and the monitoring of performance measurements. Trying to find out the basic or root *cause* of the limitation to performance is termed performance *assessment analysis* and the set of tools which can be used for this task are described subsequently, in Section 1.5.

1.4.1 Simple Performance Assessment Tools

Several novel graphical tools have been devised for the preliminary analysis and problem identification stage of the performance assessment exercise. In this section the so-called spider chart, the performance matrix and trend charts are presented.

Spider Charts
A spider chart can be used to make a comparison of several sets of measured performance data. Normally the chart would display up to (say) five performance variables across two or three sets of performance data. The procedure to be followed is now given.

Procedure 1.2 Spider chart construction

Step 1 Optimum or benchmark performance values are used to normalise each of the performance data sets to a common scaling. Typically the data is scaled to a range [0,1] where '1' represents optimum performance.

Step 2 A set of concentric circles is drawn to represent a circular grid of scale intervals, typically, 0.25, 0.5, 0.75, 1.0, 1.25.

Step 3 From the common origin of the circles a set of axis rays are drawn; one axis ray for each performance variable. Together the circles and the axis rays look like a spider's web and this gives the chart its name.

Step 4 Each complete set of scaled performance data is marked up on the axis rays. These points are then joined up to form a figure; one figure for each data set and one for the optimum performance values (this is a joining up of all the '1' points).

Step 5 The closer the figure of the performance data set resembles the optimum performance figure, the better the performance.
Procedure end

A process industry example follows.

Example 1.1
This is a comparison of the performance of three process units, E34, F62 and G22. The selected performance variables, and the numerical data for each of the three units are given in Table 1.5.

Table 1.5. Process unit data - spider chart

	Rating	Unit E34	Unit F62	Unit G22
Throughput (tonne/wk) Normalised	750 1	700 0.93	256 0.34	600 0.80
Energy Used (kW/tonne) Normalised	6000 1	4621 0.77	5217 0.87	8217 1.37
Operator Performance Normalised	1 1	Excellent 1	Good 0.75	Good 0.75
Maintenance Used (hr/tonne) Normalised	30 1	10 0.33	43 1.43	21 0.7

The spider chart is given in Figure 1.10. In the chart, the bold rectangle represents the rated (benchmark) process unit performance. The performance of the three process units can then be viewed with reference to this rated (benchmark) performance. Clearly, for some reason, Unit E34 is under-utilizing the maintenance services. Unit F62 is not attaining anything like the expected rated throughput yet it is still using the rated energy consumption and Unit G22 is using excessive energy to achieve near rated throughput.

Performance Matrices
The performance matrix is a tool that can be used to obtain a focus on what is important in a performance assessment. It is a method of comparing quantitative

performance data against qualitative attributes. The quantitative performance data can usually be normalised on a scale [0,1], where the '0' region corresponds to Poor Performance the '1' region corresponds to Good Performance. The qualitative variable is then the ranking of the importance of a variable for performance attainment. A numerical ranking scale [0,1] could be used to indicate the subjective range from Low Importance (the '0' region) to High Importance (the '1' region). A rectangular grid is then used with the performance data on the vertical scale and the importance ranking on the horizontal scale as shown in Figure 1.11.

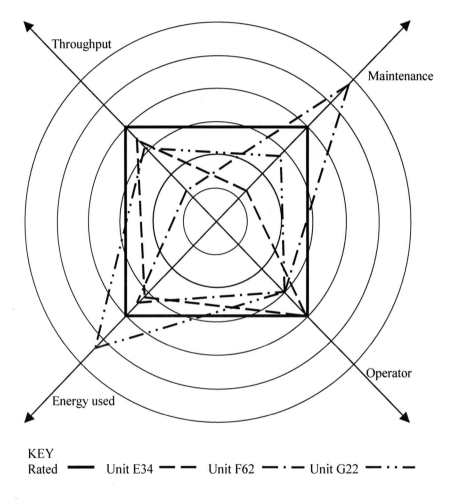

KEY
Rated ━━━ Unit E34 ━ ━ Unit F62 ━ · ━ Unit G22 ━ · · ━

Figure 1.10. Spider chart of process unit performance data

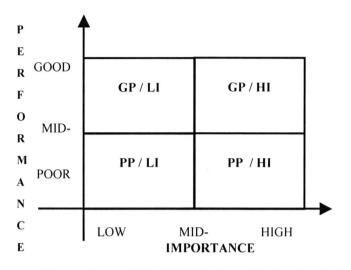

Figure 1.11. Generic performance matrix

The major squares of the grid can be categorised and interpreted as follows:

Box GP/HI Data falling in this area of the matrix is good because good performance is being achieved on the variables of importance.

Box PP/HI Data falling in this square shows that important variables are only achieving poor performance indicating that effort should be devoted to improving the performance of these variables.

Box GP/LI Data falling in this square reveals that the low importance variables are achieving good performance values. This implies that effort and energy is being expended on enhancing low priority variables.

Box PP/LI Data falling in this square is not critical for these are variables of low important and low performance is acceptable.

Example 1.2 Process Unit Comparison

Table 1.6 shows the performance assessment data for two different examples of a specific process unit. The measured performance values have been normalized on a scale of [0,1] and the actual values are given in the two right hand columns. A column ranking 'importance' is given in the middle where the scale [0,1] has been used with '1' representing 'High Importance'.

The performance matrix of Figure 1.12 clearly identifies two highly important variables achieving low performance, viz., entry temperature in Unit E34 and throughput in Unit F62 and several low importance variables that are being over-polished in performance.

Table 1.6. Process unit data

Performance Measures		Normalised Importance Rating	Normalised Actual Performance	
			Unit E34	Unit F63
Throughput	1	0.7	0.9	0.3
Surface finish	2	0.2	0.3	0.7
Energy Used	3	1.0	0.7	0.8
Entry temperature	4	0.8	0.4	0.9
Exit temperature	5	0.3	1.0	0.8

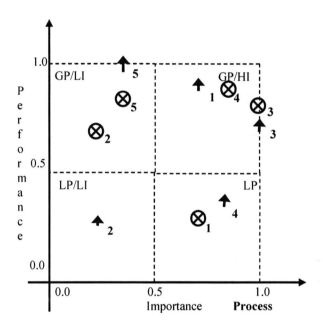

Figure 1.12. Performance matrix – Process unit data

Trend Charts
Many performance measures evolve with time and the trend chart is a straightforward graph of time-series data. The performance measure is plotted on the vertical axis and the time index occupies the horizontal axis. For multiple data sets, then several trend plots can be drawn on the same chart.

In terms of analysis, the trend chart shows how performance is evolving chronologically. The direction of the trend plot indicates whether performance is improving or worsening over time and the slope of the trend plot shows how fast these changes are taking place. Multi-trend plots can be used as a comparison tool perhaps showing the comparative performance of several company sites or the performance of the same process type across several different companies.

If the underlying process is stochastic and the trend component is removed from the data, then the residuals remaining can be subjected to a statistical analysis to test whether performance is changing with time. This method of *monitoring* performance measures has been developed into the important technique of Statistical Process Control (SPC); this method is described in the next section.

1.4.2 Statistical Process Control

Statistical Process Control is an important tool for monitoring performance measurements, especially product quality [Montgomery and Runger, 2003; Ryan, 2000]. Statistical Process Control (SPC), also referred to as Statistical Quality Control (SQC), has been widely used for quality control in discrete-parts manufacturing and in the process industries. Broadly speaking, SPC and SQC refer to a collection of statistically-based techniques that rely on *quality control charts* to monitor product quality.

The major objective in SPC is to use experimental data and statistical techniques to determine whether the process operation is normal or abnormal. The underlying assumption is that normal operation can be characterized by random variations about mean values. If this situation exists, the process is said to be *in a state of* statistical *control* (or *in control*) and the control chart measurements tend to be normally distributed about the mean value. By contrast, frequent control chart violations would indicate abnormal process behaviour or an *out-of-control* situation. Then a search would be initiated to attempt to identify the root cause of the abnormal behaviour. The root cause is referred to as the *assignable cause* or the *special cause* in the SPC literature while the normal process variability is referred to as *common cause* or *chance cause*. From an engineering perspective, SPC is more of a monitoring technique than a control technique because no automatic corrective action is taken after an abnormal situation is detected.

Quality Control Charts
An example of the most widely used type of control chart, a *Shewhart Control Chart*, is shown in Figure 1.13. A measured variable, x, or a sample mean, $\bar{x}$, are plotted versus sample number or time. The *target* represents the desired value while the *upper control limit (UCL)* and *lower control limit (LCL)* define the normal operating region. A data point beyond these limits is considered to be an abnormal situation (i.e., the process is not in a state of statistical control).

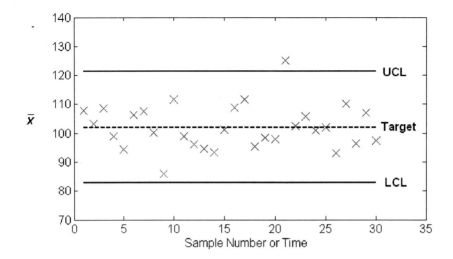

Figure 1.13. Shewhart control chart

The LCL and UCL are calculated from a set of representative data for a period of normal operation. Typically, they are specified as,

$$UCL \triangleq T + c\hat{\sigma}_{\bar{x}} \tag{1.2}$$

$$LCL \triangleq T - c\hat{\sigma}_{\bar{x}} \tag{1.3}$$

where $\hat{\sigma}_{\bar{x}}$ is an estimate of the standard deviation for the sample mean, $\bar{x}$. This estimate is calculated from the representative data. Typically, the constant c is chosen as $c = 3$ in order to provide "3σ limits".

Other types of quality control charts are also available For example, control charts can also be used to monitor the sample standard deviation as well as the sample mean, if more than one measurement is made at each sampling instant [Montgomery and Runger, 2003; Ryan, 2000].

Process Capability Indices
Process capability indices provide a measure of whether an "in control" process is meeting its product specifications. Suppose that a measured variable x must have a value between an *upper specification limit (USL)* and a *lower specification limit (LSL),* in order for a product to meet its specifications. The C_p *capability index* is defined as,

$$C_p \triangleq \frac{USL - LSL}{6\sigma} \tag{1.4}$$

where σ denotes the standard deviation of x. Suppose that $C_p = 1$ and x is normally distributed. Then we expect that 99.73% of the measurements would satisfy the specification limits, or equivalently, we would expect that only 2700 out of one million measurements would lie outside the specification limits. If $C_p < 1$, the product specifications are satisfied; for $C_p > 1$, they are not.

A second capability index C_{pk} is based on average process performance ($\bar{x}$), as well as process variability (σ). It is defined as:

$$C_{pk} \triangleq \frac{Min\,[\bar{x} - LSL, USL - \bar{x}]}{3\sigma} \tag{1.5}$$

Although both C_p and C_{pk} are useful, C_{pk} has a significant advantage over C_p for the following reason. If $\bar{x} = T$, the process is said to be "centred" and $C_{pk} = C_p$. But for $\bar{x} \neq T$, C_p does not change, even though the process performance is worse, while C_{pk} decreases. For this reason, C_{pk} is preferred. If the standard deviation σ is not known, it is replaced by an estimate, $\hat{\sigma}$.

In practice a capability index of 2.0 is often the objective while a value greater than 1.5 is considered to be acceptable [Shunta, 1995]. If the C_{pk} value is too low, it can be improved by making a change that either reduces process variability or that causes $\bar{x}$ to move closer to the target. These improvements can be achieved in a number of ways that include: tighter process control, improved process maintenance, reduced variability in raw materials, improved operator training, and process changes.

Three important points should be noted concerning the C_p and C_{pk} capability indices:
(i)   The data used in the calculations do *not* have to be normally distributed.
(ii)   The specification limits, USL and LSL, and the control limits, UCL and LCL are *not* related. The specification limits denote the desired process performance while the control limits represent actual performance during normal operation when the process is *in control*.
(iii) The numerical values of the C_p and C_{pk} capability indices are only meaningful when the process is *in a state of control*. However, other *process performance indices* are available to characterize process performance when the process is not *in a state of control*. They can be used to evaluate the incentives for improved process control [Shunta, 1995].

1.5 Performance Assessment Analysis

The performance assessment activities of modelling, performance measure selection, data collection and performance limitation identification have been covered so far, now it is the turn of analysis. This is perhaps the most difficult part of the journey and the task is to uncover the real cause of the identified limitation to performance. Fortunately, there are a number of tools of varying complexity, which can be used to assist in this analysis task.

1.5.1 Critical Incident Analysis

The preliminary assessment of the selected performance measures is designed to indicate those areas of a process that are, or likely to be, restricting the desired performance levels to be achieved. Once these areas are known, a critical incident analysis might be a tool for a fuller investigation. The critical incident analysis is outlined in Procedure 1.3.

Procedure 1.3 Critical incident analysis
Step 1   For the identified area of performance limitation, draw together a personnel team who are involved with this part of the process. Then, from brainstorming and interview techniques construct a shortlist of possible process events considered to be contributing to the problem.

Step 2   For fixed period of time, institute a simple data collection exercise using a checksheet to generate data on the frequency of occurrence of the process events listed. It is also possible that this data could be derived from historical operating data stored in the quality control database.

Step 3   The collected data are cast in cumulative tabular form, or displayed as a frequency histogram. Incidents occurring most frequently will be candidates for further investigation.
Procedure end

Within this procedure are the activities of people conducting a data collection exercise and then displaying the results using a frequency of occurrence table or a frequency histogram. Spreadsheet software such as the Microsoft Excel package could facilitate the collection and display of the required data.

Pareto Chart
A useful technique for the analysis of the critical incident data is a Pareto chart. This is an ordered histogram designed to portray the assertion that 80% of the problems arise from 20% of the possible causes. This is a paraphrase of the original interests of Pareto which was to demonstrate that 80% of the wealth of society was owned by 20% of its members. Instructions for constructing a Pareto chart are given in Procedure 1.4 and an example is shown in Figure 1.14.

Procedure 1.4   Pareto chart construction
Step 1   The list of process critical incident values should be converted to a common unit. For example all the data could be frequency of occurrence of incident but a true picture of importance only possibly appears when a monetary value is put on its occurrence or it is multiplied by an importance rating.

Step 2   Compute the percentage figure for the contribution from each cause or critical incident. Order the causes by decreasing percentage value.

Step 3   Construct an ordered histogram where the incidents or causes are placed along the horizontal axis in order of descending percentage value. On the left hand vertical axis can be marked the common comparison values, and on the right hand vertical axis the percentage contribution values.

Step 4   Draw the cumulative contribution curve and identify the 80% horizontal. This will reveal those critical incidents or causes of most significance. **Procedure end**

Example 1.3
Data for equipment faults on a rolling stand gave the figures in Table 1.7.

Table 1.7. Equipment fault data

Cause of Fault	Incident Downtime	No. of Occurrences	Total downtime	%
A Tensiometer	1.0	2	2	2
B Gaugemeter (Exit)	4.0	10	40	38
C Temperature sensor (Entry)	0.5	3	1.5	1
D Temperature sensor (Exit)	7.0	7	49	47
E Water pump	6.2	2	12.4	12

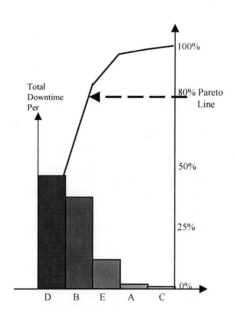

Figure 1.14. Pareto chart for equipment fault data

The Pareto chart (Figure 1.14) shows that causes D (Exit temperature sensor) and B (Exit gaugemeter) give rise to significant plant downtime.

1.5.2 Relational Diagrams

There are two types of relational diagram, qualitative and quantitative versions. Both are used to try to identify the main cause of a performance limitation [Rolstadas, 1995]. These tools can be very usefully employed with input from the personnel operating the process being investigated.

Qualitative Relational Diagram
The qualitative relational diagram uses input from a discussion group to try to construct a block diagram that identifies main causes or root causes of a problem. A procedure is outlined next [Andersen, 1999] and is shown in Figure 1.15.

Procedure 1.5 Qualitative relational diagram construction
Step 1   Assemble a task team of personnel directly involved with the process and the problem.
Step 2   Have the team list all the possible causes of the problem being investigated. Initiate a discussion of the causes and their consequences directed towards establishing the causal relationships between the various causes.
Step 3   Use the cause and consequence relationships to draw up a qualitative relational diagram. Concentrate on identifying the main causes and difficulties that create all the dependent problems.
Step 4   Use the identified main causes as the basic of a performance improvement programme.
Procedure end

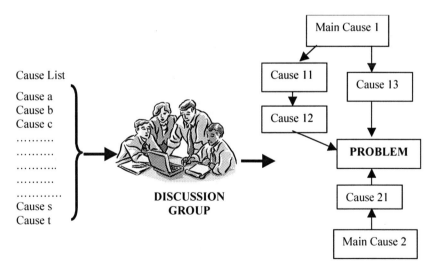

Figure 1.15. Qualitative relational diagram construction

Quantitative Relational Diagram

The quantitative relational diagram has a procedure similar to that for the qualitative relational diagram except that a scorecard system is used to highlight important features of the relationships displayed. Procedure 1.6 is illustrated in Figure 1.16.

Procedure 1.6 Quantitative relational diagram construction
Step 1   Assemble a task team of personnel directly involved with the process performance problem. Have the team list the main factors and causes of the performance limitation.
Step 2   Sketch a diagram putting the causes in an approximate circle layout. Initiate a discussion within the tasks team to determine the cause and effect relationships between the causes displayed.
Step 3   Once agreement has been reached concerning the proper cause and effect relationships, draw impact arrows. The arrow direction goes from a cause to a consequence or impact of the cause.
Step 4   For each cause on the diagram, sum a scorecard of arrows arriving (IN) and arrows leaving (OUT) each factor or cause.
Step 5   Identify *performance drivers* as those causes with the highest number of departing arrows and identify as *results indicator* those causes/factors with the highest number of arriving arrows.
Procedure end

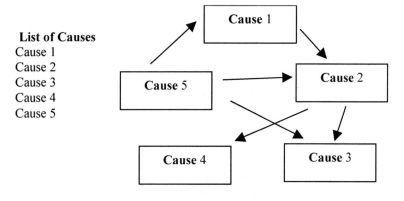

Scorecard		
	IN	OUT
Cause 1	1	1
Cause 2	2	2
Cause 3	2	0
Cause 4	1	0
Cause 5	0	3

Outcomes
Performance driver – Cause 5
Results indicator – Cause 3

Figure 1.16. Quantitative relational diagram

The scorecard can be used to identify the performance drivers and the result indicators. A performance driver can be equated to a main cause since it is a cause or factor that is impacting or driving a significant number of consequences or subsidiary effects in the problem. A performance driver has far more OUT arrows than arriving or IN arrows. The results indicator or results effect is the recipient of numerically more IN arrows than departing OUT arrows. Thus, it is the consequence or outcome of other factors or causes. The results indicator might be used as a diagnostic indicating that a particular condition has arisen.

1.5.3 Cause and Effect Charts

The cause and effect chart is a classical process control trouble-shooting tool. It has two functions; first, it can be used to analyse a performance problem and second, once the analysis is completed, it can be used to document how various performance problems arise. In this second role, the cause and effect chart fulfils the role of an archive and also a diagnostic tool that can be consulted. All these features of the cause and effect chart make it a powerful and flexible tool for performance assessment analysis.

The main cause-and-effect chart is a fishbone chart, where the name fishbone relates to the shape of the chart (see Figure 1.17). The differences in the way the chart is constructed appear to relate only to the people interaction mechanisms used rather than any distinct differences in the chart itself. Andersen [1999] details two separate routes:

Dispersion analysis. In this approach, the main arrow is drawn to represent the restriction to performance effect and then a discussion group draws in the main groups of problem areas and cause relationships one by one. This might be classed as a top-down approach.

Cause enumeration. This might be considered a bottom-up approach for it begins from an all-encompassing list of possible causes of performance restriction or limitations. The discussion group then classifies the causes into smaller cause and effect groups each focusing around a main cause. Finally, they are then drawn on the fishbone cause and effect chart once final agreement had been reached within the discussion group.

The determination of the 'Main Category' or 'Main Cause' items depends on context. In a general manufacturing process control environment, these items might be Operator, Equipment, Materials, Methods, Maintenance and so on. If the chart was for a specific piece of process equipment the list might read: Power Supply, Oil Supply, Water Supply, Sensors, Actuators, Controller Unit, and so on. Each category would then be subjected to close critical analysis to uncover the cause of limitations to achieving the desired performance.

The cause and effect chart can be used in different ways and appears in different formats. If the causes associated with a main category are sequential in relationship, this performance can be indicated by the inclusion of directional arrows on the fishbone branches. Some authors discuss a *process* cause and effect chart. This is simply the provision of a cause-and-effect chart for each step or stage in a process sequence. In such a context a decomposition of the process into stages is first required, followed by the cause-and-effect analysis archived as a fishbone chart for each stage.

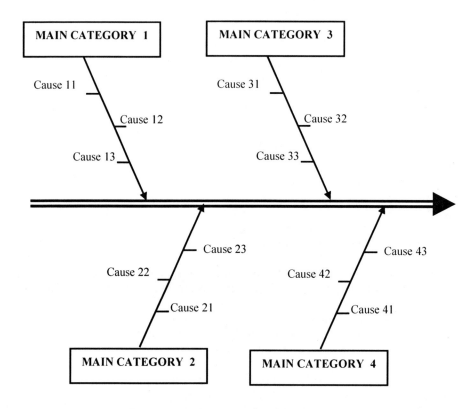

Figure 1.17. Cause and effect or fishbone chart

1.5.4 Root Cause Analysis

A different route to establishing the underlying causes of limitations or problems in attaining desired performance levels is that of root cause analysis. The technique involves persistently asking the question 'why' until there are no more 'whys' to ask and the root cause of a problem has been uncovered. This methodology is sometimes referred to as the 'five whys' method since it usually takes about five sets of 'whys' to achieve a result and the diagram used is referred to as a 'why-why chart'.

Procedure 1.7   Root cause analysis
Step 1   Identify the performance problem to be analysed.

Step 2   Form a discussion group with appropriate expertise for the problem. Have the group establish an agreed set of main cause problem areas.

Step 3   For each cause set up a recursive sequence of why questions: 'Why is this a cause for the previous problem?' Repeat the why questions until the root cause is found.

Step 4   The full analysis can then be archived on a fishbone cause and effect chart, a hierarchical block diagram or a tree diagram.
Procedure end

A typical example of a five-why chart is given in Figure 1.18 and a hierarchical block diagram for a root cause analysis document is shown in Figure 1.19.

Customer dissatisfied with output product quality

Why? *Too much is out of specification*

　　Why? *Reference changeover is not good*

　　　　Why? *Controller specification is wrong*

　　　　　　Why? *Damping ratio set at 0.707*

　　　　　　　　Why? *Design error – Damping*

　　　　　　　　should be critically damped, value = 1

Figure 1.18. Five-why chart

1.6 Performance Improvement Tools

The tools of performance assessment analysis were devised to try and locate the real source of any problem that is preventing optimum performance from being attained. In the previous section where these tools were explained, the presentation was algorithmic and prescriptive. However, when reading the literature of performance assessment it soon becomes clear that many of these tools are embedded in a wider cultural approach to the problem of deriving enhanced performance from processes, particularly the so-called business processes. A significant component of these different methodologies is concerned with people interaction tools. Tools designed to make groups of people, often from diverse expert, company or organization backgrounds, work together on enhancing the performance of a process that passes through their individual departmental domains. In this book, this aspect of performance assessment is not addressed in any significant detail but it does seem to be important to give brief definitions of the broader performance improvement tools that might be of value to the process industry and control engineer.

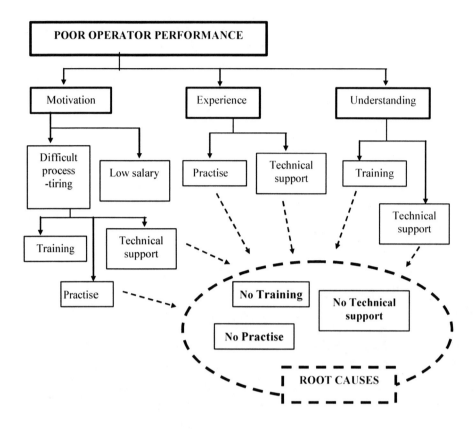

Figure 1.19. Root cause diagram

Idealizing

In some ways, idealizing is like constructing a benchmark process against which to compare the existing installed process. Idealizing involves freely constructing the ideal process model for a specified task. It is usual practice to construct the ideal process from the input of a carefully selected group of people. For an industrial process, this input will be expert technical input but for a commercial process the input is likely to be multi-departmental and hence multi-skilled. The ideal process that emerges from this idealizing exercise is then closely compared to the existing process. The target is to eliminate extraneous operations to improve performance.

Streamlining

Streamlining works in reverse to idealizing because the basic input is a model of the process as it currently exists. This model is then subjected to close analysis to remove unnecessary process steps and superfluous activities. In the business domain a close look at existing bureaucratic structures and a value-added analysis might be used to enhance the business process performance. In the industrial process domain, a material-energy balance calculation might be performed to eliminate unnecessary waste or energy lost in the process. The target in both areas is to enhance efficiency and improve the process performance.

Business Process Re-Engineering
Business process re-engineering is a fully developed methodology for re-shaping a company's business culture and *modus operandi*. Business process re-engineering has been defined as [Hammer and Champy, 1995] a '… fundamental rethinking and radical redesign of business processes to achieve dramatic improvements in critical contemporary measures, such as cost, quality, service and speed'.

Common refrains in the literature of business process re-engineering are (a) a reworking the business process to do things differently and (b) the achievement of a large step change in business performance. An interesting table of the characteristic features of business process re-engineering has been given by [Bogan and English, 1994]. Some of these features are given here as Table 1.8 to provide a flavour of the movement.

Whilst business process re-engineering has something in common with idealizing in that both have a clean-slate starting point, business process benchmarking has commonality with streamlining since both of these have a starting point in the existing business process structure. The concept of business process benchmarking is described briefly next.

Business Process Benchmarking
Like business process re-engineering, business process benchmarking is a tool for reshaping the business process culture but the procedure is quite different and probably less dramatic. Basically benchmarking is an exercise to find those practices or features that make a particular process 'best in a class'. The key objectives of a benchmarking exercise are (i) to understand, define and measure the performance of a selected process, (ii) to select and assess a group of benchmarking partners and/or processes from which to derive the benchmark standard and finally (iii) to introduce into the benchmarker's processes the best features and practices so as to enhance the business process performance. Benchmarking has considerable depth and promise as a methodology and for this reason a more detailed presentation is given in the next section.

1.7 The Art of Benchmarking

The general context of benchmarking was covered in Section 1.6 where it was explained that benchmarking is a completely self-contained methodology for process performance assessment and improvement. The method was widely disseminated in the business process domain where it is now well established as a tool of some considerable utility. However its basic principles are applicable to other domains and in particular, the process control field. As has been mentioned before, the presentations of benchmarking available in the literature (and there are many) use many of the individual tools devised for performance assessment.

However in this section, the presentation will emphasize the special features and advances of the methodology of benchmarking *per se*.

Table 1.8. Characteristics of business process re-engineering (after Bogan and English, 1994)

ITEM	CHARACTERISTICS
Time factor	Day one change over in business procedures Long term time requirement Effects of change measured in years
Risks	Moderate-to-high
People factors	Top-down participation Integration of high level teams Top-down, bottom-up communication
Primary enablers	Senior management support Business process model concept Vision of future state Best practice identification
Range	Clean-slate starting point Broad, cross functional scope Improvement change through radical breakthrough
Tools	Data collection techniques Process mapping techniques Teamwork Breakthrough thinking techniques Information technology Best practice benchmarking

1.7.1 Benchmarking – An Introduction

A benchmark process or quantity is considered to be the 'best in a class'. Thus a simple classification of benchmarking methods is based on how the 'in a class' aspect of the method is defined; this leads to four main benchmarking types.

Internal benchmarking. This is a comparison of processes against the best from within the same organization, whether it be one factory site or across many sites within the same organization.
Competitive benchmarking. This is a comparison made across a group of competing organizations. The benchmark process could be the best from all of the participating organizations or it could be the process from the organization deemed to be the competitive leader within the group.
Functional benchmarking. This is a comparison made across organizations whose processes have similarities and belong to the same technological domain. The organizations need not necessarily be competitors but should be world-class in their own field.

Generic benchmarking. In this form of benchmarking, an organization compares its processes and practices with the international leaders irrespective of whether or not the organizations are in the same commercial or industrial area.

Internal benchmarking is quite common for companies that have the same process being run and operated at several different (international) sites. By completing an internal benchmarking study a company can hope to have all its processes using the same procedures, the same best practices, and achieving the same level of internal company performance. To do internal benchmarking means a company-wide sharing of internal information but for competitive benchmarking, directly competing organizations have to agree to share what could be highly sensitive commercial information. The transfer of information necessary for competitive benchmarking may be difficult to achieve but the advantage to the laggards in the field should not be underestimated. However, it should also be noted that competitive benchmarking is ideal for fields where competition in the commercial sense is not an issue. For example, not-for-profit organizations like hospitals and government departments with services like the post, pension distribution, library services and so on would find great benefit from competitive benchmarking.

In cases where competitive benchmarking is limited by proprietary information, a more realistic exercise could be functional benchmarking on similar technological processes. For example, many large scale commercial and manufacturing organizations run supply chains, logistics for material supply, furnace and heating operations, and other similar operations. Functional benchmarking is ideal for these types of common processes. Finally, generic benchmarking is best for those higher-level operations and decisions relating to corporate issues rather than the processes active lower down in the hierarchy of company operations.

Having defined the 'class' over which benchmarking can be pursued there is also a classification of what process aspects can be benchmarked. Generally three areas are defined for the benchmark exercise.

Performance benchmarking. This is a comparison of performance measures for a specific process. The measures are usually numerical metrics describing how well a process is achieving an objective quality level.

Process benchmarking. This is benchmarking at one level up in the operational hierarchy and looks at the complete process sequence. It is a performance assessment of how to operate a process and from such an exercise best practice procedures and techniques emerge. The human element in the shape of process operators and process support staff is often an important aspect of process benchmarking.

Strategic benchmarking. This is up one level above process benchmarking and enters the managerial and corporate domains within an enterprise's activities. The comparison in strategic benchmarking concerns the higher-level decision-making processes and the information flows that support these decision activities.

In summary, Table 1.9 shows typical combinations of benchmarking types, and the next section provides an overview of the kind of activities that take place in a benchmarking study.

Table 1.9. Combinations of benchmarking types

	Internal	Competitor	Functional	Generic
Strategic	N/A	√	√	√
Process	√	√	√	N/A
Performance	√	√	N/A	N/A

1.7.2 Activities in a Benchmarking Study

In a survey of typical benchmarking steps, Bogan and English [1994] listed the Motorola Five-Step process, the Bristol-Myers *et al.* Seven-Step process, the Xerox Twelve-Step process, and the AT & T Nine-Step process as examples of company proven benchmarking cycles. It is not surprising that with so many recipes for success, the U.S. Strategic Planning Institute (SPI) Council on Benchmarking proposed a standardized benchmarking cycle based on the five-steps: Launch, Organise, Reach Out, Assimilate and Act. It was suggested that this SPI five-step cycle be taken as a template to be tailored to individual company requirements. With the requirements of the process industry in mind, the cycle described here is a minor modification the five-step benchmarking wheel due to Andersen and Pettersen [1996] as shown in Figure 1.20.

A brief explanation of the various stages in the cycle follows next.

Step 1 Plan and Model
The planning part of this activity simply covers the logistic and organization of setting up and running a benchmarking study. An early decision is to decide the type of study to be performed ranging from process – internal to strategic-generic (Table 1.9). The very nature of the benchmarking study selected will impact on the administrative steps necessary to gain study support (from higher management and others) and the necessary permissions to perform the site visits, data collection activities and to initiate the level of communication and information exchange needed for a successful study. Setting up an appropriately skilled benchmarking study team is also part of this planning activity. The detailed focus of the benchmarking study will emerge as the process model is documented and decisions are taken on the key performance assessment measures to be used in the study. Andersen [1998] estimates that around 50% of the total project time will be spent in this first step.

Step 2 Search
The extent of the search portion of the benchmarking study depends on the type (viz. internal, competitor, functional or generic). Whatever the benchmarking ambitions of the study instigator, they have to be tempered against the very real problems of accessing possibly commercially sensitive information. This activity

then is all about acquiring suitable partners for the benchmarking exercise and books like those due to Bogan and English [1994] and Andersen and Pettersen [1996] have helpful guidelines on selecting benchmarking partners.

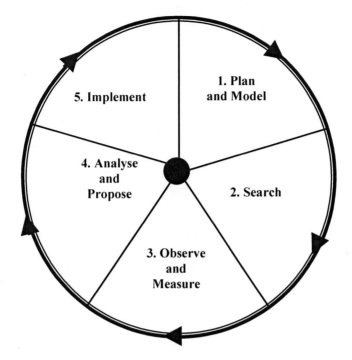

Figure 1.20. Benchmarking wheel (after [Andersen and Pettersen, 1994])

Step 3 Observe and Measure
The establishment of a process model along with a set of performance assessment metrics should make the tasks of understanding similar process installations, and measuring the performance relatively straightforward. Thus, the common process model has a normative effect on terminology, data collection, and the comparisons to be performed.

Step 4 Analyse and Propose
The analysis exercise involves locating and understanding the barriers to performance achievement. In section 1.5, a toolbox of techniques was described for the task of analysing the performance data collected in Step 3. Less clear is the task of proposing methods to remove or eliminate the identified barriers to performance achievement. In the main, the solution proposed will depend largely on the context and type of process under investigation. Nonetheless out of Step 4 should come recommendations for change and performance improvement.

Step 5 Implement
The last step of the benchmarking cycle is to implement the improvement recommendations. However, depending on context this will range from

straightforward acceptance and process upgrade to 'not possible' and then devising a compromise or trade-off implementation.

Finally, why is it a benchmark *wheel?* The assertion is that performance assessment and improvement is never finished and that companies should always have in place mechanisms to advance performance achievements continuously.

1.7.3 Benefits of Benchmarking Studies

In a recent book, Camp [1998] collected together a number of international benchmarking studies and described the benefits that were obtained. This set of case studies provided a valuable insight to the way different companies performed benchmarking exercises. Although the studies are not all documented to the same level of detail, some of them do permit a methodological analysis. Table 1.10 shows the type of benchmark studies performed by six organisations. Also listed are the types of analysis tools used and the concomitant benefits obtained.

A common (but surprising) refrain from most of the studies reported in Camp [1998] is the benefit gained from a benchmarking exercise of learning exactly how the process investigated actually works. Process and control engineers would probably be quick to comment that this is also invariably the outcome of a process control modelling activity. It was noticeable that those studies that produced a detailed process flowchart or process block diagram were more likely to use rigorous process performance assessment tools like the cause and effect diagram and cite quite precise performance improvement outcomes.

An assessment of the benchmarking literature leads to the following benefits and outcomes of company benchmarking studies.

Local and Immediate Benefits
- ♦ Leads to improved values for performance metrics and lower cost positions.
- ♦ Provides a better understanding of how the actual process operations work in the plant.
- ♦ Introduces new best practice ideas and work methods.
- ♦ Tests the rigour of established internal performance target values and operational procedures.

Wider and Longer Term Benefits
- ♦ Introduces new concepts to the process owners and operators and presents a collaborative learning opportunity.
- ♦ Opens up dialogue channels within and between organizations, departments and the process owners and operators.
- ♦ Improves employee satisfaction through involvement and empowerment.
- ♦ Creates an external business view of process operations.

Table 1.10. Analysis of benchmarking study characteristics (Data sourced from Camp, 1998)

Company	Process	Benchmark Type	Analysis Tools Used	Benefits Claimed
Sector - Manufacturing				
Allied Domecq	Information Delivery	i) Strategic ii) Functional	Matrix diagrams Team Discussions	Critical Review of success factors
NITT, India	Hardware Procurement	i) Process ii) Competitor	Brainstorming Tabular data analysis	Cycle time reduction. Overruns elimination. Better organization
Ingwe Coal Corporation	Continuous coal mining operation	i) Process ii) Internal	Cause + Effect Chart Histograms of Performance data	Process better understood Able to remove operational obstacles to performance
Sector – Not-for-Profit				
Northern New England Cardiovascular Disease Study Group	Cardiac Surgery	i) Process ii) Competitor	Cause + Effect Chart Tested Improve Process Sheet	Improved healthcare procedures
Sector - Government				
Post Office, UK	Supply Chain	i) Process ii) Generic	Site Visits Data Tables Streamlining	Understanding of Process + Holdups
Sector - Education				
Babson College	Student Enrolment	i) Process ii) Competitor	Questionnaires CSF Analysis	Understanding of Process. Best practices identified

It should be fairly clear that the local and immediate benefits of benchmarking studies are mostly quantitative in nature whilst the longer-term outcomes are more broadly qualitative and cultural.

1.8 Concluding Discussion

This review of the prevailing paradigm of business process performance monitoring, assessment and improvement reveals an interesting picture of a toolbox of techniques that are used to underpin some holistic performance improvement methods and philosophies. A diagrammatic representation of this is given in Figure 1.21, where the toolbox contains 'People Techniques' and 'Technical Methods'. In this chapter, although reference is often made to the 'people' dimension in planning and performance matters, no in-depth presentation of this special area has been made. One interesting observation is that two important techniques in this toolbox, statistical process control and cause and effect analysis are well known to process and control engineers. Finally the concept of the business process and the consequences that follow should provide inspiration for the process and control engineering community to devise a similar toolbox for technical process performance improvement.

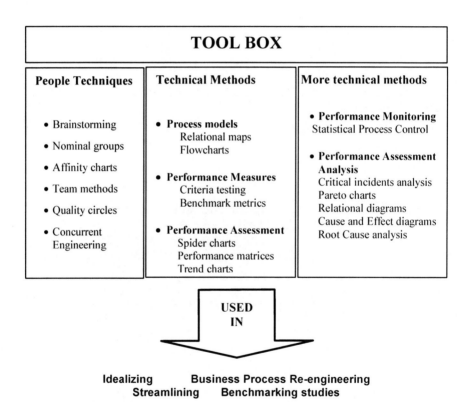

Figure 1.21. Performance improvement toolbox

Economic Auditing of Control Systems

Damien Uduehi [1], David Laing [2] and Andrzej Ordys [3]

[1] Strategy and Planning Coordinator, (West Africa) BG Group , Reading, UK,
[2] Principal Control Engineer, Marathon Oil, Aberdeen, UK,
[3] Professor of Automotive Engineering, Faculty of Engineering, Kingston University, London, UK.

2.1 Chapter Outline

In this chapter, a method for integrating the use of the benchmarking and optimisation algorithms with economic process control auditing is discussed. The focus of the methodology is to selectively target process control loops with economic importance for benchmarking and optimisation. The method is a step by step approach to prioritising control loops according to economic importance and then benchmarking and optimising the necessary loops.

Section 2.2 discusses a framework for process control benchmarking at the different layers of the process hierarchy and reviews some of the properties and characteristics of performance assessment metrics at each layer. Section 2.3 discusses the motivation for integrating process control benchmarking and optimisation with process economic control auditing and provides an integrated control and process revenue and optimisation (ICPRO) framework as a template for conducting process control audits. In Section 2.4, the integrated control and process revenue and optimisation framework is used to evaluate an industrial case study example. The case study example involves three offshore oil production platforms. The results and recommendations from this industrial case study are presented. In Section 2.5, some of these results are used to optimise a sub-system on one of the oil production platforms. Conclusions are presented in Section 2.6.

2.2 Formal Framework for Process Control Benchmarking Metrics

In complex control systems, such that can be encountered in living organisms or in large international organizations, goals are typically arranged in a hierarchy, where the higher level goals control the settings for the subsidiary goals. Such hierarchical control can be represented in terms of the process control schemes above level 0, as in Figure 2.1. The goals at the lower levels of the hierarchy become the result of an action, taken to achieve the higher level goals. In general, in the presence of a stochastic disturbance, a control loop will reduce the variability of the loop output, but will not be able to eliminate all the variations. Adding a control loop on top of the original loop may eliminate the residual variety. Therefore, the required number of levels in the control hierarchy will depend on the regulatory ability of the individual control loops. On the other hand, increasing the number of levels has a negative effect on the overall regulatory ability, since the more levels the feedback and control action signals have to pass through, the more they are likely to suffer from noise, corruption, or delays. As each device in the control hierarchy impacts composite performance of the units below it in the hierarchy, the more layers of hierarchy in a control scheme the greater the possibility that a degradation in performance of a device at the top of the hierarchy will result in a substantial reduction in the performance of the process. Because of this, control professionals have always sought to maximize the regulatory ability of layers 1 and 2 and thus minimize the number of requisite layers required to achieve the overall process objective.

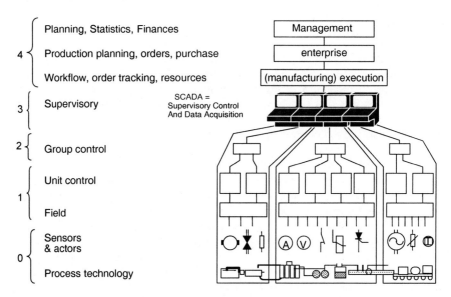

Figure 2.1. Hierarchy of Process Control

This may explain why predominantly the development and use of control performance assessment and benchmarking applications have centred around Levels 1 to 2 of the Process control hierarchy. The applications for use in Regulatory Loop Control (Level 1) assessments are by far the most commonly available commercially and have been the core of research and developments efforts over the decade. Because the characteristics of Levels 1 to 4 are different, some of the factors governing benchmarking and performance considerations at each of these levels are also different. Fundamental to the appropriate application of benchmarking applications and to effective utilisation of the results from any benchmarking exercise for process and product improvements, is to have an understanding of the different properties of benchmarking and performance assessment criteria required at each level in the control hierarchy and how these criteria relate to each other.

From Figure 2.1, the process control can be partitioned into a top level where process units are globally coordinated, a unit level where a complex process unit is operated seamlessly within the global process line and a sub-unit level where the intra-unit regulator operates autonomously. Overall process control itself can be represented as a combination of levels within the layers of an organisation's business process. The process control hierarchy in Table 2.1 describes the technical processes which intersect with the business processes at the lowest three levels (Process, Information and Economic) of the business organisation hierarchy.

Table 2.1. Business/Process Control intersection

		Layers in Business Process/Organisation		
		Cultural Level	The Company goal	
		Strategic Level	The Company strategy	
Social Level	Staff relations, Teams	**The Industrial Control Hierarchy**		
Economic Level	Profitability, Resource usage	**Level 4**	Load management	Process line interface
Information Level	Information flow system	**Level 3**	Set-point optimization	Process unit top level management
		Level 2	Dynamic set-point changeover	Automated unit level procedures
Process Level	Process instrumentation, Technical system	**Level 1**	Regulator loop control	Low level control structure and controllers
		Level 0	Process	Actuators, process equipment, sensors

Table 2.2 shows a framework for classifying the benchmarking requirements at the different layers of the business process. At the "Process" and "Information" levels, the benchmark and optimisation process is dominated by the definition of local performance metrics, technical optimisation criteria and controller design and performance, and is less influenced by the social-psychological interactions of operators and/or team work groups. At the Economic level, the benchmarking and optimisation process is dominated by definitions of global performance metrics, process objectives, business operation strategies and optimisation procedures. Performance metrics are of two types:

- *Product Performance Metrics* : These are quality variables of the process product or output.

- *Process Performance Metrics*: Those variables which indicate if the process is operating in a desired way when manufacturing the product or output.

Table 2.2. Framework for control benchmarks

INDUSTRIAL CONTROL BENCHMARK FRAMEWORK	
LEVEL	FEATURES
Economic Level	1. Discrete event characteristics 2. Dependence on operator interaction 3. Social-psychological factors 4. Process unit interaction and inter-dependence 5. Qualitative/Quantitative performance 6. Global economics
Information Level	1. Quantitative performance 2. Some qualitative performance factors 3. Performance depends less on operator skills 4. Technical and design factors important 5. Market demand and supply and economic factors
Process Level	1. Quantitative performance dominates 2. Little operator dependence 3. Performance has a high dependence on technical and design factors

The key features of a performance metrics should be:
1. The performance metric should be physically and technically meaningful for the process being assessed.
   Thus the metric may capture and measure the presence of a desirable physical property or measure an economic dimension of the process.
2. The performance metric should preferably be amenable to an optimisation analysis to enable the full achievable optimised performance be computed.
   The extension of this is that the achievable optimised performance in the presence of structured design and implementation constraints should be calculated.

2.2.1 Goals of Benchmarking

The first thing in considering the application of benchmarking and the appropriate strategy for the potential optimisation of the system under test, is to set out the goals of the levels to which the system is associated. The goals of the most prominent of these levels can sufficiently be summarised as:

1. Company: To continuously generate a healthy and increasing profit from the production and sale of the range of products.

2. Engineering process: To realise the company goal, by creating a continually improving technical environment for the efficient manufacture of the required products.

3. Control system: To implement the company and engineering goals by ensuring a safe and optimal means of increasing /maintaining a consistent production rate and product quality while simultaneously decreasing operational costs, plant downtime and maintenance costs.

2.2.2 Principles of Benchmarking

Goldratt [1993] considered the problem of optimising the performance of the entire manufacturing process, which may be made up of numerous control loops. That work is useful in developing a summary of principles to ensure that in conducting any benchmarking exercise, the exercise is structured in such a way as to actually result in a routine for performance improvements. Some of these principles include:

1. In a multivariable process, where interaction exists between the process loops, optimising each loop independently of all others does not ensure that the overall process is optimal. ("*A system of local optimums does not necessarily translate to a globally optimal system*").

2. Benchmarking the performance of individual loops in the process gives a measure of how far from a local optimum an individual loop may be, it does not say anything about the overall performance of the process and how far the process is from a global optimum.

3. The global performance of the process will be predominantly determined by the performance of constrained loops. Constrained loops are loops that have some physical, environmental or user imposed limitations applied.

4. To reach a global optimum, the ideal working point for loops with bottlenecks identified as key to the process objectives will most likely be at the constraints, the operation and performance of all other loops must allow for these limitations.

5. For global process-wide optimisation to be achieved, the control objectives must be derived from the management and process objectives.

2.3 Framework for Integrated Control and Process Revenue Optimisation

The traditional literature on benchmarking [Codling 1992, Andersen and Pettersen, 1996], has been mainly concerned with business processes rather than the problems of operating and controlling physical/mechanical/chemical process lines or factories. On the other hand, the conventional literature on process optimisation [Huang and Shah, 1998; 1999, Desborough and Harris, 1993] has been more concerned with technical performance metrics and control loop performances but not with the financial and economic aspects of the physical process.

There is a link between the economic performance of a business and the control performance of the technical process related to this business. The existence of this link has been documented by Rolstadas [1995] and Ahmad and Benson [1999]. In trying to establish and understand what exactly the relationship between economic performance and the control performance is, and how it works, a high level analysis of how the performance of the control system in an oil production facility influences the financial returns of the business will be done. The analysis will be conducted with the aid of the Return on Net Assets (RONA) business benchmark model as documented in a review of integrated performance measurement systems [CSM, 1997] in Figure 2.2.

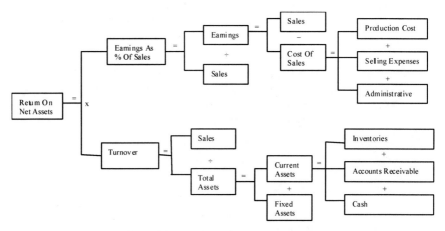

Figure 2.2. RONA performance benchmark

The oil production facility belongs to a hydrocarbon exploration and production petroleum company whose business process can broadly be defined as the production and sale of crude oil and associated products from recovered reservoir fluids. To achieve this goal the company has designed and built a number of reservoir fluid processing platforms, whose aim is to separate the commercial product in the reservoir fluid from the waste products. Each platform has a specific daily processing capacity.

To demonstrate the effect of process control on the economics of the business, consider how the process control of the platform directly influences some of the elements on the RONA tree as described in Figure 2.2

1.   Production Cost

The performance of the process control system on the oil platforms affects the efficiency of the overall process. The more efficient the production, the greater the ratio between the profit obtained from the products of the process and the production cost. In addition, in some cases, it can be shown that the efficiency of production has a direct impact on production costs.

2.   Selling Expenses

Some of the expense involved in selling the derived crude oil and associated product come from the transportation and processing tariffs the company pays for sending the crude oil to export terminals through third party pipelines and for onshore processing of its gas products. The charges of these tariffs are calculated per km of pipeline used and per tonne of gas. The process control system ensures efficient separation of commercial products from waste products ensuring that additional transportation and processing charges / or penalties are not incurred for sending unwanted waste products down the pipelines to the processing facilities.

3.   Sales

The amount of commercial product sold by the company is directly related to the production rate of its platforms. One of the functions of the process control system is to try and maintain the rate of production at a level specified by the design of the platforms. An optimal control system will ensure a rate of production that is close to the designed capacity.

4.   Earnings

The revenue received by the company from the sale of its products depends on the quality and quantity of these products. The quality of the products depends on the process units on the platform operating to the specification to which they were designed. The efficiency at which these process units operate and hence the product quality depends, in some measure, on the process control system.

5.   Current Assets

The major assets of the company are its reservoirs, oil wells and platforms. The task of maintaining these assets and ensuring maximum recovery of reservoir fluids from the wells, inherently rests on the process control and fault and condition monitoring systems.

Clearly the control system performance is interwoven with the economic performance of the business. Over the last 20 years there has been substantial progress in control system design and applications, some of which has great potential for improving process performance. However the capital expenditures required to implement these advanced technology solutions (installation, commissioning and support) are high. Therefore, what is required is a method/systematic procedure for selecting those processes where implementation of the new control technologies would have the greatest impact. This would limit the capital expense whilst maximising the revenue generated. An integrated approach to process revenue optimisation using advanced control or knowledge based expert systems can then be used to directly target specific areas in a production process where optimisation of the process unit will have substantial financial benefits in the revenue received.

2.3.1 Integrated Control and Process Revenue Optimisation (ICPRO) Method

The integrated control and process revenue optimisation approach is a means for identifying areas (bottlenecks) where advanced control and optimisation technology can have a marked effect on process revenue. This method requires that before any benchmarking analysis or control redesign is undertaken, either an in-depth plant auditing involving management, process and control objectives is carried out or information resulting from such an audit is available.

The approach identifies five steps as being critical to determining the sector of a process that not only has the necessary degrees of freedom for optimisation but also has direct impact on the financial returns from the process. These five steps can be defined as:

Step 1: Profile and Operations Assessment
The control engineer who wants to practise benchmarking must have a thorough understanding of:

- The critical business processes and products.
- The critical engineering factors for product objectives.
- The best measurements that will provide information on key performance indicators.

The linkage of the business process to the engineering process is critical to effective benchmarking. The process of control performance benchmarking must fit into an economic revenue improvement framework. The idea is that by using information about financial impact it is possible to detect the critical engineering processes and related control loops that are worth investigating. The results from the Profile and Operation Assessment should be used to design the scope and requirements for the actual benchmarking project.

Considering the multifaceted set of skills required to conduct a successful top down benchmarking and optimisation project, it is best to approach the benchmarking as a team effort. Team members need access to sensitive information on company production and operational targets and it is sometimes useful for the project to have a sponsor with a high level of seniority within the company and involve the staff with substantial knowledge of financial, engineering and the process dynamics. The benchmarking team needs to:

- Understand the critical processes and how they are measured.
- Decide what kind of data is needed and how this data will be collected.

The Profile and Operations Assessment provides insight into key company financial objectives and the engineering processes in the organization that address those objectives. At the Profile and Operations Assessment stage, the procedure involves understanding the company's business strategy. Next, it should be decided what measurements are required from those areas of company's operation from which financial benefits of the process accrue and capital expenditure or losses occur. Prime factors are:

- Product quality
- Production rate
- Raw material acquisition
- Plant operability
- Plant availability
- Power consumption
- Maintenance cost

Global benchmarks should be created and analysed for the entire process. A set of metrics for each of the objectives that the entire process aims to achieve (quality, economics and security) should be defined. Using present business and operating conditions, a set of values for the global metrics should be stored. The type of measurements (or metrics) chosen have to be useful and easily calculated e.g. production rates, hours of continuous operation, quality specifications.

Step 2: Process and System Assessment
The Process and System Assessment stage is where the benchmarking team profiles the underlying engineering process. A key step in the Process and System Assessment stage is using process and instrumentation diagrams so that the benchmarking team understands the processes and how they can be controlled and performance measured, both in the control terms and in management terms.

The purpose of Process and System Assessment is to:

- Identify processes as candidates for benchmarking.
- Establish the metrics to be used.
- For the chosen metric collect baseline data of the process variables that can be used as a calibration point for comparing the performance of the system before and after any retuning.

Identifying potential processes for optimisation is another step in the Process and System Assessment stage. It is always best to develop a list of three to five potential process units for benchmarking. Some of the potential process units may not be feasible for benchmarking on closer inspection, or may not fit within the allotted time-frame, others might not have the right sensors and instrumentation to gather data about the necessary variables.

This stage involves the identification of the important sub-processes, process goals, major control loops and control objectives. The bottlenecks existing within the process units that limit efficiency and productivity should be clearly identified and where possible, the sub-processes and control loops involved should be noted for measurement and data analyses. It is essential to obtain substantial knowledge about the company's process and control model, objective and strategies. This information can be acquired directly from staff with substantial knowledge of process and control operations and dynamics. A review of plant piping and instrumentation diagrams, operations chart and reports and maintenance reports can also help to provide a very clear picture of the physical process.

Before collecting a lot of data for an extensive benchmarking and analysis exercise, the benchmarking team needs to collect baseline data about the processes. This data can be current or archived records that show an extended period of normal plant operation with acceptable performance limits. Collecting this data will refine the measurement process and help develop the final set of metrics and application to be used in the benchmarking effort. The kinds of benchmark application and metrics chosen have to be compatible with the dynamics of the process and the performance to be assessed. For instance, there is no point in choosing a benchmarking application which relies on variance in a process to compute performance indicators if the process is relatively noise-free.

These local baseline benchmarks may sometimes be obtained by analysing the levels/units inside the process and finding a set of metrics that measure the performance of each level/unit. Using current operating conditions, a set of values for local metrics should be recorded. Also control loops within sub-processes that are either problematic, inefficient or that could be optimised should be noted.

Step 3: Correlation of Financial Benefits and Control Strategy
The Financial Benefit and Control Strategy Correlation stage is where the benchmarking team begins the process of linking control objectives and controller tuning to the organization's strategic goals. The benchmarking effort should be focused on those control loops that are most important. At this stage the correlation between subsets from which revenue accrues and sub-processes or groups of sub-processes within the system should be established. One way to determine the relative importance of loops in process units is to develop a list using the information already obtained from the previous stages:

Correlation List

1. State the mission, purpose or goal of the process or manufacturing operation.

2. List the process units associated with each of the above.

3. Identify major process units by the value or volume of their outputs.

4. Identify which processes add the most value and which add the most cost.

5. List the major enablers, bottlenecks and constraints for: production, quality and availability.

6. Identify which control loops affect these enablers, bottlenecks and constraints.

When an opportunity to enhance a company's financial objectives is identified, the engineering processes that can directly fulfil that objective can be considered as critical processes. The idea is to only benchmark critical processes, identifying weak critical processes that can give the most leverage when improved. Once this correlation exercise is done, a mapping between the related control loop and process groups should be produced. It is essential to analyse the control loops within these sub-processes, to determine if the provided control structure or algorithm is suitable.

Step 4: Optimality Assessment
At the Optimality Assessment stage the focus is on checking the process variables to determine if there exist any additional degrees of freedom by which the control action can be improved. An evaluation of the optimisation potentials at the regulatory, multivariable and supervisory levels of control hierarchy should highlight the optimisation strategy required.

Clearly defining how the evaluation process will be done, helps to define the data required and using lessons learned during collection of data for the baseline should help to refine the measurement process and develop the final set of metrics to be used in the benchmarking effort. There are measurement pitfalls to avoid as well. The benchmark team needs to have consistent collection methods (sampling rates, quantisation and compression methods for similar types of loops). The proper aggregation levels for data must be specified and the data units and intervals should also be specified to make comparison easier during analysis.

Although benchmarking stresses the use of the "best in class", often this has to be tempered with other factors, such as process dynamics, obtainable data, costs (interruption of normal process operation, model development, etc), time, and multidimensional process relationships. Analysing the benchmark performance for each identified loop or group of loops can be done as an isolated event or as an event trended over a period of time. Either method (or both) may be appropriate for the process being studied. When cost, productivity or quality is the metric under study, sometimes it is useful to look at the historical trend as well as the current performance. The benchmark metrics obtained should be used to determine if improving control action will influence/improve revenue. Note that benchmarking and optimisation criteria may be mathematical or intuitive in nature.

Step 5: Control System Adaptation
Benchmarking is about improving processes, and as such it requires a structured approach to discussing, assessing and implementing any change to the system that may be necessary as a result of the benchmarking analysis. The benchmarking team must be aware of this, before the adaptation phase is commenced, the following change management techniques should be employed:

- Communicate the benchmark findings widely.
- Involve a broad cross-functional team of employees (production, process, control and management).
- Translate the findings into a few core principles.
- Work down from principles to strategies and to action plan.

Each process has a process "owner," and process owners and other stakeholders need to have a voice in the changes recommended. Before developing control strategies, it is important to communicate with all who might be involved in the change. Communication can follow the following change management pattern [McNamee, 1994]:

- Identifying the need for change.
- Getting stakeholders to voice their opinions about the change.
- Providing a forum for all to discuss the methodology, the facts, and the findings from the benchmarking effort.
- Communicating the expectations about the changes.
- Building commitment for the change.
- Getting closure; celebrating the change.

In reaching a recommendation for a change of control strategy or design, the analysis of the collected benchmark data should expose the gap between the process performance level and the optimal level as suggested by the benchmark metric, and predict where the future gaps, constraints, and bottlenecks are likely to be. From the analysis of the benchmark results a decision on the need for retuning or redesign of the control strategy must be reached. The benchmark application used will determine the optimisation criteria that will enable full achievement of any benchmarking objective.

This means that, because of technical or business constraints, it is possible that a re-tune of the existing controllers might not result in the performance desired and more advanced solution involving process re-design might be required. Note that the decision to use an advanced control design involves the use of process models which involves additional costs. Where possible the use of simulations to compute the improvement in performance between present control strategy and the proposed strategy is most desirable. The results for the simulated global and local metrics obtained using the proposed strategy should be compared against the stored baseline metrics. The benefit of the proposed strategy must be clearly visible before any decision to change the current system setup is implemented.

The five steps in the ICPRO audit process should be considered adaptable and are intended to act as a guideline only. When applying this or any other the performance auditing /improvement method it is important to remember that the benefits are only obtained if the procedure is repeated at regular intervals.

2.4 Case Study: Oil Production Platform

To illustrate the above concepts and to place the controller performance assessment within the framework of plant wide productivity audit, the results and analysis from an industrial feasibility study conducted by Strathclyde University on the financial benefits of implementing advanced control on an oil platform [Grimble and Uduehi, 2001] are utilised. The company at the centre the study is involved in oil and gas exploration and production. The aim of the project was to examine the operation of the company offshore production platforms and determine if implementing some form of advanced control system would improve production, and therefore result in a significant revenue increase.

The feasibility study was divided into two stages. Stage one comprised an economic control audit and benchmarking exercise to include:

- Reviewing the company financial strategies as regards the offshore oil production platforms and their products,
- Reviewing the production platform process and control operation from an economic perspective to determine if there exist any financial gain in introducing advanced control.
- Identifying areas within the process that can be optimised using advanced control to yield some financial benefit.

Depending on the results of economic control audit in stage one, stage two would be a quantification and implementation exercise that would include:

- Quantify any financial gain from the identified list of potential opportunities,
- Derive any change management strategy that might be required,
- Review the advanced control optimisation packages, and recommend those packages that are offering the best application fit for building advanced control systems.

The benchmarking team was sponsored by the Production Manager and included staff members from each of the following divisions in the company: Process, Control, Production and Finance. There were three additional members of the team with benchmarking and control optimisation expertise from a university and a consulting company in charge of the feasibility study. The economic control audit was performed and the information about the company and its engineering process and the resulting recommendations was obtained by using the ICPRO approach. Some additional insight was developed from meetings and briefings by various company staff members from the Reservoir Management, Production, Control, and Process and Forecasting departments.

2.4.1 ICPRO Step 1: Profile and Operations Assessment

The company's prime concern is the production and sale of crude oil and associated products. The company has three oil platforms called here: Platform A, Platform B and Platform C. These platforms manage the production of crude from sub-sea oil wells. The crude oil and associated products are then transported by pipeline to onshore terminals for processing before being sold. The company is charged a tariff per km for using other operator pipelines to export their products. The Raw products from the company platforms can be classified as:

1. Black oil
2. Natural gas liquids (NGL)
3. Condensate

4. Gas

The Company generates revenue by the sale of its products, the quantity and quality of the products thus influencing the amount of revenue received. The finished products are:

1. Stabilised crude oil
2. NGL
3. Sales Gas

(a) Stabilised Crude Oil

Black oil is produced on the company's platforms and processed at onshore processing facilities. It is sold by the barrel, as stabilised crude oil. The price of stabilised crude on the world market and the quality of the crude determines the price received for each barrel. Its base sediment and water (BS&W) content determine the quality of the stabilised crude. There is no regulation/restriction on the amount of stabilised crude the company can sell in any given month.

(b) Natural Gas Liquid

The Natural Gas Liquid produced by the company is sold by the tonne. The price received per tonne of NGL is determined by the price of its components on the world market and the quality (composition) of the NGL for the month. There is a regulatory procedure for the sale of Natural Gas Liquid. This procedure can be summarised as follows:

1. 100% of monthly production of NGL must be lifted (i.e. sold).
2. Lifting is based on forecast production of NGL.
3. If there is under lift (less than 100% of production lifted), then the excess is stored and sold based on next month's prices.
4. The forecast production and actual production may differ.
5. The NGL component prices are released on the first day of every month.
6. The NGL is sold/lifted on the 15th of every month.

(c) Sales Gas

The Sales Gas produced by the company is sold by the tonne. The price received per tonne of NGL is determined by the price of sales gas on the world market and its quality (Gross Calorific Value) for the month. There is a sales contract in place that regulates the sale of Sales Gas. This contract can be summarised as follows:

1. Carbon Dioxide content less than 1 mol %
2. Gross Calorific value: $36.9 < GCV < 42.9$ MJm^{-3}

A substantial percentage of the monthly revenue comes from the sale of stabilised crude oil. This is produced in greater quantities and provides a higher financial return than the other company products. All three platforms A, B and C are designed to process crude oil, gas and liquids. Amongst the three platforms, A, B,

C, Platform A produces the largest quantity of stabilised crude oil, and Platform B produces the largest quantity of Gas, NGL and condensate.

2.4.2 ICPRO Step 2: Process and System Assessment

The platforms are designed to produce and process reservoir fluids. Each of the platforms is uniquely associated with a number of wells /reservoirs from which reservoir fluids are recovered and processed into black oil, NGL, sales gas and condensate. The process system can be divided into two subsystems.

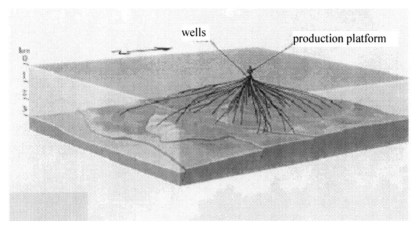

Figure 2.3. Production reservoir

1.   Reservoir system

The reservoir system is depicted in Figure 2.3. It consists of the Reservoir, Production wells and re-injection wells. The reservoir system provides the raw materials (reservoir fluids) that are processed in the topsides system. Three reservoirs and their uniquely associated production wells and gas injection wells service the platforms. Although the reservoirs are distinct, there is a level of inter-connectivity between them provided by the underlying rock formation. This introduces a level of multivariable interaction into the reservoir system. The reservoir and the well characteristics depend not only on the temperatures and pressures existing within the wells and reservoirs but also on the nature and geological topography of the underlying rock formations that surround them. There is a level of interaction and recycling between the reservoir system and the topsides system. The result of this interaction/recycling is that disturbances or events in the reservoir system affect the dynamic operation of the topside system and *vice versa*.

2.   Topside system

The topside process system provided on the platforms can be divided into four basic groups:
   a)   Wellhead system

b)   Separation systems
c)   NGL systems
d)   Gas compression systems.

Figure 2.4. Production platform christmas tree and wellhead assembly

a)   Wellhead System

The Wellhead system enables the management of the reservoir. It has associated with it a number of production wells and gas injection wells. The Wellhead system is designed to provide a safe means of producing reservoir fluids and re-injecting processed gas back into the reservoir. The 'Christmas tree' provides the facility for safe shut-off of the wells. It is an assembly of master valves and wing valves as shown in Figure 2.4. The master valves being used to shut in the wells and the wing valves to isolate the wellheads from the production manifold or gas injection manifold. On production wells, the reservoir fluids flow up the production tubing via the surface controlled sub-surface safety valve, to the wellhead and 'Christmas tree'. From the Christmas tree, the fluids flow through a choke valve which is used to control the rate of flow of reservoir fluid. From the choke valve the fluids flow through wellhead flow lines to the production manifold. Not all production wells associated with a given platform may be in operation at a particular time. The gas injection wells are used to maximise black-oil recovery by minimising reservoir pressure decay.

b)   Separation System

The separation system is designed to process reservoir fluids. Black oil, flash gas and produced water are separated in a separation train comprising the following four stages

1.   Feed and expansion system
2.   High pressure (HP) separator
3.   Medium pressure (MP) separator
4.   Low pressure (LP) separator

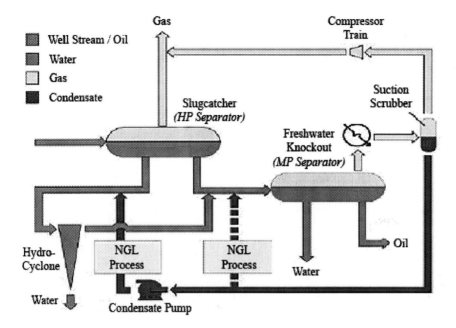

Figure 2.5. Simplified separation system

On the Platform A the operation and setup of the original system has been modified and the effective (simplified) view of the resulting system is show in the line diagram of Figure 2.5. The simplified separation process effectively consists of two tanks in series, the *High Pressure Separator* is setup as a *Slugcatcher* vessel and the *Medium Pressure Separator* is set-up as a *Free-Water Knock Out* vessel. The function of this plant is to remove gas and water from the crude oil flowing into the plant and pump this 'cleaned' crude oil to other plants down stream in the installation operation. The level of crude oil in both tanks has to be maintained between an upper and lower limit, for the Slugcatcher plant to function effectively. The level is also used as surge capacity to ensure a continuous and constant flow of crude oil downstream to other units.

c)   NGL System
A typical NGL refrigeration process is depicted in Figure 2.6. Unstable condensate and gas from the HP separator are processed within the NGL system to recover those hydrocarbons which may be exported in liquid form through the main oil export system. The unstable condensate and gas streams enter the system separately and are cooled by heat exchangers and mixed. This mixture is further cooled using liquid refrigerant in the gas chillers. The cooled mixture is then routed to the cold condensate separator. NGL is recovered from the base of the column, cooled, metered and then introduced into the black oil export pipeline. Platform B

uses an enhanced NGL recovery system. The system dehydrates and recovers NGL from the vapours of the inlet gas scrubber and HP separator in its separation system. The system returns the recovered NGL to the HP separator for subsequent export with the black oil.

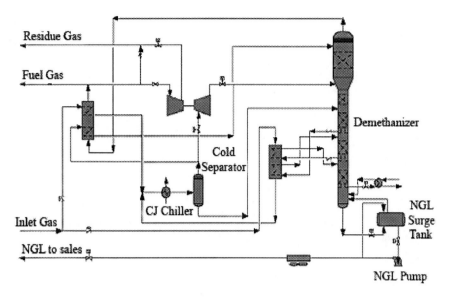

Figure 2.6. NGL refrigeration system

d)   Gas Compression System

The Gas compression and re-injection system is shown in Figure 2.7. The purpose of this system is primarily to compress gas for export and sale or for re-injection into the reservoir. Separated gas is compressed through three parallel compression trains, each with an MP separator and Export Compressor. Compressed gas is exported via pipeline and gas for re-injection is taken directly from the export header upstream of gas metering and compressed. The re-injection compressor is a two-stage, gas-turbine driven machine with dedicated anti-surge and performance control. Gas re-injection is important for increasing gas throughput. It enables more liquids to be produced from the gas.

Remarks on Platform Processes

The topside process is very interactive because it contains a number of recycle loops. There is full inter-connectivity between all the sub-systems on the platforms. This results in a highly interactive multivariable system. The critical process parameters are: pressure, temperature and level. Although the process is in general a slow one, disturbances to any part of the system can produce fast acting ripple effects (transients) that are typically amplified as they move downstream from the source. This occurs because of the interactions within the process and its multivariable nature.

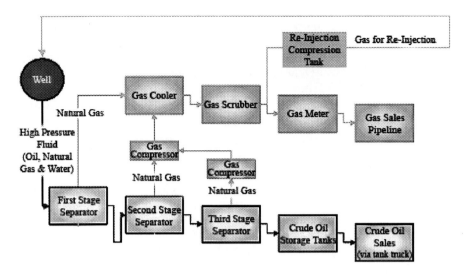

Figure 2.7. Gas compression and re-injection system

Process Control Overview

The control systems on the platforms serve two main purposes:

1. To provide a safe and efficient means of control for the production process and associated support services.

2. To provide a means for monitoring platform/system status and to initiate the necessary (shutdown) actions to preserve platform/system integrity and safety of personnel

All the primary control loops associated with the process system are controlled using PID controllers. There are three basic control loops:

1. Level Control

2. Pressure Control

3. Temperature Control

Although the process is highly interactive and contains a number of recycle loops, each control loop is tuned independently with limited consideration of the interaction with other loops or recycle effects. The platforms use the Honeywell TDC 2000 and 3000 (Total Distributed Control) system as the main platform control and data acquisition system. No supervisory control strategy or set-point optimisation is implemented, except in Platform A where the TDC 2000 is used to provide supervisory control for the gas compression system. On Platform B and Platform C, the export/re-injection gas compressors are controlled using Compressor Control Corporation (CCC) designed controllers. All the other PID controllers are located within the platform DCS system. There are no other local controllers on the platform.

2.4.3 ICPRO Step 3: Financial Benefits and Control Strategy Correlation

The information and product data obtained by the benchmark team from the first two stages of the ICPRO procedure was analysed using the correlation list discussed in Section 2.3. The objective was to determine the relative importance of loops in process units as well as the importance of the process units themselves, and to create a rational hierarchy of the various optimisation potentials that might exist. A summary of the results is presented according to correlation list.

1.   State the mission, purpose or goal

Continuous production, transportation and sale of crude oil, natural gas liquids, gas and condensates in line with established environmental policy and limits of the production facilities.

2.   List the process units associated with each of the above

The major system components that together facilitate the goals of the company are:

a) Reservoir system
b) Wellhead system
c) Separation systems
d) NGL systems
e) Gas compression systems.

3.   From all the process units identify the major units by operations

From the analysis of the operations data the following units were identified as the major operating units:

a) Separation systems
b) NGL systems
c) Gas compression systems
d) Reservoir system

4.   From the shortlist of key units, identify which processes add the most value or cost

Analysis of the production, maintenance and cost data showed that the following units contributed either the highest percentage of revenues or losses from the platform operations:

a) Separation systems
b) NGL systems
c) Gas compression systems.

5.   List the major production, quality and availability, enablers, bottlenecks and constraints

For this feasibility study the benchmark team were able to identify a number of candidate cases which could be either potential enablers or bottlenecks. These cases are presented below.

- *Analysis of Present Reservoir and Well Management Strategy*

The Company employs gas lifting, gas (re-) injection and water and gas injection (WAG) on certain wells to boost well pressures and increase reservoir fluid recovery. These techniques are used to manage the wells and limit their decline. The company also employs well scheduling. It has a detailed and accurate simulation model for their reservoirs. These models are used to simulate reservoir and well behaviour under varying circumstances. These reservoir models are however stand alone models, as they do not include either the production flow-line or the topsides process models. At present the analyses for WAG injection and the amount of gas to be injected and the rate of injection are being done as open loop calculations with no direct feedback information and without the interaction of the flow-line and platform processes. These calculations are not done online and there is a substantial time delay between analyses. This approach does not ensure optimal results and as such the resultant benefit of the whole operation is not maximised.

- *Increase In Raw Material Financial Yield*

There are two issues involved in increasing the financial yield of the raw material (reservoir fluid). One aspect of this is to increase the amount of finished product extracted per tonne of reservoir fluid processed on the platforms. The other aspect is increasing the revenue received from the finished products; this essentially involves the quality or composition of the products, since the prices per tonne/barrel of the products depend on their quality or composition.

Black-Oil Yield: measured against the company standard, black oil extraction from reservoir fluids seems to be efficient. The base sediment and water (BS&W) content determines the quality of Black oil. The efficiency of the separation process, reservoir fluid residence times in separators, interface level and the efficiency of the chemical injections affect this index. The lower the BS&W content of the black-oil, the higher its market value and the less amount of water being exported down the pipeline. Since the company is charged a transportation tariff for exporting the black-oil from the platform, reducing the BS&W should improve market value of the product and maximise returns on transportation tariff. At present company targets for BS&W are set at 0.25%. This projected target is being achieved at the Platform B, and Platform C. On the Platform A hardware problems (problems with the electrostatic coalescer) and chemical formation and injection problems (problems with the formation of solid calcium napthanate) are currently affecting the BS&W target. However, company representatives believe that they have determined the source of the problem and can bring it under control.

NGL Yield: the efficiency of product extraction or recovery from reservoir fluids cannot be claimed to be optimal in the case of NGLs. The quality of the NGL is determined by its chemical composition (the proportion of propane, butane, dry gas, etc., and waste carbon dioxide). The fractions of each of these NGL components recovered from the gas stream are influenced by the temperature and pressure conditions on the platform (particularly in the NGL / Refrigeration systems). The NGL recovered on the platforms is exported by pipeline to onshore processing plants. There is a transportation tariff per km of pipeline as well as a processing tariff per tonne of NGL sent to the processing plant. There is also a penalty charge for carbon dioxide contents exceeding a certain level. Each of these NGL fractions has a unit price that may vary from month to month. These prices become known at the 1st of each month and the NGL produced for a given month is sold on the 15th of each month.

The composition of gas re-injected into the reservoir also influences the composition of the NGL stream leaving the reservoir and entering the platform. There is about a 30-day delay (approximate) before the effects become apparent. The Company's economic department at present produces forecasts for likely prices for various NGL components in the near future and then appropriate steps are taken by the reservoir engineers to try and influence the composition of the NGL in the reservoir. There might be room for an expert system with predictive forecasting and filtering ability to improve this aspect of the operation.

- ***Reduction Of Losses Due to Plant Downtime***

A significant portion of the Company's loss of revenue from operations is due to non-availability of different platforms. Some of these losses are also due to process or control problems. The data from the monthly production report was analysed and the loss in production due to process problems trended. Figure 2.8 to Figure 2.11, show the most prominent causes of losses in production due to process problems for all three platforms over the eight month evaluation period. These process problems shown in the chart have affected the production quota more than six times. Some areas have been identified as recurring problems with a substantial contribution to losses.

Problem Areas on Platform A

1. Water treatment and handing facilities
2. Gas lift system
3. NGL system
4. LP and MP Separator control system on A and B train
5. Plant start up control system

Problem Areas on Platform A (Joint Development (J/D) Zone)

1. Slug-catcher control system
2. Booster pump control and monitoring system
3. Chemical injection and monitoring system
4. Plant start up control system
5. Water treatment and handing facilities
6. Riser pressure control system

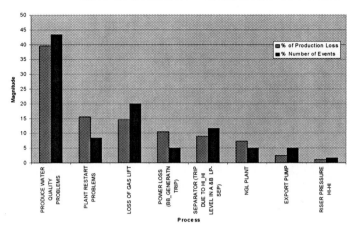

Figure 2.8. Platform A, production loss chart

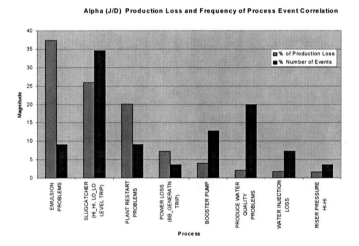

Figure 2.9. Platform A (Joint Development Zone), production loss chart

Problem Areas on Platform B

1. Refrigeration system
2. NGL plant
3. Power generation control and monitoring system

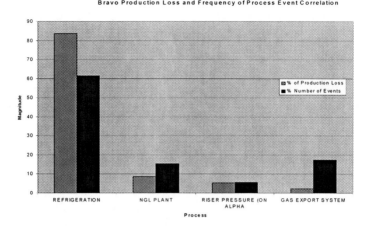

Figure 2.10. Platform B, production loss chart

Problem Areas on Platform C

1. HP Separator
2. LP and MP Separator control system on A and B train
3. Compressor control and monitoring system.

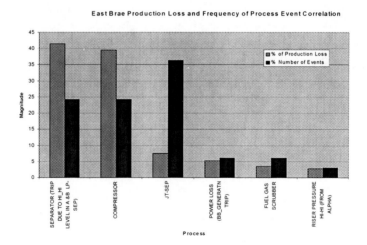

Figure 2.11. Platform C, production loss chart

- _**Analysis Of Present Control System and Strategy**_

Although the present control strategy is adequate in providing a safe, and to some extent, efficient means for production of the required product, it cannot provide the kind of efficiency or optimisation that the Company is looking for. PID controllers are by nature corrective controllers, they do not act until the system has been disturbed and a deviation from set-point has occurred. Sometimes the controller performance can be improved by including feed-forward or cascade action. These performance improvement measures are not in place at present. Analysis of the process and disturbance dynamics, showed that implementing a feed-forward or cascaded PID control strategies will not improve the control performance to the level required.

The PID controllers are tuned quasi-independently. The process is interactive, multivariable and has recycles, therefore, once all the controllers are switched to automatic, there will be some interaction between control loops. Independent tuning of PID controllers may limit the effectiveness of the required control actions and reduce the consistency of operation. As it is well known, these difficulties could be overcome by implementing an advanced (multivariable) control strategy. However, the motivation for implementation of advanced control is financially based. Therefore, changes to the existing system, where required, must be justified not only by the improved consistency of operation but, mainly, by the economics.

2.4.4 ICPRO Step 4: Optimality Assessment

The results from the audit make it possible to conclude that:

- Present control strategy is adequate but not optimal.
- Present strategy is purely based on classical control and cannot be easily adapted to meet and respond to future company process goals and operational efficiency.
- Present PID controllers are too sluggish in responding to disturbances to platform operating points.

Research on the application of model predictive control to industrial processes has shown that advanced process control techniques can enhance the performance of complex processes in petrochemical and process plants as found on the production platforms [Clarke, 1988 and 1991; Cutler and Ramaker, 1980; Richalet 1993; Schley *et al.* 2000]. Given the dynamics of the processes on the platform, then to improve platform operation and optimise revenue flow, a new supervisory level of integrated control structure is needed.

An advanced controller can be designed to continuously optimise plant operation on an economic basis according to operating conditions that prevail at any point in time. An advanced controller will reduce trips through improved disturbance rejection. From analysis and evaluation of the Company's economic targets and platform operation, a number of process areas can be targeted for an enhancement in control operation. This enhancement in the form of an upgrade from the present control strategy to an advanced one will provide positive financial benefits.

- *Recommendation For Enhanced Reservoir Management and Improved Reservoir Fluid Recovery*

With advanced control strategies and modelling techniques, the reservoir models can be integrated with the flow-line and topside models. Once an integrated model is obtained, a mathematical representation of the full multivariate system can be deduced. Using this mathematical representation, the aim would be to develop a criterion for optimising well scheduling, WAG injection, gas lift and gas re-injection such that the reservoir fluid recovery is maximised while minimising well decline. From this criterion an integrated predictive control system plus an expert decision making system can be designed. Such a control system will ensure that at any time the re-injection rates and well schedules will be optimal and recovery of reservoir fluids and input flow rates maximised. There exist a number of reservoir and well process parameters that are being monitored in real-time at present. These parameters can be used for control feedback, and other required parameters that cannot be measured directly can be inferred. Although such an advanced control and decision making system would provide large financial benefits, designing and implementing it would require a substantial amount of time, engineering and research effort plus many hours of input from the Company's personnel. Therefore it is not a strategy that can be implemented in the short term, but can be considered as a long term control development strategy.

- *Recommendations For Enhancing Black Oil and NGL Financial Yield*

There might be some advantages in introducing some form of advanced control and monitoring into the crude oil separation process. Using advanced control strategies, the efficiency of the separation process can be improved by optimising control set points and improving the control action to ensure constant production efficiency. The Company spends a substantial amount of money each year in procuring the chemicals needed for the chemical injection process. Chemical injection is necessary to deal with the situation that arises due to the nature of the process. Some of these chemicals and the situations that necessitate their use include:

- Emulsifiers: used to help with the de-emulsification process. The formation of emulsion during the separation process affects black oil quality,
- Scale Inhibitors: used to prevent scaling in separators and pipelines. Scale formation can partially/completely block pipelines hindering the flow of black oil and possibly causing plant shutdown,
- Acetic acid: on Platform C, acetic acid is used to counter the effect of sodium napthanate.

Monitoring targeted variables or indicators and setting up a control procedure to handle the rate and amount of chemical injection could significantly enhance the chemical condition monitoring and injection process. The incremental revenue that would accrue from implementing such strategies would be gained from improved product quality and reduction of lost production time. Lost production time occurs due to faults associated with the problem. Additionally, revenue would be saved through reduced chemical acquisition costs. However, since the company targets for the BS&W are already very low and are mostly being met, the amount of revenue generated by optimising these processes with advanced control will not be substantial. For the NGL using advanced control strategies, the 15 day window between price determination, production, extraction and sale can be used to optimise temperature and pressure control set-points (once the individual NGL component prices are known) to recover the optimal proportion of NGL components that maximise the revenue.

A spin-off from such a set-point optimisation strategy will be more efficient NGL recovery that should result in minimal carbon dioxide content; this should:

- Maximise returns on transportation tariff
- Maximise returns on processing tariffs
- Reduce the amount paid out as carbon dioxide penalty charges.

An example of the optimisation strategy is given below. The revenue can be calculated from the equation:

$$R = I \times a + J \times b + K \times c - C \qquad (2.1)$$

where:
a = unit price of Propane per tonne
b = unit price of Butane per tonne
c = unit price of C_5 per tonne
I = % proportion of Propane in recovered NGL
J = % proportion of Butane in recovered NGL
K = % proportion of C_5 in recovered NGL
C = Cost of operating the refrigeration system.
R = Total revenue received per tonne of NGL

I, J and K depend on the temperature (T) and the pressure (P). To obtain a formula that can be used in deriving the optimal set points for the process controllers I, J and K should be expressed in terms of T and P and substituted into Equation (2.1). Given a, b and c and the process constraints (not listed here), Equation (2.1) can be optimised for temperature and pressure set-point values that maximise R. Using such a simple optimisation criterion, an advanced controller can ensure that the recovery of NGL fractions is optimal at any time once the individual prices are known. This places the operating point of the NGL system in the optimal region.

- ***Recommendation on reducing plant downtime***
The process control audit showed that production platform trips due to separator control were responsible for over 60% of the combined production loss due to down time. A review of the PID control set-up for the level control of the separators indicated that a re-tune was necessary. Further evaluation of the separator systems showed that the process trips could be attributed to the level controller in the separator. Because the audit highlighted the Separator's level PID

control loop as a key target for reducing production losses, the loop was chosen as candidate for performance benchmarking analysis. The results of the analysis on the crude oil separation system are presented in the rest of this section.

2.4.5 ICPRO Step 5: Control System Adaptation

As discussed earlier, simulation is used to assess effects of improved control action. Firstly, this example highlights the consequences and problems that result when a proper plant audit is NOT carried out before benchmarking and optimising plant control loops (snapshot optimisation). Secondly, the function of plant auditing in prioritising the control loops for optimisation, in order to attain management level objectives is recalled, and the exercise is repeated, this time leading to performance improvement

2.4.6 Process Characteristics

The inflow of reservoir fluids into the separation train can be described as oscillatory with high amplitude and can be modelled as a sinusoidal disturbance. The PID control system associated with each separator is tasked with keeping the level in the vessel constant. However because of the sinusoidal nature of the input flow of reservoir fluid, an existing PID solution did not meet the requirements. The fluctuations in level and pressure within the first stage (HP) separator resulted in trips and shutdown of the entire platform. A MATLAB® / Simulink® model of the first two stages of the separation system (high pressure (HP) and medium pressure (MP) separators) was developed and validated with real plant data. A scaled down model from the process characteristics was obtained by linearizing this model around normal operating conditions and using balanced model reduction techniques.

   Figure 2.12 shows a simplified schematic of the process, from which a simplified mathematical model can be developed using equations for conservation of mass and pressure balance. The generic process transfer function in Equations (2.2) to (2.5) were developed from that model on the basis of the relationship between the valve position (manipulated variable) and the level of crude oil in the vessel (controlled variable).

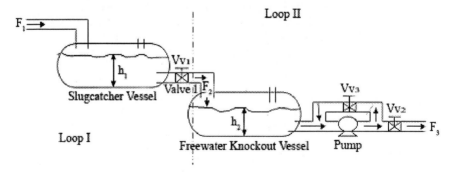

Figure 2.12. Schematic diagram for simplified 2 stage separation process

The key process assumptions, used in the model derivation are as follows:
- The gas entering the vessel along with the crude does not affect the equilibrium balance of the system.
- The vessel is rectangular with a flat base, it has a constant cross sectional area.
- Flow of crude from the vessel is laminar and the friction in the valve and pipes is negligible.
- The valves have linear characteristics.

  Notation:
  A_1 and A_2 = cross sectional area of vessel
  F_1 = Input flow into vessel 1 (slugcatcher vessel)
  F_2 = Output flow from the slugcatcher into freewater knockout
  h_1 = height of crude in slugcatcher
  F_3 = Output flow from the freewater knockout
  h_2 = height of crude in the freewater knockout vessel
  Vv_2 = hydraulic conductance of valve
  M = pump characteristics
  Vv_3 = constant valve position
  Vv_1 = hydraulic conductance of valve
  ρ = density of crude
  g = acceleration due to gravity

- Open Loop Diagram for Loop 1

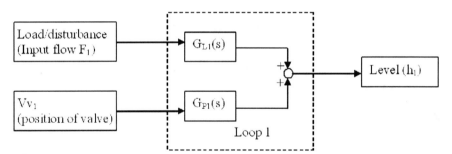

Figure 2.13. Block diagram for Loop 1: slugcatcher vessel

$$G_{P1} = \frac{\rho g h_1}{A_1 s + \rho g V v_1} \tag{2.2}$$

The input flow to the system (F_1) appears as a load/disturbance variable to the system. The transfer function between inlet flow as input and the level of crude in the vessel as output is:

$$G_{L1} = \frac{1}{A_1 s + \rho g V v_1} \tag{2.3}$$

- Open Loop Diagram for Loop 2

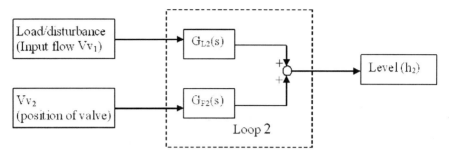

Figure 2.14. Block diagram for Loop 2: freewater knockout vessel

$$G_{P2} = -\frac{\rho g C h_2}{A_2 s + \rho g C V v_2} \qquad (2.4)$$

The valve position (Vv_1) appears as a load/disturbance variable to the system .
The transfer function between inlet flow as input and the level of crude in the vessel as output is:

$$G_{L2} = -\frac{\rho g h_1}{A_2 s + \rho g C V v_2} \qquad (2.5)$$

From the generic equations described above, a process model was built in Simulink® and optimised and calibrated using the peak input flows data collected from the real plant. The input flow disturbance to the separators is shown in Figure 2.15 and the response of the level loops in the separator is shown in Figure 2.16 and Figure 2.17. From Figure 2.16, it can be observed that the level in Loop 1 does not meet the high-level trip constraint (dotted line). This resulted in a number of plant shut downs and revenue loss.

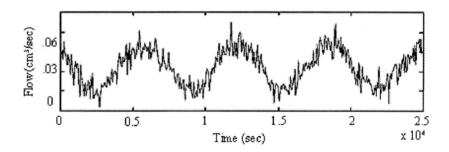

Figure 2.15. Crude oil input flow into separation system

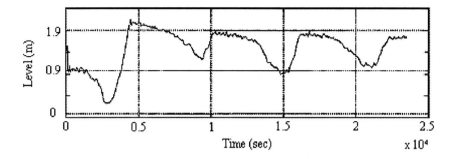

Figure 2.16. HP separator level

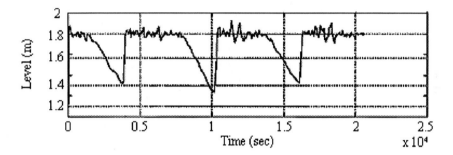

Figure 2.17. MP separator level

2.4.7 Snapshot Benchmarking and Optimisation

This approach is identical to taking a snapshot of the process at a given time. The performance of the control loops is then analysed independently, using a suitable benchmark index. In this approach no consideration is given to the overall process and management goals. Also, heuristic and knowledge based information about the process, acquired over time, is not considered.

Benchmark Analysis
A local loop performance analysis of the individual level control loops using 1000 samples was undertaken without taking the interaction between the loops and the overall process goals into account. This analysis was performed by using the normalised minimum variance control benchmark index [Desborough and Harris, 1993], to determine the performance of the control loops in the validated process model. The details of this performance index will be explained in Chapter 3. The minimum variance controller was used as a benchmark and for this, the plant performance index is given by:

$$J = \sigma_y^2 = J_0 + J_{min} \qquad (2.6)$$

$$\sigma_y^2 = E[y^2(t)] \qquad (2.7)$$

where: J is the actual output variance, J_0 is the part of the output variance which could be affected by selection of control algorithm, and J_{min} is the minimum-variance obtainable for the given plant. The normalised minimum variance index:

$$\eta = \frac{J_0}{J} = 1 - \frac{J_{min}}{J} \tag{2.8}$$

lies between 0 and 1, where 1 indicates minimum variance control (excellent control performance) and 0 indicates a very poor control. The graphical results for the two loops are shown in Figure 2.18 and Figure 2.19. From these graphs it can be deduced that Loop 1 is very far from minimum-variance (optimal) performance and therefore poorly tuned while Loop 2 is performing better.

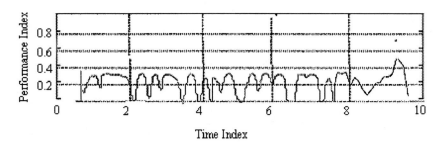

Figure 2.18. MV benchmark index for HP separator control (Loop 1)

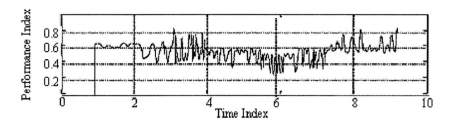

Figure 2.19. MV benchmark index for MP separator control (Loop 2)

Loop Tuning
Because both level loops are first order systems, an analytic solution to arrive at the proportional (Kc) and integral (Ti) PI controller parameters to meet the specification was used. The controller parameters were calculated using the following assumptions:

1. The damping coefficient to be used is $\zeta = 1$. The aim is to make the response of the system critically damped, so that the system response is not oscillatory and the disturbance introduces as little effect as possible during the transient period.

2. The expected maximum deviation in inlet flow is $\Delta F_{MAX} = 0.06$ m³/sec (3.6m³/min).

3. For Loop 1, the steady state level is specified as 0.91m. A 10% deviation (conservative) from specified steady state operating points during transient disturbances is assumed, hence $\Delta h_{MAX} = \pm 0.091$ m

4. For Loop 2, the steady state level is specified as 1.71m. A 5% deviation (conservative) from specified steady state operating points during transient disturbances is assumed, hence $\Delta h_{MAX} = \pm 0.0855$ m

5. The cross section areas of the vessels are specified as $A_1 = 1.75$ and $A_2 = 3.4$

This resulted in the PID parameters,

$$\begin{aligned} K_{C1} &= 29.1165 \quad and \quad T_{I1} = 0.2335 \text{ min} \\ K_{C2} &= 15.4947 \quad and \quad T_{I2} = 0.8777 \text{ min} \end{aligned} \qquad (2.9)$$

Results of Re-tuning the System
Following the findings from the previous section, the system was retuned and the new benchmark results for the tuned system, as well as the levels in the vessels can be observed in Figure 2.20 to Figure 2.23.

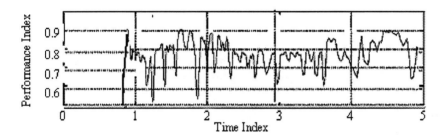

Figure 2.20. MV benchmark index for re-tuned HP separator (Loop 1)

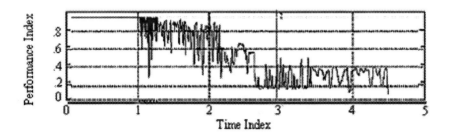

Figure 2.21. MV benchmark index for re-tuned MP separator (Loop 2)

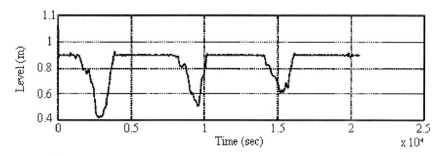

Figure 2.22. Re-tuned HP separator level (Loop 1)

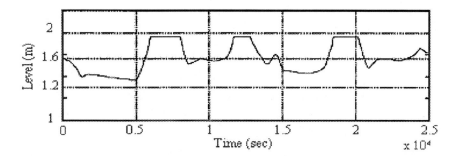

Figure 2.23. Re-tuned MP separator level (Loop 2)

From Figure 2.20 and Figure 2.21 which represent the performance index for the re-tuned loops, it can be observed that while the performance of Loop 1 has improved, the performance of Loop 2 has deteriorated. This is due to the fact that, since the control performance of Loop 1 is improved, the oscillatory dynamic disturbance of the input flow into the system is amplified and transmitted downstream to Loop 2. The effect of this can be seen from Figure 2.22 and Figure 2.23. It can be observed that Loop 1 is now meeting the high-level trip constraint while Loop 2 is breaking it. Thus the overall effect of the controller tuning effort was to shift the cause of the process trips from Loop 1 to Loop 2, with the net result that revenue will still be lost due to plant downtime in periods of peak disturbance.

2.4.8 Integrated Plant Auditing and Benchmarking

The integrated control and process revenue optimisation approach requires that before any benchmarking analysis, control design etc, is done, in-depth plant auditing involving management, process and control objectives should be carried out. The results of the audit presented earlier are now applied to the problem with the following observations:

- Management Objectives
  - The main management objective is to maximise production. The input flow oscillatory disturbance must be controlled and not transmitted downstream to other process units. A high process up time as well as the optimum separation conditions within the vessel is the target.
  - Reduction in plant down time.
    The problems caused by this sinusoidal disturbance are not only related to process control of the level, and pressure loops, but also to the platform revenue. These fluctuations in the level and pressure in the separation cause the entire plant to trip resulting in lost production and hence loss of revenue.
  - Increase in Production Rate.
    By improving set-point tracking of controlled variables (i.e. controller performance), set-points can be optimised and the process operating conditions moved closer to the constraints and production rates safely increased.
- Process and Control Objectives
  - Stabilise the flow of crude oil downstream of the HP separator.
  - Maintain the pressure and volume of crude oil in the separators at a level that ensures efficient separation.

In deciding on a criterion that will best achieve these goals, an analysis of the problem and process characteristics led to the following conclusions.

1. To stabilise the flow down stream of the separators, the volume of crude oil in the HP separator should be used as a buffer/surge control.

2. As long as the level in the HP separator is maintained within the constrained limits, adequate separation will be ensured. With the flow stabilising downstream of this vessel, the other separators will be able to perform better.

An obvious solution is to design two flow optimizing cascade controllers for the two individual loops. The cascade controllers will take measurements of the outlet flow from the loops and adjust the set points of the level controllers in the loops to compensate for any disturbance in the desired value of outlet flow. The level set points for both controllers will act, as the manipulated variables while the outlet flow from both loops will be the controlled variables. The level in both vessels will be used as surge capacity to compensate for periods of very low or very high flow rates. There are reasons however why such a design will not achieve the required performance objective and that necessitates a foray into the uses of more advanced control strategies.

Limitations of Standard PI Cascade Control Strategy.
In the cascade control structure the level loops will act as the secondary control system with the levels in both vessels serving as the measured secondary variable, while the primary system will be the flow loop.

Table 2.3 summarises the time and frequency domain characteristics for the primary and secondary systems.

Table 2.3. Loop characteristics

Time Domain Characteristics	Primary System		Secondary System	
	Loop 1	Loop 2	Loop 1	Loop 2
Rise time	120.46 sec	226.98 sec	12,455.28 sec	6455.28 sec
Settling time	205.67 sec	405.60 sec	17,850.47 sec	8000 sec
D.C. gain	0.426	0.0651	-1027	-872.30
Time constant	52 sec	114 sec		

Next, the standard design criteria for a cascade control system are examined:

a)   There must be a causal relationship between the manipulated variable and the secondary variable.
There is a causal relationship between the measured levels in the tanks and the manipulated variables (which are the level set-points). Thus, this criterion is satisfied.

b)   The secondary variable must indicate the occurrence of an important disturbance.
The measured secondary variables in the process are the levels in both vessels. The major disturbance to level Loop 1 is the sinusoidally changing input flow. Changes in the measured level in Loop 1 give an indication that the disturbance has changed value. While in Loop 2 the major disturbance is the valve position Vv_1. Changes in this valve position cause oscillations in the magnitude of crude oil flow into Loop 2, resulting in changes in the level of the vessel in Loop 2. Thus, this criterion is satisfied.

c)   The secondary variable dynamics must be much faster than the primary variable dynamics.
From Table 2.3 it can be observed that the measured secondary variables fail to meet this condition. Since both vessels are effectively integrators at steady state, changes in the input flow to the system are almost immediately reflected in the output flow. The dynamics of these secondary variables are not much faster than of the primary variables.

Implementing a cascade control structure under this conditions will not yield any really meaningful improvement in the process control performance and hence will not result in an optimized value of output flow. Since the standard cascade PI control structure is not suitable for this process, there was a need to explore more advanced control strategies. A better solution is to use a model predictive controller to implement the cascade solution since for cases when the dynamic response of the secondary system is not substantially faster than the primary, the predictive primary cascade controller offers a distinct advantage. The benefit of the predictive cascade arises because the feedback signal in a model predictive control system is the sum of the model error in the primary loop and the primary loop disturbances along with the fact that the secondary disturbances that cause deviations in the secondary measurement, appear in both the measured and predicted primary variable at about the same time and with the same magnitude (if the model is accurate), then as a result, the secondary disturbances have little or no effect on the feedback signal The model predictive controller chosen to design the cascade structure was the Internal Model Controller (Figure 2.24) developed by Morari and Garcia [1982]. The appealing feature of the IMC is that it provides a systematic approach for designing robust controllers that provide good control performance while compensating for modelling errors and usually involves only a single tuning parameter, which can be related to the desired closed loop time constant. The controllers are easy to design and are sometimes realizable in standard PID forms.

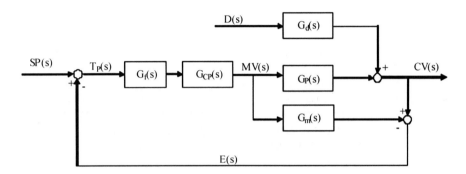

Figure 2.24. Structure of the IMC controller

Key:
SP(s)	Reference set-point	CV(s)	Controlled variable
D(s)	Disturbance variable	T$_P$(s)	Target Value
MV(s)	Manipulated variable	E(s)	Feedback signal
G$_{CP}$(s)	Predictive controller transfer function	G$_P$(s)	Process transfer
G$_d$(s)	Disturbance transfer function		function
G$_m$(s)	Process model	G$_f$(s)	Filter

The feedback signal E(s) is the difference between the measured and predicted controlled variable values. The variable E(s) is equal to the effect of the disturbance $G_d(s)D(s)$, since if the model is perfect $G_m(s) = G_p(s)$. This means that if the model is perfect then the predictive control would acts predominantly on the disturbance for feedback correction and not the combination of disturbance and errors due to model mismatch. In single loop IMC design, the convention is to use low pass filter of the form

$$G_f(s) = \left\{\frac{1}{\tau_r s + 1}\right\}^N \tag{2.10}$$

The filter time constant, τ_f is the only parameter that has to be tuned to achieve any performance specification. Increasing the filter time constant modulates the manipulated variable fluctuations and increases robustness at the expense of larger deviations of the controlled variable from its set-point. From the given process transfer functions, the IMC filter time constants for Loops 1 and 2 can be calculated as:

$$\tau_{f1} \geq 10.4 \text{ secs}$$
$$\tau_{f2} \geq 22.8 \text{ secs}$$

Because inevitably there must be an error in the determined models, a safety margin is included in the realization of the filter time constant, so:

$$\tau_{f1} = 10.4 \times 2 = 20.8 \text{ sec}$$
$$\tau_{f2} = 22.8 \times 2 = 45.6 \text{ secs}$$

Analysis of IMC Controller Performance
The performance of the IMC controllers in a supervisory role was simulated and the data obtained used as a performance benchmark. It is usually a good practice to compare system performance benchmarks before and after process optimisation. However, this does not apply to this particular exercise. Because the high level objective is to ensure stable and nearly constant flow rate, benchmarking the performance of the separator level controllers will not provide a useful indicator. From the results and data obtained from the simulations, as shown in Figure 2.25 and Figure 2.26, the estimate is that, a 15-25 % reduction in the variation of crude oil flow rate downstream of the HP and MP separator units, can be achieved by using model based control systems in a supervisory mode.

The reduction in flow rate variations will decrease the amplitude of the disturbance experienced in the level loops of other separator units. The number of process trips caused by variations in flow rate should also be reduced. This is because the flow rate trip set-point has a value of 0.057 m³/sec and as can be observed from Figure 2.25 and Figure 2.26, the IMC controller keeps the flow rate between the bounds of 0.03 ± 0.01 m³/sec.

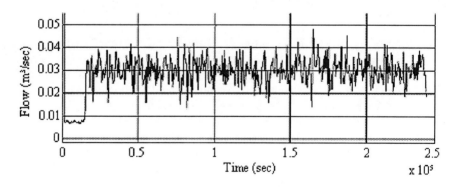

Figure 2.25. Crude oil output flow from HP separator

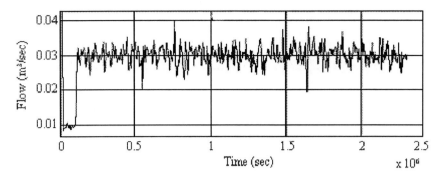

Figure 2.26. Crude oil output flow from MP separator

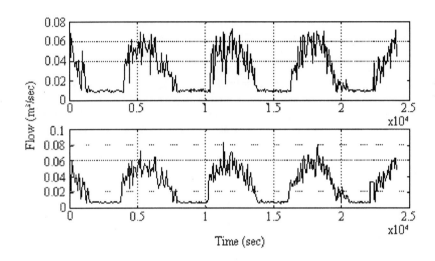

Figure 2.27. Initial crude oil output flow from HP and MP separators

For separator vessels downstream of the HP separator, a reduction in level control variations of about 10 % was recorded as shown in Figure 2.28. It is estimated that this reduction will not only reduce process shutdowns, due to the separator level trips, but also introduce the possibility of pushing the operating conditions of the separators closer to their physical constraints. This should have a significant impact in reducing the revenue lost due to plant downtime and should also enable production rates to be increased.

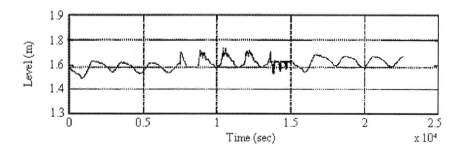

Figure 2.28. MP separator level

2.5 Conclusions

In this Chapter, the connection between process control performance and the revenue derived from industrial processes was highlighted. A method for integrating the use of process control benchmarking and optimisation algorithms with the optimisation of process revenue by means of an economic process control audit was developed. The goal of the method was to selectively benchmark and optimise process control loops in such a way as to derive maximum revenue from the process. The method was demonstrated by means of an industrial feasibility study. For large scale processes such as an integrated crude oil and gas production facility, it was possible to use this method to highlight potential areas where advanced control optimisation could be of substantial financial benefit. The method was also able to identify some process control loops which had substantial impact on process performance and revenue. For large scale processes and processes with interaction, it was shown that benchmarking and optimising individual loops without consideration for the wider objectives of the entire process can in some circumstances have a negative impact on overall process performance. As demonstrated the ICPRO approach can help focus process control optimisation and benchmarking, in conjunction with the aim of improving overall process performance for financial gain.

3

Controller Benchmarking Procedures – Data-driven Methods

Damien Uduehi [1], Andrzej Ordys [2], Michael Grimble [3], Pawel Majecki [4] and Hao Xia [5]

[1] Strategy and Planning Coordinator (West Africa), BG Group, Reading, UK,
[2] Professor of Automotive Engineering, Faculty of Engineering, Kingston University, London, UK,
[3] Professor and Director, Industrial Control Centre, Department of Electronic and Electrical Engineering, University of Strathclyde, Glasgow, UK
[4] Research Engineer, Department of Electronic and Electrical Engineering, University of Strathclyde, Glasgow, UK
[5] Research Fellow, Department of Electronic and Electrical Engineering, University of Strathclyde, Glasgow, UK

3.1 Introduction

The industrial importance of individual controller benchmarking is underpinned by the fact that there are often thousands of loops in a typical process plant and many of these will not have been tuned adequately. The market competition makes it very desirable for everyone to obtain the best from their process control system. Also, the investment in Supervisory Control and Data Acquisition (SCADA) systems, which companies may make, only makes sense if the individual plant controllers are properly tuned.

Desborough and Harris [1992, 1993] considered the assessment of control loop performance for both feedback and feedforward control using minimum variance as the benchmark cost measure. Huang and Shah [1999] summarised the state of the art in a monograph, mostly focusing on the minimum variance cost index as the performance assessment measure. Following that, over the last decade, there has been growing interest in controller performance benchmarking, often learning from the business process benchmarking community [Codling, 1992; Rolstadas, 1995; Levine, 1996; Anderson and Petersen, 1996; Ahmad and Benson, 1999]. Benchmarking techniques are also important in other industrial sectors, such as the power generation and transmission industry [Calligaris and Johnson, 1999, 2000].

In this chapter, the concept of Minimum-Variance benchmarking is introduced. Following the discussion of the basic ideas, extensions of the minimum-variance benchmark will be presented. These include Generalized Minimum-Variance and Model Predictive Control benchmarks.

3.2 Introduction to Minimum Variance Control Benchmarking

The minimum variance criterion has some value as a benchmark since the economic performance of a process is often governed by how close a set-point can be moved to an operating boundary (such as a temperature limit). This idea is illustrated in Figure 3.1: if, for example, the operating boundary represents a raw material boundary, then a decrease in the variance may result in material savings and increased revenue.

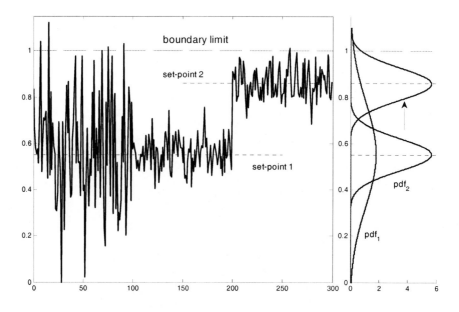

Figure 3.1. Moving the set-point closer to the boundary as a result of the reduction in variance

Historically, Minimum Variance Control (MVC) has been introduced as a simple, practical control strategy resulting from the application of linear stochastic control theory. The control algorithm was first formulated in [Åström, 1967]. Åström used the MVC technique to minimize the variance of the output signal for control of paper thickness in a paper machine. The control objective was to achieve the lowest possible variation of the paper thickness in the presence of stochastic disturbances acting on the process. Since then, the MVC technique has attracted many practical applications [e.g. Flunkert and Unbehauen, 1993; Kaminskas *et al.*, 1992] and has gained significant theoretical development leading eventually to a wide range of control algorithms known as Model Predictive Control techniques.

   Harris and Desborough [1992] introduced a normalised performance index for the assessment of controller performance against a benchmark of minimum variance control. This index can be estimated using the least squares method. The implementation of the controller performance index (CPI) based on approach of Harris and Desborough consists of fitting an auto-regressive (AR) time series model to closed-loop plant data. The theory of minimum variance control and benchmarking can be summarized as follows.

3.2.1 Minimum Variance Control

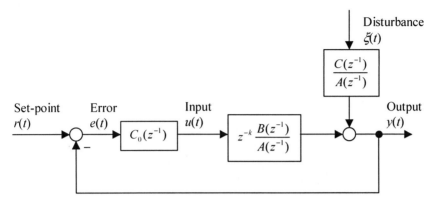

Figure 3.2. Generic stochastic feedback control system structure

Considering the generic control system shown in Figure 3.2, the minimum-variance control assumes a single-input, single-output (SISO), linear, time-invariant stationary stochastic process described by:

$$A\left(z^{-1}\right)y\left(t\right)=z^{-k}B\left(z^{-1}\right)u\left(t\right)+C\left(z^{-1}\right)\xi\left(t\right) \tag{3.1}$$

where $y(t)$ represents variation of the output signal around a given steady-state operating point, $u(t)$ is the control signal, $\xi(t)$ denotes a disturbance which is assumed to be zero mean, Gaussian white noise of variance σ^2. The constant set-point $r(t)$ is for simplicity assumed to be zero, with the more general case considered in Section 3.2.4. $A(z^{-1})$, $B(z^{-1})$, $C(z^{-1})$ are polynomials in the backward shift operator z^{-1} and the roots of $C(z^{-1})$ and $B(z^{-1})$ are assumed to lie within the unit circle, where:

$$A\left(z^{-1}\right)=1+a_1z^{-1}+a_2z^{-2}+...+a_nz^{-n_A} \tag{3.2}$$

$$B\left(z^{-1}\right)=b_0+b_1z^{-1}+b_2z^{-2}+...+b_nz^{-n_B} \qquad b_0\neq0 \tag{3.3}$$

$$C\left(z^{-1}\right)=1+c_1z^{-1}+c_2z^{-2}+...+c_nz^{-n_C} \tag{3.4}$$

$$z^{-1}\cdot x\left(t\right)=x\left(t-1\right) \tag{3.5}$$

Thus z^{-k} in Equation (3.1) represents a k-step delay in the control signal ($k\geq1$). This means that the control signal starts to act on the system after k time increments. The plant is controlled by a linear feedback controller $C_0(z^{-1})$:

$$u(t)=-C_0y(t) \tag{3.6}$$

where the argument z^{-1} has been omitted for convenience.

The output at time $t+k$, using the process model can be written as:

$$y(t+k) = \frac{B}{A}u(t) + \frac{C}{A}\xi(t+k)$$
(3.7)

The following Diophantine equation can be defined to, effectively, split the polynomial C into two time frames, one related to the past values of ξ and the other related to the future values of ξ:

$$C = AF + z^{-k}G$$
(3.8)

where F and G are polynomials defined as:

$$F(z^{-1}) = 1 + \sum_{i=1}^{k-1} f_i z^{-i}, \qquad G(z^{-1}) = g_0 + \sum_{i=1}^{n_G} g_i z^{-i}$$
(3.9)

with $n_G = \max(n_C - k, n_A - 1)$. Substitution into (3.7) yields:

$$y(t+k) = \frac{BF}{C}u(t) + \frac{G}{C}y(t) + F\xi(t+k)$$
(3.10)

resulting in the k-step ahead predictor:

$$\hat{y}(t+k \mid t) = \frac{BF}{C}u(t) + \frac{G}{C}y(t)$$
(3.11)

The cost index to be minimised is the variance of the output at time $t+k$ given all the information up to time t:

$$J(t) = E\left[y(t+k)^2 \mid t \right]$$
(3.12)

where $E[.]$ denotes the expectation operator. It is assumed that the control selected must be a function of information available at time instant t, i.e. all past control signals, all past outputs and the present output:

$$Y(t) = \left[u(t-k-1) \quad u(t-k-2) \quad \ldots \quad y(t) \quad y(t-1) \quad \ldots \right]$$
(3.13)

Note that the first two terms on the right-hand side of Equation (3.10) are uncorrelated with the third term as they refer to mutually exclusive time frames. Therefore, the variance of the output may be obtained as:

$$\sigma_y^2 = E[y^2(t+k)] = E[(\frac{BF}{C}u(t) + \frac{G}{C}y(t))^2] + E[(F\xi(t+k))^2]$$
(3.14)

$$J_{\min} = E[(F\xi(t+k))^2] \quad \text{and} \quad J_0 = E[(\frac{BF}{C}u(t) + \frac{G}{C}y(t))^2]$$
(3.15)

$$J = \sigma_y^2 = J_{\min} + J_0$$
(3.16)

where the variance J_{min} is independent of the control action and the variance J_0 is dependent on the process control law. To achieve the minimum variance of the output variable, set:

$$\frac{BF}{C}u(t) + \frac{G}{C}y(t) = 0 \tag{3.17}$$

$$u(t) = -\frac{G}{BF}y(t) \tag{3.18}$$

Substituting the minimum variance controller in Equation (3.18) into Equation (3.10) yields

$$y(t+k) = F\xi(t+k) \tag{3.19}$$

Minimum variance control may be viewed as a stochastic equivalent of deadbeat control in the sense that it places all closed-loop poles in the origin. This is achieved effectively by cancelling out the process dynamics, which explains why minimum variance control is usually aggressive and sensitive to process-model mismatch. Also, it needs a minimum-phase system for stability.

3.2.2 Minimum Variance Benchmarking

The output variance under minimum variance control can be calculated from Equation (3.15) as:

$$J_{min} = \sum_{i=0}^{k-1} f_i^2 \sigma^2 \tag{3.20}$$

For most systems, the controller performance does not match that of a minimum variance controller. That is,

$$\frac{BF}{C}u(t) + \frac{G}{C}y(t) \neq 0 \tag{3.21}$$

Therefore, in the expression defining the future output, $y(t+k)$, there is an extra term whose variance is nonzero. This term is a result of the controller not being a minimum variance controller:

$$y(t+k) = F\xi(t+k) + \hat{y}(t) \tag{3.22}$$

where $\hat{y}(t)$ is the part of the output due to suboptimality of the controller (function of time range $[-\infty, t-k]$), and $F\xi(t+k)$ is the k-step ahead prediction error (function of time range $[t-k+1, t]$). Because there is no correlation between $\hat{y}(t)$ and $F\xi(t+k)$, the variance of the output under non-minimum variance control can be computed as

$$\sigma_y^2 = E[(y(t+k))^2] = E[(F\xi(t+k))^2] + E[(\hat{y}(t))^2] \tag{3.23}$$

Denoting $E[(\hat{y}(t))^2] = \sigma_{soc}^2$, then

$$\sigma_y^2 = \sigma_{soc}^2 + \sigma_{mvc}^2 \tag{3.24}$$

where *soc* is the short term for suboptimal control and *mvc* for minimum variance control. The variance is increased by σ_{soc}^2 as compared with the minimum variance control case. The term σ_{soc}^2 vanishes under minimum variance control. In order to obtain a universal tool for comparing different systems, define the following controller performance index (CPI):

$$\eta(k) = \frac{\sigma_{mvc}^2}{\sigma_{soc}^2 + \sigma_{mvc}^2} \in [0,1] \tag{3.25}$$

The main advantage of this definition is that it provides a performance indicator that is normalised and bounded. This allows the engineer to see how close to minimum variance the system is controlled. The value $\eta(k) = 1$ indicates an ideal case of minimum variance control and $\eta(k) = 0$ the case of the worst control.

3.2.3 Estimation of the Controller Performance Index

In the previous section, the normalised performance index $\eta(k)$ was introduced. This section reviews a simple way to estimate $\eta(k)$ from routine closed loop process data using linear regression methods. This approach eliminates the necessity of solving a Diophantine equation or performing polynomial long divisions.

From Equation (3.10), the process output under feedback control is given by

$$y(t) = F\xi(t+k) + z^{-k}\left(\frac{G - BFC_0}{C}\right)y(t) \tag{3.26}$$

Since the closed loop is presumed stable, the second component in the sum in (3.26) can be approximated by a finite length autoregressive model, as in Equation (3.28):

$$y(t) = F\xi(t) + \alpha_1 y(t-k) + \alpha_2 y(t-k-1) + \ldots + \alpha_m y(t-k-m+1) \tag{3.27}$$

$$y(t) = F\xi(t) + \sum_{i=1}^{m} \alpha_i y(t-k-i+1) \tag{3.28}$$

where *m* is the autoregressive model length.

Running *t* over a range of values and stacking up similar terms yields:

$$\mathbf{y} = \mathbf{X}\alpha + \mathbf{F}\xi \tag{3.29}$$

with

$$
\mathbf{y} = \begin{bmatrix} y_n \\ y_{n-1} \\ \vdots \\ y_{k+m} \end{bmatrix} \quad \mathbf{X} = \begin{bmatrix} y_n & y_{n-k-1} & \cdots & y_{n-k-m+1} \\ y_{n-k-1} & y_{n-k-2} & \cdots & y_{n-k-m} \\ \vdots & \vdots & \ddots & \vdots \\ y_m & y_{m-1} & \cdots & y_1 \end{bmatrix} \quad \alpha = \begin{bmatrix} \alpha_1 \\ \alpha_2 \\ \vdots \\ \alpha_m \end{bmatrix} \tag{3.30}
$$

where n is the ensemble length.

Using linear regression to estimate autoregressive parameters $\{\alpha_i\}$ by fitting the recorded closed-loop data $\{y_1, y_2, \cdots, y_n\}$ to the model (3.28), the least squares solution results as:

$$
\hat{\alpha} = (\mathbf{X}^T \mathbf{X})^{-1} \mathbf{X}^T \mathbf{y} \tag{3.31}
$$

From Equation (3.29), the estimate of the minimum variance can be calculated as the residual variance:

$$
\hat{\sigma}_{mvc}^2 = \sum_{i=0}^{k-1} f_i^2 \sigma^2 = \frac{1}{n-k-2m+1}(\mathbf{y} - \mathbf{X}\alpha)^T (\mathbf{y} - \mathbf{X}\alpha) \tag{3.32}
$$

while the actual variance follows as

$$
\hat{\sigma}_y^2 = \frac{1}{n-k-m+1}\mathbf{y}^T\mathbf{y} = \hat{\sigma}_{soc}^2 + \hat{\sigma}_{mvc}^2 \tag{3.33}
$$

The least squares estimate for the normalised performance index $\eta(k)$, given in Equation (3.25), can finally be written as:

$$
\hat{\eta}(k) = \frac{n-k-m+1}{n-k-2m+1} \cdot \frac{(\mathbf{y} - \mathbf{X}\alpha)^T (\mathbf{y} - \mathbf{X}\alpha)}{\mathbf{y}^T\mathbf{y} + (n-k-m+1)\bar{\mathbf{y}}^2} \tag{3.34}
$$

where the mean square error was used rather than the variance, hence penalizing nonzero steady state errors.

It is also desirable to know how reliable this estimated value is. Therefore, looking at the variance, $E[\hat{\eta}^2(k)]$ gives an idea of the confidence levels of the estimated stochastic variable. A correct selection of n can be verified in a different way. If n is too small, then the estimation cannot be reliable and will be subject to the effect of outliers. In addition, the stochastic distribution of $\hat{\eta}(k)$ will not be normal as it is analytically predicted by assuming a large value for n. Therefore, firstly n is given a reasonable value and $\hat{\eta}(k)$ is estimated online. If the performance index has a high variance, then its plot will show a wide deviation around the average. In this case, n should be increased until the distribution has a reasonably low deviation.

However, for practical implementation, a proper value of n, large enough ($n \gg m$) for better averaging out the noise and small enough not to overload the computing power, should be chosen by the engineer to obtain a reliable estimation of $\hat{\eta}(k)$. It is imperative that the plant data used for control loop benchmarking is unfiltered and uncompressed, i.e. the data must be exactly identical to the feedback information utilised by the loop controller.

Based on the above results, the following procedure for controller performance assessment using the minimum-variance control benchmark can be detailed:

Procedure 3.1 Applying the Minimum-Variance Benchmark
Step 1:   Collect open-loop or closed-loop output data from the plant. The data should be raw, i.e. not be compressed or filtered.

Step 2:   Select a data set of $n = 500$ samples or more (preferably $n \geq 1000$) containing representative disturbances present in the system.

Step 3:   From prior knowledge, an open-loop experiment or a closed-loop estimation technique [Harrison and Qin, 2004] estimate the time delay in samples.

Step 4:   Choose the autoregressive model length m for parameter estimation. The typical values are $m = 5$ up to 30. In general, the model length depends on the sample time and should be chosen such that the system impulse response is fully captured.

Step 5:   Construct the vector **y** and matrix **X** according to Equation (3.30).

Step 6:   Fit the model to the data to obtain α_i parameters.

Step 7:   From Equation (3.34), calculate the Controller Performance Index.

Procedure End

Example 3.1 Minimum Variance Benchmarking
A simple example will be used to illustrate the benchmarking procedure. Consider a continuous-time, first order plant with the time constant $T = 2$ seconds and the time delay $\tau = 2$ seconds, whose transfer function is given by:

$$G(s) = \frac{e^{-2s}}{2s+1}$$

The process is subject to stochastic disturbance modelled by adding coloured noise to the plant output. Only the regulatory performance is considered and hence the set-point is assumed to be zero.

The plant is initially controlled by a PI controller with proportional gain $K = 2$ and integral gain $K_i = 0.25$ ($T_i = 4s$). At time $t_0 = 1000$ sec. the proportional gain is reduced to $K = 1$. The simulation results at the neighbourhood of the switching instant t_0 are presented in Figure 3.3. Also plotted in that figure are the autocorrelation functions computed based on 1000 samples collected separately before and after the controller change, with the sample time chosen as $T_s = 1$ sec. Note that this corresponds to the discrete time delay of $k = 1 + \tau/T_s = 3$ samples.

Clearly, for the initial controller settings, there are significant correlations beyond the time delay, which indicates poor performance (recall from Equation (3.19) that in the case of MV control there should not be any correlations beyond the time lag of $k = 3$ samples). After retuning the controller the autocorrelations die out quickly, indicating performance close to the optimum.

The performance can be assessed quantitatively using the MV benchmarking algorithm. Setting the autoregressive model length to $m = 10$, the minimum achievable variance has been estimated as 0.22, and the Controller Performance Indices for the two cases follow as $\eta(PI_1) = 0.31$ and $\eta(PI_2) = 0.81$. These values are relative to the minimum-variance controller of the same sampling time as data collection.

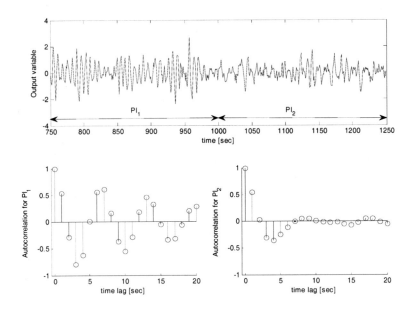

Figure 3.3. Control error and its autocorrelation function

The index of 0.81 indicates very good performance. It may still be possible to achieve a small improvement by retuning the controller, however it is hard to say if we have already achieved the best that can be achieved with the current controller. It is because the minimum variance control is usually unachievable by simple controllers like PID, particularly in the presence of a deadtime and when a plant is of higher order than in this simple example. Besides, it is often undesirable to achieve MV control ($\eta = 1$) as this can lead to excessive control action and poor robustness of the control loop. For these reasons alternative benchmarks have been proposed for assessing the performance of control loops.

3.2.4 The Case of Non-zero Set-point

In the discussion so far it has been assumed that the benchmarking data comes from regulatory loops where the control objective is to maintain the controlled variable at a certain constant set-point and reject any disturbances. Thus in the assumed linearized system model the set-point was implicitly set to zero, and only deviations from this nominal operating point were considered.

It may however be the case that the set-point changes occasionally during the data collection for the benchmarking experiment. In this case the error $e(t) = r(t) - y(t)$ can be used instead of the output, and Procedure 3.1 can be applied with no further changes. If the data set contains a large number of such set-point changes, then they will effectively be treated as an additional disturbance and the CPI will then reflect the combined regulatory and tracking performance. In general, it is important to ensure that the data used to calculate the benchmark contains an adequate representation of the typical disturbances acting on the system.

3.3 Introduction to Generalized Minimum Variance Control Benchmarking

The minimum variance criterion is valuable as a benchmark, since it gives the absolute limit on the performance achievable by any linear controller. Hence, if a loop has scored a high MV benchmark index, and the output signal variations are still unacceptable, performance cannot be improved by simply retuning the controller – changes in the controller structure (e.g. introducing feedforward action) are then necessary.

However, there are real problems in implementing minimum variance controllers, since they often result in wide bandwidth, large noise amplification and lead to very aggressive control action. Under normal circumstances such controllers would result in saturation and excessive wear and tear on the actuators. Another limitation is the inability of MV control to handle non-minimum phase systems.

3.3.1 Generalized Minimum Variance Control

Clarke and Gawthrop [1975b, 1979] developed the Generalised Minimum Variance Controller (GMVC) for self tuning control applications by introducing the control signal and dynamic weighting functions into the performance index.

Assume plant description as for the MV case (Equation (3.1)) where the set-point $r(t)$ is for simplicity assumed to be zero. The GMV controller minimises the variance of an auxiliary output of the form $\phi(t) = -P_c y(t) + F_c u(t)$ as shown in Figure 3.4.

The following performance index is therefore minimised subject to the process dynamics:

$$J(t) = E\left[\phi^2(t + k|t) \right] \tag{3.35}$$

where $P_c = \dfrac{P_{cn}}{P_{cd}}$, $F_c = \dfrac{F_{ck}}{F_{cd}}$ are stable dynamic weighting functions. Normally the weighting P_c should be chosen as an integral term, which leads to integral action in the controller. The control weighting can be chosen as a constant or as a lead term to ensure the controller rolls-off appropriately.

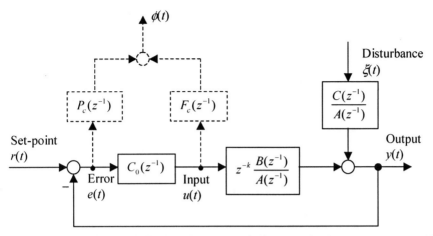

Figure 3.4. Conceptual Generalised Minimum Variance stochastic system structure

The control action $u(t)$ affects the output with a k step delay, and hence the control weighting is defined to include the delay, i.e. $F_c = z^{-k}F_{ck}$. The 'generalised' output k steps into the future can then be represented as:

$$\phi(t + k) = -P_c(z^{-1})y(t + k) + F_c(z^{-1})u(t) \qquad (3.36)$$

A simple way of interpreting the GMV control problem is to reformulate it as an MV problem for a generalized plant. The last equation can be rewritten as follows:

$$\phi(t + k) = -P_c y(t + k) + F_c u(t + k) = P_c(-\frac{B}{A}u(t) + \frac{C}{A}\xi(t + k)) + F_{ck}u(t)$$

$$= \frac{P_c C}{A}\xi(t + k) + (F_{ck} - \frac{P_c B}{A})u(t) \qquad (3.37)$$

Introducing new symbols for the generalized plant $T_G = F_{ck} - P_c\dfrac{B}{A}$ and the weighted disturbance model $N_G = P_c\dfrac{C}{A}$, obtain

$$\phi(t + k) = T_G u(t) + N_G \xi(t + k) \qquad (3.38)$$

which is in the form equivalent to (3.7).

By analogy with the MV control law derivation, the weighted disturbance term can be split into unpredictable and predictable components using the Diophantine identity:

$$N_G = F + z^{-k}R. \qquad (3.39)$$

Equation (3.38) can now be rewritten as:

$$\phi(t) = z^{-k}\left(T_G u(t) + R\xi(t)\right) + F\xi(t) \tag{3.40}$$

The expression for the optimal controller clearly follows by setting the term in brackets to zero. This gives the optimal control:

$$u_{gmv}(t) = T_G^{-1} R\xi(t) \tag{3.41}$$

and the generalized output under optimal control:

$$\phi_{gmv}(t) = F\xi(t) \tag{3.42}$$

By some straightforward algebraic manipulations, the explicit expression for the controller can be obtained as:

$$u_{gmv}(t) = \left(FR^{-1}T_G + F_{ck}\right)^{-1} P_c y(t) \tag{3.43}$$

Note that by setting $P_{cd} = F_{cd} = 1$ and $F_{ck} = 0$, the basic MV controller is obtained. The weighting P_c may be used to influence the magnitude of control input signal. In particular, choosing $P_c = P_{c0}/(1 - z^{-1})$ introduces integral action in the controller. The weighting F_{ck} may be used to adjust the speed of response of the controller and hence prevent actuator saturation. The closed-loop transfer function may be derived by substituting for $u(t)$ to obtain,

$$y(t) = \frac{FB + F_{ck}C}{P_c B + F_{ck} A}\xi(t) \tag{3.44}$$

Solving the characteristic equation:

$$F_{cd} P_{cn} B + F_{cn} P_{cd} A = 0 \tag{3.45}$$

gives the closed loop poles. The weightings P_c and F_c may be selected such that the poles are located in a desired position on the complex plane.

Error and Control Weightings for GMV Control Design

A good starting point for GMV weightings is the rule of thumb documented by Grimble [2000] where the error weighting can be chosen to have integral action at low frequencies and to roll off at high. The control weighting can be chosen as a lead term to ensure the controller rolls-off appropriately (in the absence of a measurement noise model). This implies that at low frequencies the error weighting should be large and control weighting small while at high frequencies it should be the reverse. Using these guidelines for weighting selection, the GMV controller will have a very similar performance to that of the usual LQG cost function design [Grimble and Johnson, 1988]. That is, by appropriate choice of weightings:

$$E\left[(P_c e(t) + F_c u(t))^2\right] \cong E\left[(P_c e(t))^2 + (F_c u(t))^2\right] \tag{3.46}$$

- At low frequencies $|P_c|$ is large and $|F_c|$ small
- At high frequencies $|P_c|$ is small and $|F_c|$ large

A detailed discussion on the selection of cost weightings for the benchmarking algorithms is presented in Section 4.5.

3.3.2 Generalised Minimum Variance Benchmarking

The GMV control algorithm minimises the variance of a 'generalised' output, which is defined as a weighted combination of the process variable (or control error) and the control signal. This is a more practical benchmark than that based on the minimum variance controller, since it introduces a penalty on excessive control action and also allows, by an appropriate choice of the weightings, the achievement of desirable features of the control loop, such as high gain in low frequencies, low gain in high frequencies and good robustness. An overview of the GMV benchmarking algorithm is presented in this section.

The minimum value of the cost function (minimum variance of the generalized output $\phi(t)$) follows from Equation (3.42) as

$$J_{min} = Var\left[F\xi(t)\right] = \sum_{i=0}^{k-1} f_i^2 \tag{3.47}$$

and depends on the disturbance and the process time delay.

It is interesting to note that the above expression for the minimum cost is equivalent to that for the minimum variance control but now it refers to $\phi(t)$, defined in (3.42) rather than to $y(t)$. The minimum variance controller would therefore lead to suboptimal control as it would not minimise the variance of $\phi(t)$. Numerically, the two minimum costs are equal if the error weighting P_c is set to unity.

The estimation of the GMV performance index is performed in much the same way as in the minimum variance case, with an autoregressive model fitted to the generalized output constructed from the measured data:

$$\phi(t) = \sum_{j=1}^{m} \alpha_j \phi(t - k - j + 1) + \varepsilon(t) \tag{3.48}$$

Assuming that the data are collected over a range of values of t, Equation (3.48) leads to the following vector-matrix formulation:

$$\varphi = X\alpha + \varepsilon \tag{3.49}$$

where:

$$\varphi = \begin{bmatrix} \phi(n) \\ \phi(n-1) \\ \vdots \\ \phi(k+m) \end{bmatrix}, \quad X = \begin{bmatrix} \phi(n-k) & \phi(n-k-1) & \cdots & \phi(n-k-m+1) \\ \phi(n-k-1) & \phi(n-k-2) & \cdots & \phi(n-k-m) \\ \vdots & \vdots & \ddots & \vdots \\ \phi(m) & \phi(m-1) & \cdots & \phi(1) \end{bmatrix}, \quad \alpha = \begin{bmatrix} \alpha_1 \\ \alpha_2 \\ \vdots \\ \alpha_m \end{bmatrix}$$

k – process time delay (in samples)
n – data length
m – autoregressive model length

Minimising $E[\varepsilon^T \varepsilon]$ leads to the least squares solution:

$$\hat{\alpha} = \left(\mathbf{X}^T \mathbf{X} \right)^{-1} \mathbf{X}^T \varphi \qquad (3.50)$$

The estimate of the minimum variance follows from the residual variance as

$$\hat{\sigma}^2_{gmv} = \frac{1}{n-k-2m+1} \left(\varphi - \mathbf{X}\hat{\alpha} \right)^T \left(\varphi - \mathbf{X}\hat{\alpha} \right) \qquad (3.51)$$

This value is compared with the actual variance of $\phi(t)$ leading to the following expression for the GMV controller performance index:

$$\hat{\eta}(k) = \frac{\hat{\sigma}^2_{gmv}}{\hat{\sigma}^2_{\phi}} \qquad (3.52)$$

where $\qquad \hat{\sigma}^2_{\phi} = \frac{1}{n-k-m+1} \varphi^T \varphi$.

The basic steps of the benchmarking procedure are summarized below.

Procedure 3.2 Applying the Generalised Minimum Variance Benchmark

Step 1: Collect closed-loop output and control data from the plant. The data should be raw, i.e. not be compressed or filtered

Step 2: Select a data set of $n = 500$ samples or more (preferably $n \geq 1000$) containing representative disturbances acting on the system

Step 3: From prior knowledge, an open-loop experiment or a closed-loop estimation technique [Harrison and Qin, 2004] determine the time delay in samples.

Step 4: Choose the autoregressive model length for parameter estimation. The typical values are $m = 5$ up to 30. In general, the model length depends on the sample time and should be chosen such that the system impulse response is fully captured

Step 5: Filter the data to obtain the generalized output signal

Step 6: Construct the vector φ and the matrix $\mathbf{X}$.

Step 7: Fit the model to the data to obtain α_i parameters

Step 8: From Equation (3.52), calculate the Controller Performance Index

Procedure End

For the controller design, the restriction imposed on the choice of the dynamic weightings is that the polynomial $D_c = P_{cn}F_{cd}B - F_{ck}P_{cd}A$ must be strictly Schur (this ensures the stability of the closed-loop system). If this is not the case, the CPI computed from the data obtained from a (stable) existing control loop will be finite but corresponding to an inadmissible benchmark, and therefore will underestimate the controller performance.

Example 3.2. Generalized minimum variance benchmarking
The plant considered in this example has already been discretized and includes a time delay of $k = 2$ samples. It can be represented by the following ARMAX model:

$$y(t) = \frac{z^{-2}}{1-0.8z^{-1}}u(t) + \frac{1}{1-0.8z^{-1}}\xi(t)$$

The dynamic weightings for GMV benchmarking have been selected as

$$P_c(z^{-1}) = \frac{1-0.2z^{-1}}{1-z^{-1}} \quad \text{and} \quad F_c(z^{-1}) = -z^{-2}(1-0.5z^{-1})$$

The magnitude frequency responses of the weightings are shown in Figure 3.5. Note that the error weighting has infinite gain in low frequencies, which penalizes any zero steady-state error. As a result, the optimal controller will contain integral action, and so should any controller that is to be benchmarked against it. Otherwise the benchmark value would be very small and not providing a useful comparison.

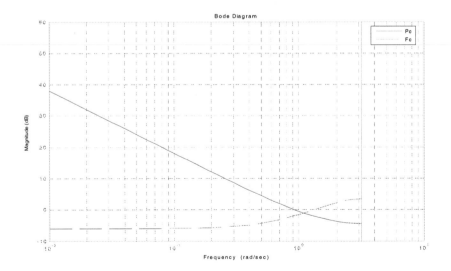

Figure 3.5. Frequency responses of the GMV weightings

The weighting choice was motivated by the fact that very often, in the process industries, it is the regulatory performance that is of real importance but set-point changes do occur occasionally. Hence the controller should include integral action. With the above choice of the error weighting any nonzero steady-state error would result in a very large value of the cost function so integral action is the basic requirement for any control law to be benchmarked against the optimal controller with the above weightings.

The assumed transfer function of the existing controller does in fact contain an integrator:

$$C_0 = \frac{1 - 0.4z^{-1}}{(1 - z^{-1})(1 + 0.5z^{-1})}$$

The stochastic performance of this controller is shown in Figure 3.6. Besides the error and control signal plots, the generalized output $\phi(t)$ was also computed and plotted, together with its autocorrelation function. Significant correlations can be seen beyond the plant time delay, and the estimated CPI is 0.49, indicating potential for improvement.

These results can be compared with those obtained from the actual GMV controller, calculated analytically:

$$C_{GMV} = \frac{2.44 - 1.44z^{-1}}{(1 - z^{-1})(2 + 1.8z^{-1})}$$

The corresponding plots are shown in Figure 3.7 and the performance is clearly better. In particular, the autocorrelations of the generalized output are negligible at and beyond the time delay, which is confirmed by the near-optimal index of 0.96. For such a choice of weightings the optimal minimum variance controller (which does not contain an integral term) would have a very low performance index.

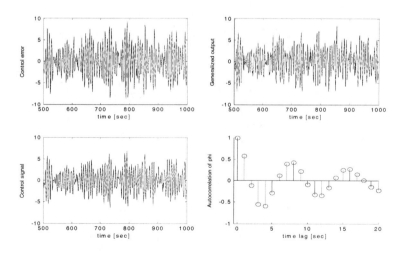

Figure 3.6. Simulation results for the nominal controller

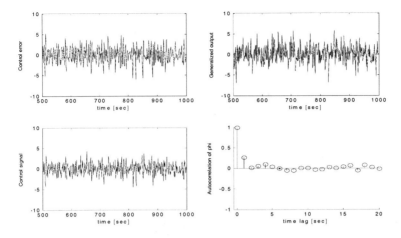

Figure 3.7. Simulation results for the optimal GMV controller

3.4 MV Benchmarking of Multivariable Systems

The SISO MV benchmark is attractive for its simplicity and minimum prior information required. It has become a standard feature in a number of commercial monitoring and diagnostic software packages. However, it has to be used on the loop-by-loop basis, ignoring the interactions between control loops. While this approximation is often valid enough to make the results useful, a typical industrial process contains at least a few interacting control loops and clearly it is desirable to assess their performance as a whole. The MIMO minimum-variance controller benchmark is therefore required.

For multivariable systems, the concept of the interactor matrix was introduced as a multivariable generalisation of the SISO time delay term [Goodwin and Sin, 1984]. Assuming full knowledge of the interactor matrix, MIMO MV benchmarking method was presented in [Harris, *et al.*, 1996, Huang, *et al.*, 1997, Ko and Edgar, 2001]. In this section, we review the basic ideas involved, discuss the similarities and differences as compared with the scalar case, and present an alternative performance assessment approach that returns the bounds on the controller performance index rather than the index itself, but requires less *a priori* information.

3.4.1 MIMO MV Control

Performance limitations due to time delays are amongst the most fundamental problems in control. Wolovich and Falb [1976] generalised the concept of time delays in scalar systems to the multivariable case. They illustrated that the analog of a time-delay term in SISO systems is an interactor matrix in MIMO systems. In [Goodwin and Sin, 1984], it was proved that for every $n \times m$ strictly proper, rational transfer-function matrix T, there exists a unique, non-singular, $n \times n$ lower left triangular polynomial matrix D, such that $| D |= z^r$ and

$$\lim_{z^{-1} \to 0} DT = \lim_{z^{-1} \to 0} \tilde{T} = K \qquad (3.53)$$

where K is a full rank constant matrix. The integer r is defined as the number of infinite zeros of T, and $\tilde{T}$ is the delay-free transfer-function matrix of T which contains only finite zeros. The matrix D is defined as the interactor matrix and can be written as

$$D = D_0 z^d + D_1 z^{d-1} + ... + D_{d-1} z \qquad (3.54)$$

where d denotes the order of the interactor matrix and is unique for a given transfer-function matrix, and D_i (for $i=0,...,d$-1) are coefficient matrices.

The interactor matrix D can assume one of the three forms described in the sequel. If D is of the form: $D = z^d I$, then the transfer-function matrix T is regarded as having a *simple interactor matrix*. If D is a diagonal matrix, i.e., $D = diag(z^{d_1}, z^{d_2}, ..., z^{d_n})$, then T is considered having a *diagonal interactor matrix*. Otherwise, T is considered to have a *general interactor matrix*.

A special type of an interactor matrix, which is called a *unitary interactor matrix*, was introduced in [Peng and Kinnaert, 1992], which satisfies

$$D^T (z^{-1}) D(z) = I \qquad (3.55)$$

As will be seen, this is a particularly useful property for controller performance assessment.

Example 3.3. Interactor matrices
Consider a MIMO process which can be described by the 2×2 transfer-function matrix:

$$T = \begin{bmatrix} \dfrac{z^{-1}}{1-0.1z^{-1}} & 0 \\ \dfrac{2z^{-2}}{1-0.3z^{-1}} & \dfrac{2z^{-2}}{1-0.4z^{-1}} \end{bmatrix}$$

It can be seen by inspection that the diagonal delay matrix $D_d = \begin{bmatrix} z & 0 \\ 0 & z^2 \end{bmatrix}$ containing the minimum time delays in each output channel is actually the system interactor matrix. This is because the matrix $\lim_{z^{-1} \to 0} D_d T = \begin{bmatrix} 1 & 0 \\ 2 & 2 \end{bmatrix}$ has full rank.

If T is changed to $T = \begin{bmatrix} \frac{z^{-1}}{1-0.1z^{-1}} & \frac{z^{-1}}{1-0.2z^{-1}} \\ \frac{2z^{-2}}{1-0.3z^{-1}} & \frac{2z^{-2}}{1-0.4z^{-1}} \end{bmatrix}$, then $\lim\limits_{z^{-1} \to 0} D_d T = \begin{bmatrix} 1 & 1 \\ 2 & 2 \end{bmatrix}$ is singular and

hence D_d is not the diagonal interactor of T.

Also when T is changed to $T = \begin{bmatrix} \frac{z^{-1}}{1-0.1z^{-1}} & 0 \\ \frac{2z^{-2}}{1-0.3z^{-1}} & \frac{2z^{-3}}{1-0.4z^{-1}} \end{bmatrix}$, $\lim\limits_{z^{-1} \to 0} D_d T = \begin{bmatrix} 1 & 0 \\ 2 & 0 \end{bmatrix}$ is a singular

matrix and the interactor matrix of T is not diagonal.

The above example illustrates that there are two possible reasons for the interactor matrix of a given system not being diagonal. That is, the singularity may arise due to either a particular coefficient combination or due to the inherent rank deficiency. There is a fundamental difference between these two cases: the singularity due to coefficient combination is very sensitive to the coefficient perturbation while small variations of coefficients have no effect on the inherent rank deficiency case. In reality, small perturbations of the system are inevitable, and as a result, singularity due to coefficients combination is unlikely. It is reasonable to conclude that, in practice, the existence of a general interactor matrix is only due to the inherent rank deficiency.

Using the unitary interactor matrix, [Huang and Shah, 1999] proposed a simple method of deriving the MIMO MV controller. This derivation is briefly summarized below.

For a multivariable process:

$$Y(t) = TU(t) + N\xi(t) \tag{3.56}$$

where $\xi(t)$ is a vector of random white noise sources with zero mean, let D be the interactor matrix of T with order, d. The linear quadratic objective function is defined as

$$J_{MV} = E\left[\tilde{Y}(t)^T \tilde{Y}(t) \right] \tag{3.57}$$

where the filtered output signal:

$$\tilde{Y}(t) = z^{-d} D Y(t) \tag{3.58}$$

Multiplying both sides of Equation (3.56) by $z^{-d}D$ yields:

$$z^{-d} DY(t) = z^{-d} \tilde{T} U(t) + z^{-d} DN\xi(t) = \tilde{T}U(t-d) + R\xi(t-d) + F\xi(t) \tag{3.59}$$

where F and R satisfy the Diophantine identity:

$$z^{-d} DN = \underbrace{F_0 + ... + F_{d-1}q^{-d+1}}_{F} + z^{-d}R \tag{3.60}$$

and R is a rational proper transfer-function matrix.

Since the last term in Equation (3.59) cannot be affected by the control action, the minimum variance control is achieved when the sum of the first two terms on the right hand side of Equation (3.59) is set to zero,

$$\tilde{T}U(t-d) + R\xi(t-d) = 0 \tag{3.61}$$

which yields $U(t) = -\tilde{T}^{-1}R\xi(t)$. After some manipulation, we have:

$$U(t) = -\tilde{T}^{-1}RF^{-1}(z^{-d}D)Y(t) \tag{3.62}$$

If D is a unitary interactor matrix, then the controller defined in Equation (3.62) is output-ordering invariant and unique. Furthermore, this controller also minimizes the LQ objective function of the original output $Y(t)$,

$$E\left[\tilde{Y}(t)^T \tilde{Y}(t)\right] = E\left[Y(t)^T Y(t)\right] \tag{3.63}$$

From the above, it is clear that the interactor matrix is vital in the computation of the MIMO MV benchmark. From Equations (3.60) and (3.65), the benchmark can have the same value for two different plants as long as they have the same interactor matrix.

3.4.2 MIMO MV Benchmarking

For an arbitrary linear multivariable or multi-loop controller, the following inequality holds:

$$tr\left[Cov(Y(t))\right] = tr\left[Cov(\tilde{Y}(t))\right] \ge tr\left[Cov(F\xi(t))\right] \tag{3.64}$$

The MIMO MV benchmark may therefore be defined as:

$$J_{MV} = tr[Cov(F\xi(t))] \tag{3.65}$$

The above value can be estimated from routine closed-loop data using a generalization of the Harris' algorithm presented in Procedure 3.1. An alternative method was proposed by Huang and Shah [1999] and is included below.

Procedure 3.3 Applying the MIMO Minimum Variance Benchmark: Filtering and Correlation (FCOR) algorithm

Step 1:   Collect the output data from the plant

Step 2:   Determine the interactor matrix of the plant D and find its order d.

Step 3:   Filter the data through the interactor matrix:

$$\tilde{Y}(t) = z^{-d}DY(t)$$

Step 4:   Estimate the noise vector sequence $\xi(t)$ (whitening process) – a common approach is to model the output signal as a VAR (Vector Auto Regressive) time series and then filter it to obtain the white noise 'innovations' sequence:

$$\xi(t) = A(z^{-1})\tilde{Y}(t)$$

Step 5:   Compute the cross-correlations between the output and the estimated noise:

$$r_{\tilde{Y}\xi}(0) = E[\tilde{Y}(t)\xi(t)^T] = F_0\Sigma_\xi$$

$$r_{\tilde{Y}\xi}(1) = E[\tilde{Y}(t)\xi(t-1)^T] = F_1\Sigma_\xi$$

$$\vdots$$

$$r_{\tilde{Y}\xi}(d-1) = E[\tilde{Y}(t)\xi(t-d+1)^T] = F_{d-1}\Sigma_\xi$$

where $\Sigma_\xi = E[\xi(t)\xi(t)^T]$. The right-hand sides follow from Equation (3.62), and the coefficients of F are thus determined.

Step 6:   Calculate the optimum covariance matrix:

$$\Sigma_{mv} = F_0\Sigma_\xi F_0^T + ... + F_{d-1}\Sigma_\xi F_{d-1}^T$$

Step 7:   Compute the controller performance index:

$$\eta = \frac{tr(\Sigma_{mv})}{tr(\Sigma_{\tilde{Y}})}$$

Procedure End

In practice all the above theoretical values are replaced by their estimates calculated from a finite data set.

Although it is possible to estimate the interactor matrix from closed loop data [Huang and Shah, 1999; Mutoh and Ortega, 1993], this makes the computation of the benchmark much more difficult. An excellent discussion of this issue has been presented in [Huang, et al., 2005]. The next subsection will explicitly deal with this problem.

3.4.3 Performance Assessment Using Reduced *a priori* Information

It is desirable to assess the controller performance based on information which is easy to obtain. The accurate estimation of the interactor matrix is rather involved, as can be seen from the algorithms for calculating a lower triangular and nilpotent interactor matrices proposed in [Wolovich and Falb, 1976] and [Rogozinski, et al., 1987] respectively. It was later shown in [Huang and Shah, 1999] that the interactor matrix could also be determined from the leading Markov parameter matrices of the plant model. Therefore, the multivariable minimum variance may not seem a practical benchmark. However, its upper and lower bounds can still be obtained which require less *a priori* information. For example, in the recent papers [Huang, et al., 2005] and [Xia, et al., 2005], it was shown that it is possible to derive such bounds based on the knowledge of the order of the interactor matrix alone or from the time delays between inputs and outputs.

As opposed to estimating the interactor matrix, it is relatively easy to determine the time delays between each input/output pair. This information can be obtained through either an open loop experiment or physical insights, and an I/O delay matrix can thus be constructed. It will be shown that the order of the interactor matrix can be found from this time delay matrix. With the order of the interactor matrix known, a simple interactor of the same order as a substitute for the original interactor matrix may be used. The benchmark figure thus computed turns out to be an upper bound of the true minimum variance. Although the result is suboptimal, the computational procedure can be greatly simplified. Furthermore, a lower bound of the MV benchmark can also be computed based on the delay matrix. These results provide the bounds on the controller performance index.

Finding the Order of the Interactor Matrix from I/O Delay Matrix

Let the Markov parameter representation of a strictly proper transfer-function matrix T be written as:

$$T = \sum_{i=0}^{\infty} G_i z^{-i-1} \tag{3.66}$$

With Markov parameter matrices given, the interactor matrix can readily be found from the algorithm presented in [Rogozinski, *et al.*, 1987]. Obviously, the order of the interactor matrix can also be obtained as a by-product of this algorithm.

For a transfer-function matrix with only its I/O delay matrix given, the structures of the first few Markov parameter matrices can be obtained even though the exact coefficients may be unknown.

Consider a plant represented by a 3×3 strictly proper, non-singular transfer matrix T with an I/O delay matrix $D_{I/O}$:

$$D_{I/O}(T) = \begin{bmatrix} z^{-1} & z^{-3} & z^{-4} \\ z^{-2} & z^{-4} & z^{-5} \\ z^{-2} & z^{-3} & z^{-4} \end{bmatrix}$$

From this structure the Markov parameters can be determined as

$$G_0 = \begin{bmatrix} * & 0 & 0 \\ 0 & 0 & 0 \\ 0 & 0 & 0 \end{bmatrix}, \; G_1 = \begin{bmatrix} * & 0 & 0 \\ * & 0 & 0 \\ * & 0 & 0 \end{bmatrix}, \; G_2 = \begin{bmatrix} * & * & 0 \\ * & 0 & 0 \\ * & * & 0 \end{bmatrix}, \; G_3 = \begin{bmatrix} * & * & * \\ * & * & 0 \\ * & * & * \end{bmatrix}, \; G_4 = \begin{bmatrix} * & * & * \\ * & * & * \\ * & * & * \end{bmatrix},$$

where $*$ indicates a nonzero number.

A procedure was given in [Mutoh and Ortega, 1993] which can be used to determine the order of the interactor matrix and obtain a reduced parameterization for the interactor. This procedure revealed that the order of the interactor matrix is in general only determined by the zero/nonzero pattern of the Markov parameter matrices, i.e. it is determined by $D_{I/O}(T)$. This indicates that any transfer-function matrix with the same I/O delay matrix has an interactor matrix of the same order. By replacing $*$ of the above Markov parameter matrices with any random nonzero number and using the algorithm of [Rogozinski, *et al.*, 1987], the order of the interactor matrix can easily be determined.

Upper Bound of the MIMO MV Benchmark

In this subsection, it will be shown that an upper bound of the MV benchmark can be estimated based only on the information on the order of the interactor matrix.

Let $J = tr[Cov(F\xi(t))]$, where F satisfies the following identity:

$$z^{-d} DN = F + z^{-d} R \qquad (3.67)$$

where $F(z) = F_0 + ... + F_{d-1}z^{-d+1}$, D is any unitary interactor matrix of order d and $N = \sum_{i=0}^{\infty} N_i z^{-i}$ is a given transfer-function matrix. Then it was shown in [Xia, et al., 2005] that J is maximised when $D = z^d I$.

Given the order of the interactor matrix is d, the upper bound of J_{MV} can be found by estimating $J_{upper} = tr(Cov(\bar{F}\xi(t)))$, where $\bar{F}$ satisfies the following identity:

$$N = \bar{F} + z^{-d}\bar{R} \qquad (3.68)$$

where $\bar{F}(z) = N_0 + ... + N_{d-1}z^{-d+1}$.

The above result implies that the upper bound J_{upper} is the minimum output variance for the plant with its interactor matrix replaced by a simple interactor of the same order. In other words, an attempt is made to estimate the first d coefficient matrices of the disturbance model (3.68), i.e. the $\bar{F}$ polynomial matrix.

It can be shown that to estimate $\bar{F}$ it is necessary to introduce additional delays to the system, for the duration of the benchmarking experiment. The reason for this can be briefly explained as follows. With the actual plant (having a general interactor matrix) in the feedback loop, and with an arbitrary controller C_0, the output is:

$$Y(t) = \bar{F}\xi(t) + z^{-d}\bar{R}\xi(t) - TC_0 Y(t) = \bar{F}\xi(t) + z^{-d}\bar{R}\xi(t) - D^{-1}\tilde{T}C_0 Y(t) \qquad (3.69)$$

Since in the general interactor case the term $D^{-1}\tilde{T}C_0$ contains delays smaller than d in at least one channel, the original polynomial matrix $\bar{F}$ cannot be correctly estimated from data. Therefore additional delays need to be introduced in those channels. The actual number of delays needed to be added to the $(i,j)^{th}$ element of the controller transfer matrix equals $d - d_{ij}$ where d_{ij} is the time delay of the $(i,j)^{th}$ element of the I/O delay matrix.

The advantage of this approach is that the obtained bound is controller invariant (depending only on the disturbance model N and the interactor order d), however the required temporary controller modification may be considered restrictive. In that case, a simple solution is to use an estimation algorithm with the original data, and it was shown in [Huang, et al., 2005] that this approach also provides an upper bound on the true MIMO MV performance. This corresponds to comparing the current controller performance with that of the d-step ahead controller [Harris, et al., 1996, Huang, et al., 2005]. Because of the use of routine data in estimation, this is the preferred approach, although the estimated upper bound depends on the existing controller and hence would be different for two different controllers.

Compared with the former bound, based on simulation studies, it tends to result in a higher value for poorly tuned controllers, and a lower value for well tuned controllers. In the limiting case of the true MV controller operating in the loop it equals the true value exactly [Huang, et al., 2005]. Both considered upper bounds could therefore be computed together, and the lower of the two selected. This will be illustrated in Example 3.4.

Lower Bound of the MIMO MV Benchmark
An upper bound of the MIMO MV benchmark was discussed in the previous section. In this section, a lower bound will be introduced.

Based on the lemmas given in [Xia, et al., 2005], the following result concerning the lower bound of the MIMO MV benchmark is obtained: Given a transfer-function matrix T, let $D_d(T)$ be its diagonal delay matrix with order d; the least conservative lower bound of J_{MV} can be found by estimating $J_{lower} = tr\left[Cov(F^*\xi(t))\right]$ where F^* satisfies the following identity:

$$z^{-d} D_d(T)N = F^* + z^{-d} R \tag{3.70}$$

where $F^* = F_0^* + ... + F_{d-1}^* z^{-d+1}$.

Remark: The above result can be interpreted as follows: the diagonal delay matrix can be considered as a simple generalization of the time delay of SISO system. Let n_i be the stochastic noise acting on the ith system output:

$$n_i = \underbrace{f_0\xi(t) + f_1\xi(t-1) + ... + f_{d_i-1}\xi(t-d_i-1)} + f_{d_i}\xi(t-d_i) + ...$$

where $\xi(t)$ is white noise and d_i is the minimum time delay of the ith row of T. Denote:

$$e = f_0\xi(t) + f_1\xi(t-1) + ... + f_{d_i-1}\xi(t-d_i-1)$$

It is evident that e is the portion of noise which happens before the feedback control can start to react and, therefore, it is independent of feedback control. Furthermore, there may be another portion of n_i which cannot be compensated due to other infinite zeros of T. This inevitably increases the achievable minimum variance.

As a summary of the results presented in this section, the following procedure is proposed for the evaluation of the MIMO MV performance based on the I/O delay matrix.

Procedure 3.4 Applying the MIMO Minimum Variance Benchmark with reduced prior information
Step 1:   Obtain the I/O delay matrix $D_{I/O}(T)$ of the plant.

Step 2:   Construct the diagonal delay matrix $D_d(T)$ and calculate

$$U = \lim_{z^{-1} \to 0} D_d(T) D_{I/O}(T)$$

Step 3:   Generate three random matrices $R_i \in R(T)$.

Step 4:   If $rank(R_i \odot U) = min(m,n)$ for all $i = 1,2,3$ go to step 7.

Step 5:   Determine the order of the interactor matrix (Section 0)

Step 6:   Estimate the lower/upper bounds of the MIMO MV benchmark including the upper bound introduced in [Huang, et al., 2005]. Select the lower of the two upper bounds. Go to Step 8.

Step 7: $D_d(T)$ is the diagonal interactor matrix of T. Estimate the MIMO MV benchmark using $D_d(T)$.

Step 8:   Compute the CPI of the existing controller from real MIMO MV benchmark or lower/upper bounds.

Procedure End

The procedure shows how to determine the upper and lower bounds of the controller performance index, given only the knowledge of the I/O delay matrix rather than of the interactor matrix itself. The order d of the interactor matrix is also needed but this can be determined from the I/O delay matrix, as discussed before. If the true interactor matrix is diagonal, which can be established in Step 4, then an estimate of the true performance index is returned. The following example will illustrate the controller performance assessment procedure.

Example 3.4. MIMO Minimum Variance Benchmarking
The example is similar to the one used in [Huang, et al., 2005] and involves a two-by-two system, as described by Equation (3.56), with the plant and disturbance transfer matrices given as:

$$T = \begin{bmatrix} \dfrac{z^{-(d-1)}}{1-0.4z^{-1}} & \dfrac{0.7z^{-d}}{1-0.1z^{-1}} \\ \dfrac{0.3z^{-(d-2)}}{1-0.1z^{-1}} & \dfrac{z^{-(d-1)}}{1-0.8z^{-1}} \end{bmatrix} \quad N = \begin{bmatrix} \dfrac{1}{1-0.5z^{-1}} & \dfrac{-0.6z^{-1}}{1-0.6z^{-1}} \\ \dfrac{0.5z^{-1}}{1-0.7z^{-1}} & \dfrac{1.0}{1-0.8z^{-1}} \end{bmatrix}$$

The white noise input $\xi(t)$ is a two-dimensional white noise sequence with the covariance matrix $\Sigma_\xi = I$.

The diagonal delay matrix D_d for this plant can be determined from the I/O delay matrix as

$$D_d(T) = \begin{bmatrix} z^{-(d-1)} & 0 \\ 0 & z^{-(d-2)} \end{bmatrix}$$

Due to the inherent rank deficiency the plant has a general interactor matrix, and its order can be found to equal d. Assuming for the moment full knowledge of the plant model, the actual unitary interactor matrix for this example can be determined as

$$D = \begin{bmatrix} 0.9578z^{d-1} & 0.2873z^{d-2} \\ -0.2873z^d & 0.9578z^{d-1} \end{bmatrix}$$

The order of the interactor matrix is equal to the largest time delay of the system, however generally this does not have to be the case. The MV criterion to be minimized is the sum of the variances of the two outputs: $J = E[Y_t^T Y_t]$, and the controller performance index (CPI) is defined as,

$$\eta = \frac{J_{MV}}{E\left[Y(t)^T Y(t)\right]}$$

In the following, it is assumed that the existing controller has a multi-loop structure:

$$K = \begin{bmatrix} \dfrac{0.5 - 0.2z^{-1}}{1 - 0.5z^{-1}} & 0 \\ 0 & \dfrac{0.25 - 0.2z^{-1}}{(1 - 0.5z^{-1})(1 + 0.5z^{-1})} \end{bmatrix}$$

The benchmark figures computed for increasing values of d are presented in Table 3.1, and also shown graphically in Figure 3.8. The rows in the table correspond to the true minimum variance J_{MV} (computed for comparison purposes assuming full knowledge of the plant and the true interactor matrix), its lower and upper bounds J_{lower} and J_{upper}, the alternative upper bound MV_d introduced in [Huang, et al., 2005], and the lower of the two upper bounds.

All these values have been calculated from theoretical formulas. The actual output variance (the denominator in the expression for the performance index) has been computed as the H_2 norm of the system $(I + TK)^{-1}N$. The upper bounds of the CPI (the lower plot) have been calculated using the lower of the two upper bounds J_{upper} and MV_d.

As can be seen from these results, the d-step ahead upper bound MV_d is lower than the J_{upper} bound for small d and higher for $d \geq 6$. The difference between the bounds and the true minimum variance decreases rapidly with the increasing interactor order - this is because, for a stable disturbance model, the Markov coefficient matrices that determine this difference become smaller and smaller.

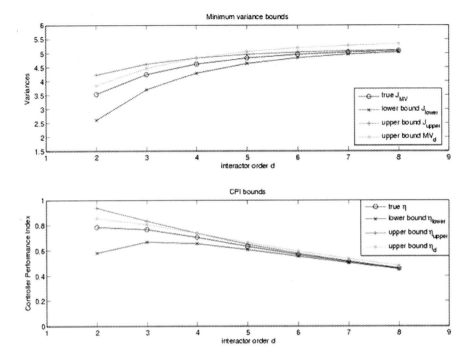

Figure 3.8. Minimum variance and CPI bounds: theoretical figures

The MV_d upper bound for large interactor orders is more conservative, which can be explained by the dependence of this bound on the controller. Since the performance of this fixed controller deteriorates with increasing time delays, this is reflected also in the bound.

Table 3.1. Minimum variance and its bounds

d	3	4	5	6	7	8	9
J_{MV}	3.53	4.24	4.61	4.83	4.95	5.03	5.08
J_{lower}	2.61	3.69	4.29	4.63	4.83	4.96	5.03
J_{upper}	4.22	4.61	4.83	4.96	5.03	5.08	5.11
MV_d	3.84	4.46	4.83	5.06	5.19	5.27	5.33
$min(J_{upper}, MV_d)$	3.84	4.46	4.83	4.96	5.03	5.08	5.11

All the above concerns the determination of the exact theoretical values of the performance bounds. In reality, these values have to be estimated from plant data.

The FCOR (Filtering and Correlation) algorithm is used here for the numerical estimation, as described in [Huang and Shah, 1999]. The algorithm involves modelling the outputs as a multivariable time series in order to estimate the white noise driving sequences, and subsequently correlating them with the outputs to obtain the coefficient matrices of $\bar{F}$ (Equation (3.69)) and F^* (Equation (3.70)). This "whitening" step is not unique and may result in different polynomial matrices. However, it is worth noting that their H_2 norm (the estimated variance) remains unchanged.

The obtained estimates (calculated using a data set of 5000 samples), together with the theoretical true values, are plotted in Figure 3.9. The estimates are reasonably close to their theoretical values, although in some cases the estimated upper bound is actually lower than the true minimum variance. Taking averages over a number of realizations (or taking a longer data set) would improve the accuracy of the estimates.

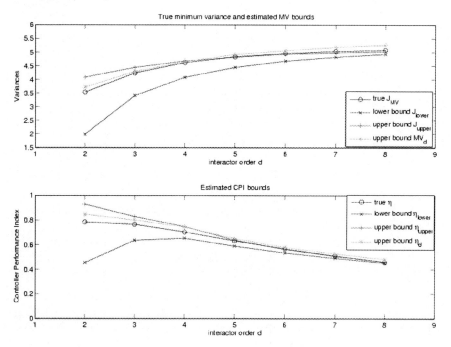

Figure 3.9. Estimated minimum variance and CPI bounds

3.5 GMV Benchmarking of Multivariable Systems

As in the scalar case, the MV benchmarking algorithm presented in the previous section can be generalized to the GMV case. Most of the developments presented in this section have equivalents in Sections 3.3 and 3.4 so the reader is encouraged to refer to these sections first.

By analogy with the Minimum Variance control law, the GMV control algorithm can be defined as one that minimizes the variance of the "generalized" output signal $\phi(t)$, which is a sum of dynamically weighted control error and control signals, and:

$$\phi(t) = P_c e(t) + F_c u(t) \tag{3.71}$$

In the multivariable case the GMV cost function has the form:

$$J = E\left[\phi(t)^T \phi(t)\right] = E\left[(P_c e(t) + F_c u(t))^T (P_c e(t) + F_c u(t))\right] \tag{3.72}$$

The multivariable dynamic weighting transfer-function matrices are defined as:

$$P_c = P_{cn} P_{cd}^{-1}, \quad F_c = F_{cn} F_{cd}^{-1} \tag{3.73}$$

These weightings can be used to frequency-shape the responses produced by the controller.

3.5.1 Generalized Plant

The GMV problem can be reformulated as an MV problem for the generalized plant (as in Section 3.3), and subsequently it can be reduced from the general to the simple interactor problem (as in Section 3.4). Recalling the process equation:

$$Y(t) = D^{-1}\tilde{T}U(t) + N\xi(t) \tag{3.74}$$

the generalized output (3.71) can be rewritten as follows:

$$\phi(t) = P_c e(t) + F_c u(t) = P_c(-D^{-1}\tilde{T}u(t) + N\xi(t)) + F_c u(t)$$
$$= (F_c - P_c D^{-1}\tilde{T})u(t) + P_c N\xi(t) \tag{3.75}$$

Introducing new symbols for the generalized plant $T_G = D_G^{-1}\tilde{T}_G = F_c - P_c T$ and the weighted disturbance model $N_G = P_c N$, obtain

$$\phi(t) = D_G^{-1}\tilde{T}_G u(t) + N_G \xi(t) \tag{3.76}$$

which is in the form equivalent to (3.74). Note that D_G is a unitary interactor matrix of the generalized plant and in general may be different from the original interactor D.

In order to simplify the problem to the case of a simple interactor matrix, consider the filtered signal

$$\tilde{\phi}(t) = z^{-d} D_G \phi(t) = z^{-d}\tilde{T}_G u(t) + \tilde{N}_G \xi(t) \tag{3.77}$$

where d is the order of D_G (the highest degree of its entries). Note that if D_G is a simple interactor matrix ($D_G = z^d I$), then $\tilde{\phi}(t) = \phi(t)$, but in general the signal $\tilde{\phi}(t)$ will be different though its spectrum remains the same. Because of the unitary property of the interactor, the original problem (3.72) is thus reduced to minimizing the variance of signal $\tilde{\phi}_1$.

Similar comments regarding the estimation of the interactor matrix apply as they did in the MV case, the difference involving additional filtering by P_c and F_c. Another approach is to constrain F_c to have a specific structure so that the interactor matrix of the generalized plant is the same as the interactor of the weighted plant P_cW .

3.5.2 Multivariable GMV Controller

The main step in the derivation of the GMV control law is to split the cost function (3.72) into two terms, one of which is independent of, and the other dependent on the controller. The control law then results simply by setting the second term to zero.

Splitting the disturbance term into unpredictable and predictable components using the Diophantine identity

$$\tilde{N}_G = F + z^{-d}R , \tag{3.78}$$

Equation (3.77) can be rewritten as

$$\tilde{\phi}(t) = z^{-d}\left(\tilde{T}_G u(t) + R\xi(t)\right) + F\xi(t) \tag{3.79}$$

The expression for the optimal controller follows by setting the term in brackets to zero. This gives the optimal control:

$$u_{gmv}(t) = \tilde{T}_G^{-1}R\xi(t) \tag{3.80}$$

and the generalized output under optimal control:

$$\tilde{\phi}_{gmv}(t) = F\xi(t) \tag{3.81}$$

In order to obtain the explicit expression for the controller, combine (3.80), (3.81) and (3.71) to obtain, after some straightforward algebraic manipulations:

$$u_{gmv}(t) = -\left(FR^{-1}\tilde{T}_G + z^{-d}D_GF_c\right)^{-1}z^{-d}D_GP_ce(t) \tag{3.82}$$

3.5.3 GMV Benchmarking

The minimum value of the cost function (minimum variance of the generalized output $\phi(t)$) follows from Equation (3.81) as

$$J_{min} = Var[F\xi(t)] = \sum_{i=0}^{d-1}trace\left(F_i^T F_i\right) \tag{3.83}$$

and depends on the combined noise model and the plant interactor matrix. This makes the minimum variance benchmark more difficult to estimate in the multivariable case than in scalar case where it was only necessary to know the noise model and the time delay of the system. For MIMO systems, the knowledge of the interactor matrix is in general equivalent to knowledge of the plant itself and although there have been attempts to relax this requirement, this is still a serious limitation in multivariable performance assessment.

The FCOR estimation algorithm can be generalized from the minimum-variance case. Recall that the idea is to estimate the coefficients of the matrix polynomial F and then use the expression (3.83) directly. This can be achieved by cross-correlating the generalized output ϕ_t with the estimated white noise input to the system (signal $\xi(t)$), as outlined in the procedure below.

Procedure 3.5 Applying   MIMO   Generalised   Minimum   Variance Benchmark: FCOR algorithm

Step 1:   Filter the error and control signals to obtain the generalized output signal:

$$\phi(t) = P_e e(t) + F_e u(t)$$

Step 2:   Estimate the interactor matrix of the generalized plant D_G and determine its order d.

Step 3:   Filter ϕ_t through the interactor matrix D_G to obtain $\tilde{\phi}_t$:

$$\tilde{\phi}(t) = z^{-d} D_G \phi(t)$$

Step 4:   Estimate the noise vector sequence $\xi(t)$ (whitening process) – a common approach is to model the output signal as a VAR (Vector Auto Regressive) time series and then filter it to obtain the white noise 'innovations' sequence:

$$\xi(t) = A(z^{-1}) \tilde{\phi}(t)$$

Step 5:   Compute the cross-correlation between the output and the estimated noise:

$$r_{\tilde{\phi}\xi}(0) = E[\tilde{\phi}(t)\xi(t)^T] = F_0 \Sigma_\xi$$

$$r_{\tilde{\phi}\xi}(1) = E[\tilde{\phi}(t)\xi(t-1)^T] = F_1 \Sigma_\xi$$

$$\vdots$$

$$r_{\tilde{\phi}\xi}(d-1) = E[\tilde{\phi}(t)\xi(t-d+1)^T] = F_{d-1} \Sigma_\xi$$

where $\Sigma_\xi = E[\xi(t)\xi(t)^T]$. The right-hand sides follows from Equation (3.79), and the coefficients of F are thus determined.

Step 6:   Calculate the optimum covariance matrix:

$$\Sigma_{mv} = F_0 \Sigma_\xi F_0^T + \ldots + F_{d-1} \Sigma_\xi F_{d-1}^T$$

Step 7:   Compute the controller performance index:

$$\eta = \frac{tr[\Sigma_{mv}]}{tr[\Sigma_\phi]}$$

Procedure End

Example 3.5 MIMO Generalised Minimum Variance Benchmarking
   The above benchmarking algorithm will be illustrated with a simple numerical example. The example comes from [Huang and Shah, 1999] and involves a 2×2 linear first-order system with time delays:

$$T = \begin{bmatrix} \dfrac{z^{-1}}{1-0.4z^{-1}} & \dfrac{K_{12}z^{-2}}{1-0.1z^{-1}} \\ \dfrac{0.3z^{-1}}{1-0.1z^{-1}} & \dfrac{z^{-2}}{1-0.8z^{-1}} \end{bmatrix}, \quad N_d = N = \begin{bmatrix} \dfrac{1}{1-0.5z^{-1}} & \dfrac{-0.6}{1-0.5z^{-1}} \\ \dfrac{0.5}{1-0.5z^{-1}} & \dfrac{1}{1-0.5z^{-1}} \end{bmatrix}$$

The set-point is assumed to be zero for both outputs (pure regulatory control). The parameter K_{12} in the plant transfer-function matrix determines the level of interaction between input 2 and output 1 and varies from 0 to 10. Irrespective of K_{12}, the plant has a general interactor matrix which can be determined as

$$D = \begin{bmatrix} -0.9578z & -0.2873z \\ 0.2873z^2 & -0.9578z^2 \end{bmatrix}$$

The existing controller is the multi-loop minimum variance controller C_0 calculated for the two single loops without considering the interactions:

$$C_0(z^{-1}) = \begin{bmatrix} \dfrac{0.5-0.2z^{-1}}{1-0.5z^{-1}} & 0 \\ 0 & \dfrac{0.25-0.2z^{-1}}{(1-0.5z^{-1})(1+0.5z^{-1})} \end{bmatrix}$$

Two choices of dynamic weighting functions have been considered in this example:

Case 1: Minimum variance weightings:

$$P_c = \begin{bmatrix} 1 & 0 \\ 0 & 1 \end{bmatrix}, \quad F_c = \begin{bmatrix} 0 & 0 \\ 0 & 0 \end{bmatrix},$$

Case 2: GMV constant (static) weightings

$$P_c = \begin{bmatrix} 1 & 0 \\ 0 & 1 \end{bmatrix}, \quad F_c = -D^{-1} \begin{bmatrix} R_1 & 0 \\ 0 & R_2 \end{bmatrix},$$

Parameters R_1 and R_2 determine the relative importance attached to both control variances with respect to each other and to the error variances. In this example they have been fixed to $R_1 = 3$ and $R_2 = 5$. Also note that the weighting F_c has been premultipled by D^{-1} and in this case the interactor matrix D_G simply equals D.

The system has been simulated for different values of K_{12} and 5000 samples of the error and control signal have been collected in each case. Then the FCOR algorithm has been applied to the interactor-filtered generalized output and the benchmark cost evaluated. The controller performance index (CPI) has also been calculated. The comparison of the error and control variances for the optimal MV and GMV controllers is presented in Table 3.2, and the benchmarking results are shown in Figure 3.10.

Table 3.2. Error and control variances

K_{12}	Controller	Var[e_1]	Var[e_2]	Var[u_1]	Var[u_2]
1	MV	1.39	1.55	0.79	0.30
	GMV	2.82	2.11	0.18	0.014
5	MV	1.39	1.55	∞	∞
	GMV	3.23	1.91	0.21	0.005
10	MV	1.40	1.55	∞	∞
	GMV	3.39	1.88	0.24	0.002

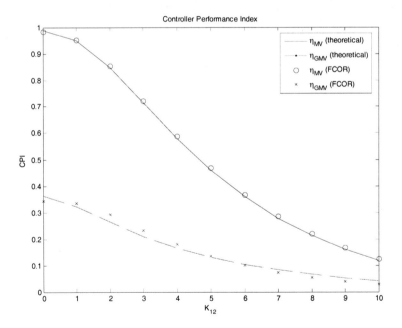

Figure 3.10. MV and GMV controller performance indices

Comments

Although rather simplistic, the static GMV weightings do provide a means of defining a benchmark with balanced error and control variances. In this particular example, the plant becomes non-minimum phase for $K_{12} > 2$, resulting in the unbounded minimum variance control, as seen in Table 3.2. The introduction of the control weighting makes the benchmark realizable.

The analysis of Figure 3.10 indicates that the performance of the multi-loop MV controller is close to unity for small interactions between loops and decreases when these interactions increase. The existing controller, however, is not so good if the control variances are also of importance. This implies that detuning the controller may be needed in practice.

3.6 Benchmarking Based on Predictive Performance Index

In the process industry, Generalised Predictive Control (GPC) algorithms have become increasingly popular and are being used both at the supervisory level and the regulatory level. However, most of the benchmarks developed so far have not tackled the problem of assessing the performance of predictive controllers. A new normalised performance index ($\Psi(HP)$) for benchmarking controller performance against the GPC performance is developed in this section. The GPC benchmarking index can also be used to assess the performance of non-predictive process controllers against a predictive controller. This can provide a measure to indicate whether it is appropriate to replace an existing classical process controller with a more advanced predictive controller. The route followed to obtain the GP SISO benchmarking algorithm is modelled on that used by Harris and Desborough [1993], to obtain their minimum variance performance index, and it is shown that the new normalised performance index can be estimated by linear regression methods.

3.6.1 Introduction to Generalised Predictive Control

The Generalised Predictive Controller (GPC) [Clarke *et al.*, 1987, Clarke and Mohtadi, 1989] is perhaps one of the best known algorithms in the field of Model Predictive Control. Moreover, the GPC algorithm can be placed within the general class of Minimum Variance (MV) algorithms, which suggests that it can be similarly suitable for benchmarking purposes. A sketch of the derivation of the GPC controller [Clarke *et al.*, 1987] now follows.

Consider an open loop process assumed to be linear, discrete time and single input, single output, with input-output relation as in Equation (3.84)

$$y(t) = \frac{z^{-d-1}B(z^{-1})}{A(z^{-1})} u(t) \tag{3.84}$$

where A, B and C are polynomials in backshift operator z^{-1} as defined in Equations (3.2) to (3.4). If the process is disturbed by a stochastic disturbance appearing at the output, then Equation (3.84) can be rewritten as:

$$y(t) = \frac{z^{-d-1}B(z^{-1})}{A(z^{-1})}u(t) + \xi(t) \qquad (3.85)$$

where:

$$\xi(t) = \frac{C(z^{-1})}{D(z^{-1})}v(t) \qquad (3.86)$$

where $v(t)$ is an external discrete white noise source with zero mean and a variance σ_v^2 driving colouring filters. In most cases the disturbance polynomials C and D are unknown and hence D is assumed equal to $\Delta A(z^{-1})$, where $\Delta = (1 - z^{-1})$, and the polynomial C is used as a design parameter to influence the parameterisation and hence the performance of the controller. Hence:

$$\xi(t) = \frac{C(z^{-1})}{\Delta A(z^{-1})}v(t) \qquad (3.87)$$

and

$$\Delta A(z^{-1})y(t) = z^{-d}B(z^{-1})\Delta u(t-1) + Cv(t) \qquad (3.88)$$

The j-step predictor for the process can be derived by using a Diophantine equation to effectively split the polynomial C into two time frames as in minimum variance control. Define the polynomials E_j and F_j such that:

$$E_j \Delta A = C - z^{-j}F_j \qquad (3.89)$$

where:

$$E_j(z^{-1}) = 1 + e_1 z^{-1} + \ldots\ldots + e_{j-1}z^{-j+1} \qquad (3.90)$$

$$F_j(z^{-1}) = f_0 + f_1 z^{-1} + \ldots\ldots + f_m z^{-m+1} \qquad (3.91)$$

where $m = \max(n_c - j, n_A)$.

Then at time $t=k$, by multiplying (3.88) by $z^j E_j$ obtains the j step prediction equation :

$$\Delta A(z^{-1})E_j(z^{-1})y(k+j) = z^{-d}E_j(z^{-1})B(z^{-1})\Delta u(k+j-1) + E_j(z^{-1})Cv(k+j) \qquad (3.92)$$

Substituting (3.89) gives:

$$y(k+j) = \frac{z^{-d}E_j B}{C}\Delta u(k+j-1) + \frac{F_j}{C}y(k) + E_j v(k+j) \qquad (3.93)$$

Notice that the last part of the sum, i.e. $E_j v(k+j)$ consists of future white noise signals. Therefore, the j-step predictor for the process output becomes:

$$\hat{y}(k+j) = \frac{z^{-d}E_j B}{C}\Delta u(k+j-1) + \frac{F_j}{C}y(k) \tag{3.94}$$

In order to split the j-step ahead predictor in Equation (3.93) into parts that are known at time k and future signals, another Diophantine equation is introduced:

$$\frac{E_j B}{C} = G_j + z^{-j+d}\frac{H_j}{C}, \qquad \text{for } j \geq d+1 \tag{3.95}$$

Using Equation (3.95):

$$\hat{y}(k+j) = G_j \Delta u(k+j-d-1) + \frac{H_j}{C}\Delta u(k-1) + \frac{F_j}{C}y(k) \tag{3.96}$$

hence:

$$\hat{y}(k+j) = G_j \Delta u(k+j-d-1) + f_j(k) \tag{3.97}$$

where: $f_j(k) = \dfrac{H_j}{C}\Delta u(k-1) + \dfrac{F_j}{C}y(k)$

The GPC cost function to be minimised is defined as the conditional expectation of the cost function J based on information available at time k and denoted as $E[J|k]$, where:

$$J = \sum_{j=N_1}^{N_2}\left[r(k+j) - \hat{y}(k+j)\right]^2 + \lambda\sum_{j=1}^{N_u}\left[\Delta u(k+j-1)\right]^2 \tag{3.98}$$

and λ represents weighting on the feedback control signal $\Delta u(k)$. Sometimes, instead of the scalar quantity λ, a vector representing dynamic filtering coefficients can be used (λ_j inside the second sum in Equation (3.98)).

The assumption must be made that a stabilising control law exist. The optimal control solution is required to be causal. The solution to this problem can be obtained by substituting the necessary parameters into the GPC criterion and solving for required control coefficients.

Substituting Equation (3.97) in Equation (3.98), obtains:

$$J = \sum_{j=N_1}^{N_2}\left[b_j(k) - G_j \Delta u(k+j-d-1)\right]^2 + \lambda\sum_{j=1}^{N_u}\left[\Delta u(k+j-1)\right]^2 \tag{3.99}$$

where: $b_j(k) = \left(r(k+j) - f_j(k)\right) \tag{3.100}$

Define:

$$\Delta U(k) = \begin{bmatrix} \Delta u(k) & \Delta u(k+1) & \cdots & \Delta u(k+N_u) \end{bmatrix}^T$$

$$\mathbf{b}(k) = \begin{bmatrix} b_{N_1}(k) & b_{N_1+1}(k) & \cdots & b_{N_2}(k) \end{bmatrix}^T \qquad (3.101)$$

then writing Equation (3.99) in vector notation obtains:

$$J = \begin{bmatrix} \mathbf{b}(k) - \mathbf{G}\Delta U(k) \end{bmatrix}^T \begin{bmatrix} \mathbf{b}(k) - \mathbf{G}\Delta U(k) \end{bmatrix} + \lambda \begin{bmatrix} \Delta U(k) \end{bmatrix}^T \begin{bmatrix} \Delta U(k) \end{bmatrix} \qquad (3.102)$$

where, $\mathbf{G}$ is a lower triangular $(Hp \times Hp)$ matrix whose elements in the row j are coefficients of the polynomial G_j.

Differentiating Equation (3.102) with respect to the vector $\Delta U(k)$, and setting $\dfrac{dJ}{d(\Delta U(k))} = 0$, obtain:

$$\Delta U(k) = \begin{bmatrix} \mathbf{G}^T\mathbf{G} + \lambda \, \mathbf{I} \end{bmatrix}^{-1} \mathbf{G}^T \mathbf{b}(k) \qquad (3.103)$$

3.6.2 SISO GPC Benchmarking

To introduce the concept of GPC control performance assessment, consider a time invariant process whose behaviour about a nominal operating point can be modelled by a linear transfer function with additive disturbance,

$$\ddot{y}(k) = \frac{z^{-d-1}B(z^{-1})}{A(z^{-1})}\ddot{u}(k) + \xi(k) \qquad (3.104)$$

where $\ddot{y}(k) = y(k) - \mu_y$ denotes the difference between the process output variable and a nominal operating point (μ_y) assumed to be the steady state mean value of the process output. And where $\ddot{u}(k) = u(k) - \mu_u$ denotes the difference between the controller output variable and a nominal operating point (μ_u) assumed to be the steady state mean value of the controller output.

It is assumed that the process disturbance $\xi(k)$ is represented by an autoregressive integrated time series of the form:

$$\xi(k) = \frac{C(z^{-1})}{\Delta A(z^{-1})} v(k) \qquad (3.105)$$

Applying Diophantine equations, in the same way as in the previous section, the j-step nominal future process output becomes:

$$\hat{\ddot{y}}(k+j) = G_j \Delta \ddot{u}(k+j-d-1) + f(k) \qquad (3.106)$$

The modified GPC cost function to be minimised (Ω), which will provide the basis of the benchmark cost is defined as the unconditional expectation of a function J:

$$\Omega = \mathrm{E}[J] \tag{3.107}$$

$$J = \sum_{j=N_1}^{N_2} [\bar{e}(k+j)]^2 + \lambda \sum_{j=1}^{N_u} [\Delta \bar{u}(k+j-1)]^2 \tag{3.108}$$

where $\bar{e}(k+j) = e(k+j) - \mu_e$ denotes the difference between the future process output error and a nominal operating point (μ_e) assumed to be the steady state mean value of the process error, and where $\Delta \bar{u}(k+j-1) = \Delta(u(k+j-1) - \mu_u)$ denotes the difference between the future incremental controller output variable and a nominal operating point (μ_u). Then, J can be written as:

$$J = \sum_{j=N_1}^{N_2} \left[\{r(k+j) - y(k+j)\} - \{\mu_r - \mu_y\} \right]^2 + \lambda \sum_{j=1}^{N_u} [\Delta \bar{u}(k+j-1)]^2 \tag{3.109}$$

It is assumed that the process set point remains constant, i.e. $r(k+j) = r(k) = \mu_r$, therefore:

$$J = \sum_{j=N_1}^{N_2} [-\bar{y}(k+j)]^2 + \lambda \sum_{j=1}^{N_u} [\Delta \bar{u}(k+j-1)]^2 \tag{3.110}$$

$$J = \left\{ \begin{array}{l} \sum_{j=N_1}^{N_2} \left[-G_j \Delta \bar{u}(k+j-d-1) - H_j \Delta \bar{u}(k-1) - F_j \bar{y}(k) - E_j v(k+j) \right]^2 \\ + \lambda \sum_{j=1}^{N_u} [\Delta \bar{u}(k+j-1)]^2 \end{array} \right\} \tag{3.111}$$

Taking the expectation of Equation (3.111) and noting that the expectation of all cross terms involving the white noise signal will be zero gives:

$$\Omega = \mathrm{E}\left[\Delta \bar{u}^T \left[G^T G + \lambda I \right] \Delta \bar{u} - \Delta \bar{u}^T (G^T b) - (b^T G) \Delta \bar{u} + b^T b + v^T E^T E v \right] \tag{3.112}$$

where: G and b were defined after Equation (3.102), and E is a lower triangular matrix whose elements in the row j are coefficients of the polynomial E_j.

Equation (3.112) can be presented as the summation of two terms, one independent of future or past process output or control action, and hence irreducible and the second dependent on future control action and hence reducible:

$$\Omega = J_{\min} + J_0 \tag{3.113}$$

where:

$$J_{min} = E\left[\mathbf{v}^T\mathbf{E}^T\mathbf{E}\mathbf{v}\right] = \sigma_v^2 \sum_{j=N_1}^{N_2} E_j^2 , \qquad (3.114)$$

$$J_0 = E\left[\Delta\tilde{\mathbf{u}}^T\left[\mathbf{G}^T\mathbf{G} + \lambda\mathbf{I}\right]\Delta\tilde{\mathbf{u}} - \Delta\tilde{\mathbf{u}}^T(\mathbf{G}^T\mathbf{b}) - (\mathbf{b}^T\mathbf{G})\Delta\tilde{\mathbf{u}} + \mathbf{b}^T\mathbf{b}\right] \qquad (3.115)$$

The term $E\left[\mathbf{v}^T\mathbf{E}^T\mathbf{E}\mathbf{v}\right]$ can be interpreted as summation of the minimum variance of the nominal process output over the prediction/minimisation interval $HP = N_2 - N_1 + 1$. The minimum performance cost Ω will be the sum of J_{min} and the minimum value of J_0. The minimum (optimal) value of the cost J_0 can be obtained by substituting the optimal control equation, derived in the previous section, to obtain:

$$J_0 = E\left[\mathbf{b}^T\mathbf{b} - (\mathbf{b}^T\mathbf{G})\left[\mathbf{G}^T\mathbf{G} + \lambda\mathbf{I}\right]^{-1}\mathbf{G}^T\mathbf{b}\right] \qquad (3.116)$$

Thus, the minimum performance cost Ω:

$$\Omega = \sigma_v^2 \sum_{j=N_1}^{N_2} E_j^2 + E\left[\mathbf{b}^T\mathbf{b} - (\mathbf{b}^T\mathbf{G})\left[\mathbf{G}^T\mathbf{G} + \lambda\mathbf{I}\right]^{-1}\mathbf{G}^T\mathbf{b}\right] \qquad (3.117)$$

At time k_l, the performance bound Ω can be interpreted as the minimum value of the performance cost function achievable in the interval of time (k_l+N_l) to (k_l+N_2) for a process described by Equation (3.104). This cost is achievable if the control signals applied in the interval (k_l) to (k_l+HP) are given by Equation (3.103). Given that the process described in Equation (3.104) is defined as linear time invariant, then this property allows us to look back in time at the performance of the process and say what the minimum value of the cost function would have been in the interval (k_l-HP) to (k_l). This cost would have been achieved over this interval if the control action applied in the interval $(k_l- N_2)$ to $(k_l- N_l)$ were given by Equation (3.103). Hence, not only the future performance of a process, but also the past performance can be assessed.

It is appropriate at this stage to consider the limiting case of this performance indicator. This case can be characterised by the following assumptions, $\lambda = 0$, $N_u = HP$, $N_2 - N_1 + 1 = HP$, $N_1 = d + 1$. Under these conditions the optimal value of J_0 reduces to:

$$J_0 = E\left[\mathbf{b}^T\mathbf{b} - (\mathbf{b}^T\mathbf{G})\left[\mathbf{G}^{-1}\mathbf{G}^{-T}\right]\mathbf{G}^T\mathbf{b}\right] = E\left[\mathbf{b}^T\mathbf{b} - \mathbf{b}^T\mathbf{b}\right] = 0 \qquad (3.118)$$

which is equivalent to the minimum variance performance cost

$$\Omega = J_{min} = \sigma_v^2 \sum_{j=N_1}^{N_2} E_j^2 \qquad (3.119)$$

3.6.3 Estimation of the Normalised SISO GPC Performance Index

Now that a theoretical minimum cost is available, it is possible to express the current process controller performance in terms of the theoretical minimum variance of the process output and the controller output. Following the lead of [Harris et al., 1996] and to account for offset in the process output, it is useful to assess the controller performance using the mean squared error (mse) of the process output and the controller output given as,

$$\sum_{k_1=k-HP}^{K} mse(y_{k_1}) + \lambda \times mse(\Delta u_{k_1-N_1}) = \left\{ \sigma_y^2 + \lambda \sigma_{\Delta \bar{u}}^2 + (\lambda \mu_{\bar{u}}^2 + \mu_y^2) \right\} \tag{3.120}$$

where

$$\sigma_y^2 = \left\{ \left[\bar{y}_{k_1-HP} \quad \bar{y}_{k_1-HP+1} \quad \cdots \quad \bar{y}_{k_1} \right] \times \left[\bar{y}_{k_1-HP} \quad \bar{y}_{k_1-HP+1} \quad \cdots \quad \bar{y}_{k_1} \right]^T \right\} \tag{3.121}$$

$$\sigma_{\Delta \bar{u}}^2 = \left\{ \left[\Delta \bar{u}_{k_1-HP-d-1} \quad \Delta \bar{u}_{k_1-HP-d} \quad \cdots \quad \Delta \bar{u}_{k_1-d-1} \right] \times \left[\Delta \bar{u}_{k_1-HP-d-1} \quad \Delta \bar{u}_{k_1-HP-d} \quad \cdots \quad \Delta \bar{u}_{k_1-d-1} \right]^T \right\} \tag{3.122}$$

and μ_y, μ_u are the mean deviation of the process output from set point and the mean deviation of the control output from steady state operating point. The normalised performance index $\Psi(HP)$, which expresses the fractional increase in the process output and controller output mean square errors that arise from not implementing a GPC controller is defined as

$$\Psi_{(HP)} = \frac{\Omega}{\left\{ \sigma_y^2 + \lambda \sigma_{\Delta \bar{u}}^2 + (\lambda \mu_{\bar{u}}^2 + \mu_y^2) \right\}} \qquad \Psi_{(HP)}, \text{ lies within } [0, 1]. \tag{3.123}$$

To obtain an estimate of performance index $\Psi(HP)$, it is necessary to estimate Ω. An estimate of Ω can be obtained from routine operating plant data. Some a priori knowledge of the process dead time and estimate of system order is required. The procedure consists of fitting a controlled autoregressive integral moving average time series. The approach eliminates the necessity of solving the Diophantine equations. From Equation (3.106) the j-step nominal future process output is given by:

$$\bar{y}(k+j) = G_j \Delta \bar{u}(k+j-d-1) + H_j \Delta \bar{u}(k-1) + F_j \bar{y}(k) + E_j v(k+j) \tag{3.124}$$

If it is assumed that $k = k_1 - j$, where k_1 is the present time index, then

$$\bar{y}(k_1) = G_j \Delta \bar{u}(k_1-d-1) + H_j \Delta \bar{u}(k_1-j-1) + F_j \bar{y}(k_1-j) + E_j v(k_1) \quad \text{for } j \geq d+1 \tag{3.125}$$

and:

$$E_j v(k_1) = \bar{y}(k_1) - \left\{ G_j \Delta \bar{u}(k_1-d-1) + H_j \Delta \bar{u}(k_1-j-1) + F_j \bar{y}(k_1-j) \right\} \tag{3.126}$$

Noting that,

$$G_j \Delta \ddot{u}(k_1 - d - 1) = g_0 \Delta \ddot{u}(k_1 - d - 1) + g_1 \Delta \ddot{u}(k_1 - d - 2) + \ldots\ldots + g_{j-1} \Delta \ddot{u}(k_1 - d - j)$$

$$(3.127)$$

$$H_j \Delta \ddot{u}(k_1 - j - 1) = h_0 \Delta \ddot{u}(k_1 - j - 1) + h_1 \Delta \ddot{u}(k_1 - j - 2) + \ldots\ldots + h_{n_h - 1} \Delta \ddot{u}(k_1 - j - n_h)$$

$$(3.128)$$

$$F_j \ddot{y}(k_1 - j) = f_0 \ddot{y}(k_1 - j) + f_1 \ddot{y}(k_1 - j - 1) + \ldots\ldots + f_{n_f - 1} \ddot{y}(k_1 - j - n_f - 1) \qquad (3.129)$$

where , $n_f = \max(n_c - j, \ n_{\Delta\theta A} - 1)$ and $n_h = \max(n_B - j + d, \ n_A - 1)$. Then, using data points k_1 to k_N, Equation (3.129) is written in matrix notation as:

$$\xi = Y_j - X_j \phi_j \qquad (3.130)$$

where $k_1 > k_N$, and:

$$\xi = \begin{bmatrix} E_j v_{k_1} \\ E_j v_{k_2} \\ \vdots \\ E_j v_{k_N} \end{bmatrix} k_N \quad \phi_j = \begin{bmatrix} g_0 & \cdots & g_{j-1} & h_0 & \cdots & h_{n_h - 1} & f_0 & \cdots & f_{n_f - 1} \end{bmatrix}^T \qquad (3.131)$$

$$\mathbf{X}_j = \begin{bmatrix} \Delta \ddot{u}_{k_1 - d - 1} & \cdots & \Delta \ddot{u}_{k_1 - d - j} & \Delta \ddot{u}_{k_1 - j - 1} & \cdots & \Delta \ddot{u}_{k_1 - j - n_h} & \ddot{y}_{k_1 - j} & \cdots & \ddot{y}_{k_1 - j - n_f - 1} \\ \Delta \ddot{u}_{k_2 - d - 1} & \vdots & \vdots & \Delta \ddot{u}_{k_2 - j - 1} & \cdots & \Delta \ddot{u}_{k_2 - j - n_h} & \ddot{y}_{k_2 - j} & \vdots & \ddot{y}_{k_2 - j - n_f - 1} \\ \vdots & \vdots & \vdots & \vdots & \cdots & \vdots & \vdots & \vdots & \vdots \\ \Delta \ddot{u}_{k_N - d - 1} & \cdots & \Delta \ddot{u}_{k_N - d - j} & \Delta \ddot{u}_{k_N - j - 1} & \cdots & \Delta \ddot{u}_{k_N - j - n_h} & \ddot{y}_{k_N - j} & \cdots & \ddot{y}_{k_N - j - n_f - 1} \end{bmatrix} \qquad (3.132)$$

Vector ϕ_j contains the coefficients of the polynomials $G_j \ H_j \ F_j$. The estimates of ϕ_j are found by minimizing:

$$\Theta_j = \left(Y_j - X_j \phi_j \right)^T \left(Y_j - X_j \phi_j \right) \qquad (3.133)$$

leading to:

$$\phi_j = \left(X_j^T X \right)^{-1} X_j^T Y_j \qquad (3.134)$$

To get the estimates of matrices $\tilde{\mathbf{G}}, \ \tilde{\mathbf{H}}, \ \tilde{\mathbf{F}}$ which are needed to obtain the minimum performance cost, the minimization problem for Θ_j, has to be solved *HP* number of times from $j = N_1$ to N_2. Noting the structure of the matrix $\mathbf{G}$, the size of the optimisation problem can reduced by using an iterative method to obtain the parameters of $\tilde{\mathbf{G}}, \ \tilde{\mathbf{H}}, \ \tilde{\mathbf{F}}$. Over the iteration interval, the residual mean squared error is given by:

$$\kappa^2 = \left(\ddot{\mathbf{Y}} - \tilde{\mathbf{G}}\Delta\ddot{\mathbf{u}} - \tilde{\mathbf{H}}\Delta\ddot{\upsilon} - \tilde{\mathbf{F}}\ddot{\mathbf{y}}\right)^T \left(\ddot{\mathbf{Y}} - \tilde{\mathbf{G}}\Delta\ddot{\mathbf{u}} - \tilde{\mathbf{H}}\Delta\ddot{\upsilon} - \tilde{\mathbf{F}}\ddot{\mathbf{y}}\right) \tag{3.135}$$

where:

$$\Delta\ddot{\mathbf{u}} = \begin{bmatrix} \Delta\ddot{u}_{k_1-N_2} \\ \vdots \\ \vdots \\ \Delta\ddot{u}_{k_1-N_1} \end{bmatrix} \quad \ddot{\mathbf{Y}} = \begin{bmatrix} \ddot{y}_{k_1-HP} \\ \vdots \\ \ddot{y}_{k_1} \end{bmatrix} \quad \Delta\ddot{\upsilon} = \begin{bmatrix} \Delta\ddot{\upsilon}_{k_1-N_2-1} \\ \vdots \\ \Delta\ddot{\upsilon}_{k_1-N_2-n_h} \end{bmatrix} \quad \ddot{y} = \begin{bmatrix} \ddot{y}_{k_1-N_2} \\ \vdots \\ \ddot{y}_{k_1-N_2-n_f-1} \end{bmatrix} \tag{3.136}$$

This provides an estimate for the summed minimum variance expressed in Equation (3.114). The vector **b** can be estimated using matrices **H**, **F** and a vector of past nominal process output and controller output signals; $\tilde{\mathbf{b}} = \left(-\tilde{\mathbf{H}}\Delta\ddot{\mathbf{u}}_{(k_1-1)} - \tilde{\mathbf{F}}\ddot{\mathbf{y}}_{(k_1)}\right)$. The estimate of the minimum GPC performance cost is obtained as:

$$\tilde{\Omega} = \kappa^2 + \left(\tilde{\mathbf{b}}^T\tilde{\mathbf{b}} - (\tilde{\mathbf{b}}^T\tilde{\mathbf{G}})\left[\tilde{\mathbf{G}}^T\tilde{\mathbf{G}} + \lambda\,\mathbf{I}\right]^{-1}\tilde{\mathbf{G}}^T\tilde{\mathbf{b}}\right) \tag{3.137}$$

The least square estimate of the normalised GPC performance index $\Psi(HP)$ is given as:

$$\tilde{\Psi}(HP) = \frac{\kappa^2 + \left(\tilde{\mathbf{b}}^T\tilde{\mathbf{b}} - (\tilde{\mathbf{b}}^T\tilde{\mathbf{G}})\left[\tilde{\mathbf{G}}^T\tilde{\mathbf{G}} + \lambda\,\mathbf{I}\right]^{-1}\tilde{\mathbf{G}}^T\tilde{\mathbf{b}}\right)}{\left\{\ddot{\mathbf{Y}}^T\ddot{\mathbf{Y}} + \mu_y^2\right\} + \lambda\left\{\Delta\ddot{\mathbf{u}}^T\Delta\ddot{\mathbf{u}} + \mu_{\ddot{u}}^2\right\}} \tag{3.138}$$

A successive approximation algorithm to obtain the least square estimate of the normalised GPC performance index $\Psi(HP)$ in the time range k_1, where k is the present time index and $k_N < k_1$, $k_1 - k_N = 1000$, is set up as follows:

Procedure 3.6 Applying the Generalised Predictive Control Benchmark
Step 1:
Obtain appropriately collected closed-loop plant data for the controlled variable and manipulated variable (process output and controller output). This means that the data should be free of missing and spurious points, should not be compressed or filtered and should be long enough. The data should be well sampled to eliminate aliasing effects.
Step 2:
Set up matrices $\Delta\ddot{\mathbf{u}}$, $\ddot{\mathbf{Y}}$, $\Delta\ddot{\upsilon}$, $\ddot{y}$ as defined in Equation (3.136)
Step 3:
For $j = N_2$ set up matrices $\mathbf{Y}_j$, $\mathbf{X}_j$ as defined in Equation (3.131) and (3.132) use the least square solution in Equation (3.134) to obtain ϕ_j. From ϕ_j extract vectors $\tilde{\mathbf{G}}_{N_2}$, $\tilde{\mathbf{H}}_{N_2}$, $\tilde{\mathbf{F}}_{N_2}$

Step 4:

From the vector $\tilde{\mathbf{G}}_{N_2}$ which contains all the coefficients of the matrix $\tilde{\mathbf{G}}$ set-up

the matrix $\tilde{\mathbf{G}}$

Step 5:

For $j = N_1$ to $N_2 - 1$, set up the modified matrices and vector $\mathbf{Y}_j$, $\mathbf{X}_j$ defined as:

$$
\mathbf{Y}_j = \begin{bmatrix} \ddot{y}_{k_1} \\ \ddot{y}_{k_2} \\ \vdots \\ \ddot{y}_{k_N} \end{bmatrix} - \begin{bmatrix} \Delta \ddot{u}_{k_1-d-1} & \cdots & \Delta \ddot{u}_{k_1-d-j} \\ \Delta \ddot{u}_{k_2-d-1} & \vdots & \vdots \\ \vdots & \vdots & \vdots \\ \Delta \ddot{u}_{k_N-d-1} & \cdots & \Delta \ddot{u}_{k_N-d-j} \end{bmatrix} \times \begin{bmatrix} g_{j-1} \\ \vdots \\ g_1 \\ g_0 \end{bmatrix}
$$

$$
\mathbf{X}_j = \begin{bmatrix} \Delta \ddot{u}_{k_1-j-1} & \cdots & \Delta \ddot{u}_{k_1-j-n_h} & \ddot{y}_{k_1-j} & \cdots & \ddot{y}_{k_1-j-n_f-1} \\ \Delta \ddot{u}_{k_2-j-1} & \vdots & \Delta \ddot{u}_{k_2-j-n_h} & \ddot{y}_{k_2-j} & \cdots & \ddot{y}_{k_2-j-n_f-1} \\ \vdots & \vdots & \vdots & \vdots & \cdots & \vdots \\ \Delta \ddot{u}_{k_N-j-1} & \cdots & \Delta \ddot{u}_{k_N-j-n_h} & \ddot{y}_{k_N-j} & \cdots & \ddot{y}_{k_N-j-n_f-1} \end{bmatrix}
$$

Step 6:

Use the least squares solution to obtain the modified ϕ_j vector defined as:

$$
\phi_j = \begin{bmatrix} h_0 & \cdots & h_{n_h-1} & f_0 & \cdots & f_{n_f-1} \end{bmatrix}^T
$$

Step 7:

Set up the matrices $\tilde{\mathbf{H}}$, $\tilde{\mathbf{F}}$

Step 8:

Obtain the minimum variance sum

$$
\kappa^2 = \left\{ \left(\ddot{\mathbf{Y}} - \tilde{\mathbf{G}}\Delta \ddot{\mathbf{u}} - \tilde{\mathbf{H}}\Delta \ddot{\upsilon} - \tilde{\mathbf{F}}\ddot{\mathbf{y}} \right)^T \left(\ddot{\mathbf{Y}} - \tilde{\mathbf{G}}\Delta \ddot{\mathbf{u}} - \tilde{\mathbf{H}}\Delta \ddot{\upsilon} - \tilde{\mathbf{F}}\ddot{\mathbf{y}} \right) \right\}
$$

Step 9:

Obtain the vector $\tilde{\mathbf{b}} = \left(-\tilde{\mathbf{H}}\Delta \ddot{\mathbf{u}}(k_1 - 1) - \tilde{\mathbf{F}}\ddot{\mathbf{y}}(k_1) \right)$

Step 10:

Obtain the least square estimate of the normalised GPC performance index $\tilde{\psi}_{k_1}(HP)$

$$
\tilde{\psi}_{k_1}(HP) = \frac{\kappa^2 + \left(\tilde{\mathbf{b}}^T\tilde{\mathbf{b}} - (\tilde{\mathbf{b}}^T\tilde{\mathbf{G}})\left[\tilde{\mathbf{G}}^T\tilde{\mathbf{G}} + \lambda \mathbf{I} \right]^{-1} \tilde{\mathbf{G}}^T\tilde{\mathbf{b}} \right)}{\left\{ \ddot{\mathbf{Y}}^T\ddot{\mathbf{Y}} + \mu_y^2 \right\} + \left\{ \Delta \ddot{\mathbf{u}}^T\Delta \ddot{\mathbf{u}} + \mu_u^2 \right\}}
$$

Procedure end

Example 3.6 Generalised Predictive Control Benchmarking
This experiment was designed to illustrate the effectiveness of the GPC benchmark
index. A discrete time SISO process with a time delay of six (6) sample periods
was used. Two controllers were used in turn to control the process and the
performances compared using the GPC benchmark index. The controllers used in
the experiment were, a PID controller which in the first instance was poorly tuned
and then later re-tuned using Ziegler-Nichols method and a GPC controller . The
process data from the plant was sampled every 1 second and the following
performance index parameters where applied; $\lambda = 0.7$, $m = 5$, $N= 1200$, $N_1 = 7$
$N_2= 15$, process delay: $d =6$, *model order* = 4, SP=55.

Plant transfer function model:

$$y_k = \frac{z^{-7}\left(1 + 0.21\, z^{-1} + 0.002z^{-2}\right)}{1 - 1.42z^{-1} + 0.45z^{-2} - 0.0026z^{-3}} u_k \tag{3.139}$$

The sensor noise was assumed to be a white noise disturbance with zero mean
and unity variance. The disturbance model used to design the GPC controller is
defined as:

$$\xi_k = \frac{1 - 1.98\, z^{-1} + 0.9801\, z^{-2}}{1 - 2.42\, z^{-1} + 1.87\, z^{-2} - 0.4526z^{-3} + 0.0026z^{-4}} v_k \tag{3.140}$$

$\lambda = 0.7$, $N_u = HP$, $N_2 - N_1 + 1 = HP$, $N_1 = d + 1$, $N_2 = 15$

Poorly tuned PID Controller:

$$u_k = \frac{0.43698 - 0.74449z^{-1} + 0.3171z^{-2}}{1 - z^{-1}} \tag{3.141}$$

Ziegler-Nichols tuned PID Controller:

$$u_k = \frac{0.14566 - 0.24816z^{-1} + 0.1057z^{-2}}{1 - z^{-1}} \tag{3.142}$$

Figure 3.11 shows the behaviour of the system using the different controllers. On
the left hand side the step responses are presented: for the poorly tuned PID
controller, the retuned PID and for the GPC controller. On the right hand side,
there are values of the GPC benchmark index. It can be observed that for the
poorly tuned PID controller the GPC index is approximately 0.33, on a scale from
0 to 1. This confirms that the controller is poorly tuned, as seen from the step
response plot. The GPC benchmark index for the process controlled by the
retunded (Ziegler-Nichols) PID controller is 0.95, which when compared to the
previous performance of the PID controller shows a remarkable improvement. The
GPC benchmark index for the process controlled by a GPC controller equals unity,
as expected.

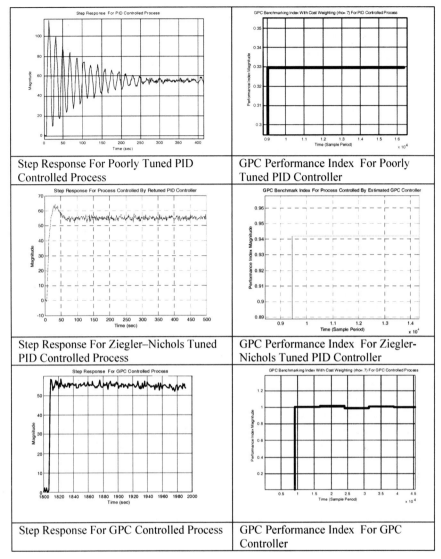

Figure 3.11. Step responses and performance indices for the two cases of PID control and for the GPC control

3.7 Summary

There has been considerable activity in the development of performance benchmarking tools based upon minimum variance criteria, that was recorded in the text by Huang and Shah. Harris showed how time series analysis techniques can be used to find an expression for the minimum possible cost from plant operating data. The main difficulty with the minimum-variance control law is that it often gives high gain, wide bandwidth and unrealistically large control signal variations. It has value as a benchmark cost though, since economic performance can often be related to the variance of the set point following error. It also has the virtue that the value of the minimum cost can be found from simple plant test results. However, the fact that in many real systems the measure of changes of the control signal is important in assessing performance, reduces the utility of the minimum variance criterion.

Although the GMV benchmark is an improvement on the minimum variance benchmarking index it does have a number of limitations. These include the fact that its stability depends on cost weightings and also that it compares the performance of the controller installed on the plant, which is usually a low order controller, against the full order optimal controller.

The controller performance assessment of multivariable systems, using the minimum variance controller as a benchmark, was also discussed. In order to avoid the problem of estimating the interactor matrix, it was proposed to estimate the upper and lower bounds of the minimum achievable variance, instead of the minimum variance itself. An upper bound of the MV benchmark can be estimated using only the known order of the interactor matrix, whereas the estimation of the lower bound requires only the knowledge of the minimum time delays between any given output and all the inputs. All these can be obtained from the I/O delay matrix of the plant. The lower bound can be estimated from routine operating data, while the estimation of the upper bound normally requires introducing additional delays to the controller. The method can be applied to evaluate the regulatory performance of MIMO industrial controllers.

The performance assessment, using the Generalised Predictive Controlled as a benchmark, was discussed for single-input, single-output systems. In this approach, the matrices needed for the predictive design are estimated from the past plant data and then used to assess the current performance.

The methods were summarised in step-by-step procedures and illustrated by examples.

4

Controller Benchmarking Procedures – Model-based Methods

Andrzej Ordys [1], Michael Grimble [2], Damien Uduehi [3] and Pawel Majecki [4]

[1] Professor of Automotive Engineering, Faculty of Engineering, Kingston University, London, UK,
[2] Professor and Director, Industrial Control Centre, Department of Electronic and Electrical Engineering, University of Strathclyde, Glasgow, UK
[3] Strategy and Planning Coordinator (West Africa), BG Group, Reading, UK,
[4] Research Engineer, Department of Electronic and Electrical Engineering, University of Strathclyde, Glasgow, UK

4.1 Introduction

There are generally two types of performance assessment and benchmarking algorithms. The MV, GMV and GPC benchmark algorithms reviewed so far only utilize normal operating records and do not need explicit models. The benchmark figures of merit are computed completely independently of the control law calculation. At the regulatory level the controller performance assessment is based on variances of the process and the controller outputs.

On the other hand, this chapter focuses on algorithms that require model information. Some of them enable limitations of equipment to be taken into consideration. Then, as a bi-product of the analysis, they also provide the optimal control law parameters, even for the case where the structure of the controller is limited to PID design. In those algorithms the benchmark performance measures are therefore provided together with the recommended controller parameters. The disadvantage is the requirement for more model information.

Section 4.2 looks at algorithms for multivariable systems derived from a predictive performance index which can evaluate steady state as well as dynamic system performance. Sections 4.3 and 4.4 focus on performance indices that can take into account the structure of the required controller.

4.2 Multivariable Benchmarking Based on Predictive Performance Index

4.2.1 Setting the Scene

For continuous processes, the performance of the controller in steady state is the most important factor. In this, the crucial key performance indicator (KPI) is the variance of a signal that can be related to process revenue.

For systems which exhibit frequent dynamic changes (for instance batch processes), the variance is still an important KPI, however there are other KPI's that directly affect the process revenue. These include:

- Rise Time
- % overshoot
- Settling time
- Steady state output error

It can be observed that the formulation of the predictive control, and especially the predictive performance index, contains explicitly the performance indicators needed for dynamic performance assessment. Indeed, consider a step change in the set-point signal, then, the predictive controller (which is designed to minimize the integral of the square of the tracking error over a finite horizon) will provide short rise time, low percentage overshoot and short settling time. Using the cost associated with the square of the input signal in the performance index will reduce the variance of this signal.

This section will present the derivation of equations which can be used to benchmark a system against a predictive performance index. In developing the benchmark index, based on the formulation of the GPC controller, one disadvantage is the fact that this formulation does not lead to the true minimum of the cost function. To reach the global minimum (in the steady state) the performance index over an infinite time period would have to be considered. This fact was first noted by MacGregor and Tidwell [1977] in relation to GMV controller. In the case of GMV controller, the corresponding optimal controller, i.e. the controller providing steady-state minimum of the GMV performance index, is the Linear Quadratic Gaussian (LQG) controller. For the predictive case, a corresponding (optimal) controller, which results in a minimum cost of its performance index, will be called Linear Quadratic Gaussian Predictive Controller (LQGPC). It will be derived in this chapter and used as a benchmarking reference. Some of the properties that make the LQGPC benchmark desirable include:

- LQG type cost functions are more robust than MV or GMV type functions
- LQG cost functions provide the lowest practically achievable performance bound.

Also, there are many situations in which the LQGPC benchmarking index would be very useful, including: benchmarking dynamic performance of process controllers, and benchmarking supervisory control systems. The LQGPC benchmark cost can used in situations where the main area of interest is a finite region / or the transient stage of the process as opposed to the steady state operating region. It is more applicable to servo, machine, automotive and robotic systems where reference tracking and transient behaviour are a priority.

Using predictive control benchmarks has the additional advantage that no separate benchmarking algorithm is required for benchmarking supervisory controllers. This characteristic of predictive controllers was investigated for SISO systems by Sáez *et al.* [2002] and for MIMO systems by Uduehi *et al.* [2003]. The results obtained show that the same algorithm used to benchmark multivariable regulatory level controllers can be utilised to benchmark the performance of supervisory controllers.

4.2.2 GPC/LQGPC Controller Derivation

Unlike in the SISO case, presented in Chapter 3, the derivation of predictive controllers in this chapter will be based on state-space models. State space models are more convenient for the representation of multivariable systems. The results of this section apply to both the MIMO and SISO cases. In the SISO case, the state-space GPC controller is equivalent to the polynomial version of GPC predictive controller, derived in Chapter 3.

Consider the system model in the state space form:

$$\begin{aligned} x_{t+1} &= Ax_t + Bu_t + Gv_t \\ y_t &= Dx_t + w_t \end{aligned}$$

(4.1)

where: x_t is a vector of system state variables of size n_x.

The initial state of the system is assumed to have Gaussian distribution with the mean value: $\mathrm{E}[x_0] = \overline{x}_0$ and the covariance matrix:

$$\mathrm{E}\left[\left(x_0 - \mathrm{E}[x_0] \right) \left(x_0 - \mathrm{E}[x_0] \right)^T \right] = X_0$$

(4.2)

u_t is a vector of control signals of size n_u,

y_t is a vector of output signals of size n_y,

A, B, G, D are constant matrices.

v_t and w_t are vectors of disturbances, whose elements are assumed to be Gaussian white noises with zero mean value and the co-variance matrix:

$$\mathrm{E}\left[\begin{bmatrix} v_t \\ w_t \end{bmatrix} \begin{bmatrix} v_t^T & w_t^T \end{bmatrix} \right] = \begin{bmatrix} V & Z \\ Z^T & W \end{bmatrix}$$

(4.3)

Notice that the correlation between state and output disturbances can be, for example, a result of converting from a polynomial description (CARMA or Box-Jenkins) into a state space description. If the disturbances in the state space formulation are correlated, i.e. Z in Equation (4.3) is non-zero, a transformation is applied which removes the correlation. This then enables a straightforward use of Kalman filter equations. Following the idea from [Anderson and Moore, 1979], denote:

$$\zeta_t = w_t - ZV^{-1}v_t \tag{4.4}$$

Then:

$$E\left[\begin{bmatrix} v_t \\ \zeta_t \end{bmatrix}\begin{bmatrix} v_t^T & \zeta_t^T \end{bmatrix}\right] = \begin{bmatrix} V & 0 \\ 0 & W - ZV^{-1}Z^T \end{bmatrix} \tag{4.5}$$

and the disturbances v_t and ζ_t are uncorrelated. Substituting (4.4) into the output equation in (4.2.1) yields:

$$\begin{aligned} x_{t+1} &= Ax_t + Bu_t + Gv_t \\ y_t &= Dx_t + \zeta_t + ZV^{-1}v_t \end{aligned} \tag{4.6}$$

The state is now expanded to incorporate the noise signal v_t:

$$\chi_{t+1} = \begin{bmatrix} x_{t+1} \\ v_{t+1} \end{bmatrix} = \begin{bmatrix} A & G \\ 0 & 0 \end{bmatrix}\begin{bmatrix} x_t \\ v_t \end{bmatrix} + \begin{bmatrix} B \\ 0 \end{bmatrix}u_t + \begin{bmatrix} 0 \\ I \end{bmatrix}v_{t+1} \tag{4.7}$$

$$y_t = \begin{bmatrix} D & ZV^{-1} \end{bmatrix}\begin{bmatrix} x_t \\ v_t \end{bmatrix} + \zeta_t \tag{4.8}$$

Equations (4.7) and (4.8) describe the system with uncorrelated state and output disturbances. The system initial state has a Gaussian distribution with the mean value:

$$\bar{\chi}_0 = \begin{bmatrix} E[x_0] \\ 0 \end{bmatrix} = \begin{bmatrix} \bar{x}_0 \\ 0 \end{bmatrix}$$

and the co-variance matrix:

$$E\left[(\chi_o - E[x_0])(\chi_o - E[x_0])^T\right] = \begin{bmatrix} X_0 & 0 \\ 0 & V \end{bmatrix}$$

The predictive control algorithms usually incorporate integral action. It is possible to introduce the integral action in the controller design by changing the model which is used for that design. The modified model will have as the control input the increment in the control action Δu_t rather than the actual control. This can be achieved by the following substitution [Hangstrup et al., 1997]:

$$\chi_{t+1} = \begin{bmatrix} x_{t+1} \\ u_t \end{bmatrix} = \begin{bmatrix} A & B \\ 0 & I \end{bmatrix}\begin{bmatrix} x_t \\ u_{t-1} \end{bmatrix} + \begin{bmatrix} B \\ I \end{bmatrix}\Delta u_t + \begin{bmatrix} G \\ 0 \end{bmatrix}v_t \tag{4.9}$$

$$y_t = \begin{bmatrix} D & 0 \end{bmatrix}\chi_t + w_t \tag{4.10}$$

where the new control signal is: $\Delta u_t = u_t - u_{t-1}$.

Remark: For the systems with so-called direct feed through, a more general state space formulation can be used:

$$\begin{aligned} x_{t+1} &= Ax_t + Bu_t + Gv_t \\ y_t &= Dx_t + Cu_t + w_t \end{aligned} \tag{4.11}$$

In this case, the state vector is extended [Ordys and Pike, 1998] as follows:

$$\chi_{t+1} = \begin{bmatrix} x_{t+1} \\ u_{t+1} \end{bmatrix} = \begin{bmatrix} A & B \\ 0 & I \end{bmatrix}\begin{bmatrix} x_t \\ u_t \end{bmatrix} + \begin{bmatrix} 0 \\ I \end{bmatrix}\Delta u_{t+1} + \begin{bmatrix} G \\ 0 \end{bmatrix}v_t \tag{4.12}$$

$$y_t = \begin{bmatrix} D & C \end{bmatrix}\chi_t + w_t \tag{4.13}$$

State estimator
A state space representation has been used to describe the MIMO system under consideration, however in the process industry it is more common to find systems using output feedback as opposed to state feedback. Hence to obtain the LQGPC/GPC controllers using the state equations described above, a state estimator is required. Assuming a known distribution for the initial state of the system (Equation (4.2)) the estimate of the system state can be calculated using standard Kalman filter equations:

$$\hat{x}_{t+1/t+1} = A\hat{x}_{t/t} + Bu_t + K_{t+1}\left[y_{t+1} - D\left(A\hat{x}_{t/t} + Bu_t\right)\right] = \hat{x}_{t+1/t} + K_{t+1}\left[y_{t+1} - D\hat{x}_{t+1/t}\right] \tag{4.14}$$

$$K_{t+1} = P_{t+1/t}D^T\left(DP_{t+1/t}D^T + W\right)^{-1} \tag{4.15}$$

$$P_{t+1/t} = AP_{t/t}A^T + GVG^T \tag{4.16}$$

$$P_{t+1/t+1} = P_{t+1/t} - P_{t+1/t}D^T\left(DP_{t+1/t}D^T + W\right)^{-1}DP_{t+1/t} \tag{4.17}$$

or, equivalent to the above:

$$\begin{aligned} \hat{x}_{t+1/t} &= A\hat{x}_{t/t-1} + AK_t\left[y_t - D\hat{x}_{t/t-1}\right] + Bu_t \\ &= \left(A - AK_tD\right)\hat{x}_{t/t-1} + AK_ty_t + Bu_t \end{aligned} \tag{4.18}$$

$$P_{t+1/t} = GVG^T + A\left[P_{t/t-1} - P_{t/t-1}D^T\left(DP_{t/t-1}D^T + W\right)^{-1}DP_{t/t-1}\right]A^T \tag{4.19}$$

with the initial conditions (assuming that the first available measurement is y_0):

$$\hat{x}_{0/-1} = E\left[x_0\right] = \bar{x}_0 \tag{4.20}$$

$$P_{0/-1} = X_0 \tag{4.21}$$

or (equivalent to the above):

$$K_0 = X_0 D^T \left(D X_0 D^T + W \right)^{-1} \tag{4.22}$$

$$P_{0/0} = X_0 - X_0 D^T \left(D X_0 D^T + W \right)^{-1} D X_0 \tag{4.23}$$

$$\hat{x}_0 = \mathrm{E}[x_0] + K_0 \left[y_0 - D\mathrm{E}[x_0] \right] \tag{4.24}$$

When the iterations described by (4.15), (4.16) and (4.17) (or, equivalently, by (4.15) and (4.19)) reach the steady state, the Kalman filter becomes stationary and it is then described by the equations:

$$\begin{aligned}\hat{x}_{t+1/t+1} &= A\hat{x}_{t/t} + Bu_t + K\left[y_{t+1} - D\left(A\hat{x}_{t/t} + Bu_t \right) \right] \\ &= (A - KDA)\hat{x}_{t/t} + (B - KDB)u_t + Ky_{t+1}\end{aligned} \tag{4.25}$$

$$K = P_{pred} D^T \left(D P_{pred} D^T + W \right)^{-1} \tag{4.26}$$

or, in the predictor form:

$$\hat{x}_{t+1/t} = (A - AKD)\hat{x}_{t/t-1} + AKy_t + Bu_t \tag{4.27}$$

where P_{pred} is a steady-state solution of the Kalman filter prediction equation:

$$P_{pred} = GVG^T + A\left[P_{pred} - P_{pred} D^T \left(D P_{pred} D^T + W \right)^{-1} D P_{pred} \right] A^T \tag{4.28}$$

If the model used in the controller design is an incremental model (including integral action), as described by Equations (4.9) and (4.10) certain simplifications to the Kalman filter equations are possible. The initial conditions can be set-up as follows:

$$\hat{\chi}_{0/-1} = \begin{bmatrix} \hat{x}_{0/-1} \\ \hat{u}_{-1/-2} \end{bmatrix} = \begin{bmatrix} \bar{x}_0 \\ u_{-1} \end{bmatrix} \tag{4.29}$$

where u_{-1} is the known (constant) level of the control signal before the Kalman filter is put into action, and,

$$P_{0/-1} = \begin{bmatrix} X_0 & 0 \\ 0 & 0 \end{bmatrix} \tag{4.30}$$

Using the definitions of block matrices as in Equations (4.9) and (4.10) and denoting those block matrices by $\breve{A}$, $\breve{B}$, $\breve{G}$, $\breve{D}$, the iterations of the Kalman filter Riccati equation as described by (4.16) and (4.17) will become:

$$\breve{P}_{t+1/t} = \begin{bmatrix} P_{t+1/t} & 0 \\ 0 & 0 \end{bmatrix} \tag{4.31}$$

where $P_{t+1/t}$ is updated using the prediction covariance Equation (4.19).

The Kalman filter gain is given by:

$$\tilde{K}_{t+1} = \tilde{P}_{t+1/t} \bar{D}^T \left(\bar{D} \tilde{P}_{t+1/t} \bar{D}^T + W_t \right)^{-1} = \begin{bmatrix} K_{t+1} \\ 0 \end{bmatrix} = \begin{bmatrix} P_{t+1/t} D^T \left(DP_{t+1/t} D^T + W_t \right)^{-1} \\ 0 \end{bmatrix} \qquad (4.32)$$

Then, the state estimate update is given by:

$$\hat{\chi}_{t+1/t+1} = \bar{A}\hat{\chi}_{t/t} + \bar{B}\Delta u_t + \tilde{K}_{t+1} \left[y_{t+1} - \bar{D} \left(\bar{A}\hat{\chi}_{t/t} + \bar{B}\Delta u_t \right) \right]$$

$$\begin{bmatrix} \hat{x}_{t+1/t+1} \\ \hat{u}_{t/t} \end{bmatrix} = \begin{bmatrix} A\hat{x}_{t/t} + Bu_t + K_{t+1} \left[y_{t+1} - D \left(A\hat{x}_{t/t} + Bu_t \right) \right] \\ u_t = u_{t-1} + \Delta u_t \end{bmatrix} \qquad (4.33)$$

Therefore, the calculations of the Kalman filter can be performed for the original system, i.e. the system described by Equations (4.1) or (4.7) and (4. 8).

Output Prediction
Following [Ordys and Clarke, 1993], the *k*-step prediction of the output signal may be calculated from the relationship:

$$\hat{y}_{(t+k)/t} = E\left[y_{t+k} | t \right] = DA^k \hat{x}_{t/t} + \sum_{j=1}^{k} DA^{k-j} Bu_{t+j-1} \qquad (4.34)$$

Collecting together the formulae for $\hat{y}_{(t+k)/t}$ when *k* changes from *1* to *N+1* obtains in a block matrix form:

$$\hat{Y}_{t,N} = \begin{bmatrix} \hat{y}_{(t+1)/t} \\ \hat{y}_{(t+2)/t} \\ \vdots \\ \hat{y}_{(t+N+1)/t} \end{bmatrix} = \begin{bmatrix} D \\ DA \\ \vdots \\ DA^N \end{bmatrix} A\hat{x}_{t/t} + \begin{bmatrix} DB & O & \cdots & O \\ DAB & DB & & \vdots \\ \vdots & \vdots & \ddots & O \\ DA^N B & DA^{N-1} B & \cdots & DB \end{bmatrix} \begin{bmatrix} u_t \\ u_{t+1} \\ \vdots \\ u_{t+N} \end{bmatrix} \qquad (4.35)$$

State-space GPC Controller
The performance index to be minimized is defined as follows:

$$J_t = E\left[\sum_{j=0}^{N} \left(y_{t+j+1} - r_{t+j+1} \right)^T \tilde{Q}_e \left(y_{t+j+1} - r_{t+j+1} \right) + u_{t+j}^T \tilde{Q}_u u_{t+j} \right] \qquad (4.36)$$

where: r_{t+j+1} represents   a   vector   of   reference   (set   point)   signals, $\tilde{Q}_e > 0$ and $\tilde{Q}_u \geq 0$ are weighting matrices. Denote:

$$R_{t,N} = \begin{bmatrix} r_{t+1} \\ r_{t+2} \\ \vdots \\ r_{t+N+1} \end{bmatrix}, \quad U_{t,N} = \begin{bmatrix} u_t \\ u_{t+1} \\ \vdots \\ u_{t+N} \end{bmatrix}, \quad Q_e = diag_N \left(\tilde{Q}_e \right), \quad Q_u = diag_N \left(\tilde{Q}_u \right) \qquad (4.37)$$

so that $R_{t,N}$ is a block vector of $N+1$ future reference signals, $U_{t,N}$ is a block vector of $N+1$ future control signals. Then, the performance index, neglecting a constant (control independent) term, can be expressed in a static (vector) form:

$$J_t = \left(\hat{Y}_{t,N} - R_{t,N}\right)^T Q_e \left(\hat{Y}_{t,N} - R_{t,N}\right) + U_{t,N}^{\ T} Q_u U_{t,N} \tag{4.38}$$

Thus, by substituting (4.35) and finding the stationary point, the vector of optimal control signals is found as:

$$U_{t,N} = \left(S_N^{\ T} Q_e S_N + Q_u\right)^{-1} S_N^{\ T} Q_e \left(R_{t,N} - F_{t,N}\right) \tag{4.39}$$

where the following matrix and vector are defined:

$$S_N = \begin{bmatrix} DB & O & \cdots & O \\ DAB & DB & & \vdots \\ \vdots & \vdots & \ddots & O \\ DA^N B & DA^{N-1}B & \cdots & DB \end{bmatrix}, \tag{4.40}$$

$$F_{t,N} = \begin{bmatrix} D \\ DA \\ \vdots \\ DA^N \end{bmatrix} A\hat{x}_{t/t} = \Phi_N A\hat{x}_{t/t} \quad \text{and} \quad \Phi_N = \begin{bmatrix} D \\ DA \\ \vdots \\ DA^N \end{bmatrix} \tag{4.41}$$

LQGPC Controller
Notice that the GPC performance index involves a "one shot" optimisation, to obtain the control action (vector $U_{t,N}$). It is therefore similar to the GMV minimisation problem. To minimise a performance index similar to LQG, but suitable for predictive formulation, the LQGPC algorithm is introduced (Ordys *et al.*, 2006). The LQGPC performance index is defined as a sum of the indices in the form of (4.38):

$$J = E\left\{ \lim_{T_f \to \infty} \frac{1}{T_f + 1} \sum_{t=0}^{T_f} J_t \right\} \tag{4.42}$$

If the vector of $N+1$ future control signals (4.2.3) is considered as an input to the system and the vector of $N+1$ future outputs (4.35) is treated as an output, then the state space Equations (4.2.1) can be rewritten in the form:

$$x_{t+1} = Ax_t + \beta U_{t,N} + \Gamma V_{t,N} \tag{4.43}$$

$$Y_{t,N} = \Phi_N Ax_t + S_N U_{t,N} + \tilde{S}_N V_{t,N} + W_{t,N} \tag{4.44}$$

where the system matrices are defined as:

$$\beta = \begin{bmatrix} B & O & \cdots & O \end{bmatrix} , \quad \Gamma = \begin{bmatrix} G & O & \cdots & O \end{bmatrix} \tag{4.45}$$

$$\tilde{S}_N = \begin{bmatrix} DG & O & \cdots & O \\ DAG & DG & & \vdots \\ \vdots & \vdots & \ddots & O \\ DA^N G & DA^{N-1}G & \cdots & DG \end{bmatrix} \tag{4.46}$$

and:

$$V_{t,N} = \begin{bmatrix} v_t \\ v_{t+1} \\ \vdots \\ v_{t+N} \end{bmatrix}, \quad W_{t,N} = \begin{bmatrix} w_{t+1} \\ w_{t+2} \\ \vdots \\ w_{t+N+1} \end{bmatrix} \tag{4.47}$$

Substituting the output Equation (4.44) into the performance index (4.42) obtains:

$$
\begin{aligned}
J &= E\left[\lim_{T_f \to \infty} \frac{1}{T_f+1} \sum_{t=0}^{T_f} \left[\left(Y_{t,N}-R_{t,N}\right)^T Q_e \left(Y_{t,N}-R_{t,N}\right) + U_{t,N}{}^T Q_u U_{t,N} \right] \right] \\
&= E\left[\lim_{T_f \to \infty} \frac{1}{T_f+1} \sum_{t=0}^{T_f} \begin{bmatrix} \left(\Phi_N A x_t + S_N U_{t,N} + \tilde{S}_N V_{t,N} + W_{t,N} - R_{t,N}\right)^T Q_e \\ \times \left(\Phi_N A x_t + S_N U_{t,N} + \tilde{S}_N V_{t,N} + W_{t,N} - R_{t,N}\right) + \left(U_{t,N}{}^T Q_u U_{t,N}\right) \end{bmatrix} \right]
\end{aligned}
\tag{4.48}
$$

Then, performing conditional expectation and neglecting control independent terms, the performance index to be minimized (4.48) takes the form:

$$\hat{J} = \lim_{T_f \to \infty} \frac{1}{T_f+1} \sum_{t=0}^{T_f} \left[\left(\Phi_N A\hat{x}_t + S_N U_{t,N} - R_{t,N}\right)^T Q_e \left(\Phi_N A\hat{x}_t + S_N U_{t,N} - R_{t,N}\right) + U_{t,N}{}^T Q_u U_{t,N} \right]$$

$$\tag{4.49}$$

It is assumed that the future reference signal values may be evaluated from the N first reference signals, as follows:

$$R_{t+1,N} = \Theta_N \cdot R_{t,N} \tag{4.50}$$

$$\text{where: } R_{t+1,N} = \begin{bmatrix} r_{t+2} \\ r_{t+3} \\ \vdots \\ r_{t+N+2} \end{bmatrix}$$

and the matrix Θ_N is selected so that the best representation is given of the expected future behaviour of the reference signal. For example:

$$\text{choice (a)} \quad \Theta_N^a = \begin{bmatrix} O & I & \cdots & O \\ \vdots & \ddots & \ddots & \vdots \\ & & O & I \\ O & \cdots & O & I \end{bmatrix} \tag{4.51}$$

denotes that the reference signal will remain constant after N steps; or:

$$\text{choice (b)} \quad \Theta_N^b = \begin{bmatrix} O & I & \cdots & O \\ \vdots & \ddots & \ddots & \vdots \\ & & O & I \\ O & \cdots & O & O \end{bmatrix} \tag{4.52}$$

denotes that the reference signal will be set to zero after N steps.

Introducing the extended state vector, comprising the state of the system and the reference signal, the extended state space equation may be written as follows:

$$\begin{bmatrix} x_{t+1} \\ R_{t+1,N} \end{bmatrix} = \chi_{t+1} = \begin{bmatrix} A & O \\ O & \Theta_N \end{bmatrix} \chi_t + \begin{bmatrix} \beta \\ O \end{bmatrix} U_{t,N} = \Lambda \chi_t + \Psi U_{t,N} \tag{4.53}$$

The prediction equation then becomes:

$$\hat{Y}_{t,N} = \Phi_N A \hat{x}_t + S_N U_{t,N} = \begin{bmatrix} \Phi_N A & O \end{bmatrix} \hat{\chi}_t + S_N U_{t,N} \tag{4.54}$$

and the error between the predicted output and the set-point is given by:

$$\hat{Y}_{t,N} - R_{t,N} = \begin{bmatrix} \Phi_N A & -I \end{bmatrix} \hat{\chi}_t + S_N U_{t,N} = L_N \hat{\chi}_t + S_N U_{t,N} \tag{4.55}$$

Substituting Equation (4.55) to the performance index (4.49):

$$\hat{J} = \lim_{T_f \to \infty} \frac{1}{T_f + 1} \sum_{t=0}^{T_f} \left[\left(L_N \hat{\chi}_t + S_N U_{t,N} \right)^T Q_e \left(L_N \hat{\chi}_t + S_N U_{t,N} \right) + U_{t,N}^T Q_u U_{t,N} \right] \tag{4.56}$$

the optimal control solution is given by:

$$U_{t,N} = -\left(Q_u + S_N^T Q_e S_N + \Psi^T \tilde{H} \Psi \right)^{-1} \left(S_N^T Q_e L_N + \Psi^T \tilde{H} \Lambda \right) \hat{\chi}_t \tag{4.57}$$

where $\tilde{H}$ is a steady state solution of the Ricatti equation:

$$\begin{aligned} \tilde{H}_j = L_N^T Q_e L_N + \Lambda^T \tilde{H}_{j+1} \Lambda - \left(L_N^T Q_e S_N + \Lambda^T \tilde{H}_{j+1} \Psi \right) \times \\ \left(Q_u + S_N^T Q_e S_N + \Psi^T \tilde{H}_{j+1} \Psi \right)^{-1} \times \left(S_N^T Q_e L_N + \Psi^T \tilde{H}_{j+1} \Lambda \right) \end{aligned} \tag{4.58}$$

Equations (4.57), and (4.58) may be further simplified. Assuming that matrix $\tilde{H}_j$ is divided into four matrix blocks of appropriate dimensions:

$$\tilde{H}_j = \begin{bmatrix} \tilde{H}_j^1 & \tilde{H}_j^2 \\ \tilde{H}_j^{2T} & \tilde{H}_j^3 \end{bmatrix} \tag{4.59}$$

and using definitions of matrices Λ, Ψ and L as implied in Equations (4.53) and (4.54) obtains:

$$S_N^T Q_e L = \begin{bmatrix} S_N^T Q_e \Phi_N A & -S_N^T \end{bmatrix} \tag{4.60}$$

$$\Psi^T \tilde{H}_j = \begin{bmatrix} \beta^T \tilde{H}_j^1 & \beta^T \tilde{H}_j^2 \end{bmatrix} \tag{4.61}$$

$$\Psi^T \tilde{H}_j \Psi = \beta^T \tilde{H}_j^1 \beta \tag{4.62}$$

$$\Psi^T \tilde{H}_j \Lambda = \begin{bmatrix} \beta^T \tilde{H}_j^1 A & \beta^T \tilde{H}_j^2 \Theta_N \end{bmatrix} \tag{4.63}$$

$$L_N^T Q_e L_N = \begin{bmatrix} A^T \Phi_N^T Q_e \Phi_N A & -A^T \Phi_N^T Q_e \\ -Q_e \Phi_N A & Q_e \end{bmatrix} \tag{4.64}$$

$$\Lambda^T \tilde{H}_j \Lambda = \begin{bmatrix} A^T \tilde{H}_j^1 A & A^T \tilde{H}_j^2 \Theta_N \\ \Theta_N^T \tilde{H}_j^{2T} A & \Theta_N^T \tilde{H}_j^3 \Theta_N \end{bmatrix} \tag{4.65}$$

Then, Equation (4.57) becomes:

$$U_{t,N} = -\left(Q_u + S_N^T Q_e S_N + \beta^T \tilde{H}^1 \beta\right)^{-1} \left[\left(S_N^T Q_e \Phi_N A + \beta^T \tilde{H}^1 A\right)\hat{x}_t + \left(\beta^T \tilde{H}^2 \Theta_N - S_N^T Q_e\right)R_{t,N}\right] \tag{4.66}$$

and Equation (4.58) may be split into two equations:

$$\tilde{H}_j^1 = A^T \left(\Phi_N^T Q_e \Phi_N + \tilde{H}_{j+1}^1\right)A - A^T \left(\Phi_N^T Q_e S_N + \tilde{H}_{j+1}^1 \beta\right)\left(Q_u + S_N^T Q_e S_N + \beta^T \tilde{H}_{j+1}^1 \beta\right)^{-1}$$
$$\times \left(S_N^T Q_e \Phi_N + \beta^T \tilde{H}_{j+1}^1\right)A \tag{4.67}$$

$$\tilde{H}_j^2 = -A^T \Phi_N^T Q_e + A^T \tilde{H}_{j+1}^2 \Theta_N - A^T \left(\Phi^T Q_e S_N + \tilde{H}_{j+1}^1 \beta\right)\left(Q_u + S_N^T Q_e S_N + \beta^T \tilde{H}_{j+1}^1 \beta\right)^{-1}$$
$$\times \left(\beta^T \tilde{H}_{j+1}^2 \Theta_N - S_N^T Q_e\right) \tag{4.68}$$

where $\hat{x}_t$ can be obtained from a standard Kalman filter equation.

The state-space GPC and LQGPC controllers, derived in this section, will be used in the subsequent sections to formulate benchmarks enabling a comparison of the performance of any controller against these two predictive controllers.

4.2.3 Calculation of the Performance in Closed-loop System

Consider the generalised description of an ($r \times m$) multivariable controller defined as

$$U(s) = \begin{bmatrix} c_o^{11}(s) & \cdots & c_o^{1m}(s) \\ \vdots & \ddots & \vdots \\ c_o^{r1}(s) & \cdots & c_o^{rm}(s) \end{bmatrix} E(s) \tag{4.69}$$

where $U(s) = \begin{bmatrix} u^1(s) \\ \vdots \\ u^r(s) \end{bmatrix}$ and $E(s) = \begin{bmatrix} e^1(s) \\ \vdots \\ e^m(s) \end{bmatrix}$

The real rational transfer functions $c_o^{ij}(s)$ are elements of the multivariable controller which, without loss of generality, can be expressed as follows:

Reduced order:

$$c_0^{ij}(s) = \frac{c_{n0} + c_{n1}s + ... + c_{np}s^p}{c_{d0} + c_{d1}s + ... + c_{dv}s^v}$$

Lead lag:

$$c_0^{ij}(s) = \frac{(c_{n0} + c_{n1}s)(c_{n2} + c_{n3}s)}{(c_{d0} + c_{d1}s)(c_{d2} + c_{d3}s)}$$

PID:

$$c_0^{ij}(s) = k_0 + k_1 / s + k_2 s$$

Then as documented by Pota [1996] a direct minimal order state space equation can be obtained for the multivariable controller in the form,

$$\begin{aligned} x^{co}_{t+1} &= A_{co}x^{co}_t + B_{co}y_t + g_{co}\tilde{r}_t \\ u_t &= C_{co}x^{co}_t + D_{co}y_t + f_{co}\tilde{r}_t \end{aligned}$$
(4.70)

If the vector of $N+1$ future reference signals r_t is considered as an input to the state space Equations (4.70) then the controller can be rewritten in the form

$$\begin{aligned} x^{co}_{t+1} &= A_{co}x^{co}_t + B_{co}y_t + G_{co}\tilde{R}_{t,N} \\ u_t &= C_{co}x^{co}_t + D_{co}y_t + F_{co}\tilde{R}_{t,N} \end{aligned}$$
(4.71)

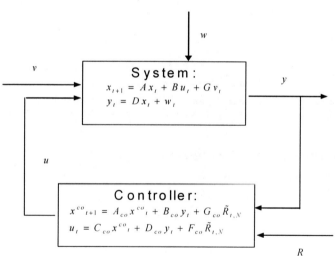

Figure 4.1. Closed loop system state space block diagram

The structure of the closed-loop system is given in Figure 4.1. The controller is described by the equations above and the system by state-space Equations (4.1), or

(4.7), or (4.9) and (4.10). A particular equation will be used depending on whether the model is extended to eliminate the noise correlation and/or to introduce integral control action (as explained in earlier sections).

Remark: It is possible to use only Equation (4.2.1) for the system description and, when designing the controller, whichever of (4.7), or (4.9) and (4.10) is appropriate. The result of such an approach would be that the model used for control design differs from the system and this fact would have to be taken into account when defining the state-space equation of the controller.

Combining together the equations for the system and for the controller

$$u_t = C_{co}x^{co}_t + D_{co}Dx_t + D_{co}w_t + F_{co}\tilde{R}_{t,N} \tag{4.72}$$

$$x_{t+1} = (A + BD_{co}D)x_t + BC_{co}x^{co}_t + BF_{co}\tilde{R}_{t,N} + BD_{co}w_t + BGv_t$$
$$x^{co}_{t+1} = B_{co}Dx_t + A_{co}x^{co}_t + G_{co}\tilde{R}_{t,N} + B_{co}w_t$$

$$\tag{4.73}$$

$$u_t = D_{co}Dx_t + C_{co}x^{co}_t + F_{co}\tilde{R}_{t,N} + D_{co}w_t$$
$$y_t = D \cdot x_t + w_t$$

obtain:

$$\mathcal{X}_{t+1} = \begin{bmatrix} x_{t+1} \\ x^{co}_{t+1} \end{bmatrix} = \begin{bmatrix} A + BD_{co}D & BC_{co} \\ B_{co}D & A_{co} \end{bmatrix}\begin{bmatrix} x_t \\ x^{co}_t \end{bmatrix} + \begin{bmatrix} BD_{co} & BG \\ B_{co} & O \end{bmatrix}\begin{bmatrix} w_t \\ v_t \end{bmatrix} + \begin{bmatrix} BF_{co} \\ G_{co} \end{bmatrix}\tilde{R}_{t,N}$$

$$\tag{4.74}$$

$$\zeta_t = \begin{bmatrix} u_t \\ y_t \end{bmatrix} = \begin{bmatrix} D_{co}D & C_{co} \\ D & O \end{bmatrix}\begin{bmatrix} x_t \\ x^{co}_t \end{bmatrix} + \begin{bmatrix} D_{co} \\ I \end{bmatrix}w_t + \begin{bmatrix} F_{co} \\ O \end{bmatrix}\tilde{R}_{t,N}$$

Note that for predictive controllers,

$$\tilde{R}_{t,N} = R_{t,N} = \begin{bmatrix} r_{t+1} \\ r_{t+3} \\ \vdots \\ r_{t+N+1} \end{bmatrix} \text{ and for non-predictive controllers } \tilde{R}_{t,N} = \begin{bmatrix} r_t \\ r_{t+3} \\ \vdots \\ r_{t+N} \end{bmatrix}$$

GPC Controller
Using Equations (4.39) and (4.41), the control action at time instant t can be extracted as the first n_u elements from the vector $U_{t,N}$:

$$u_t = \begin{bmatrix} I & O & \cdots & O \end{bmatrix}(S_N^T Q_e S_N + Q_u)^{-1} S_N^T Q_e (R_{t,N} - \Phi_N A\hat{x}_{t/t})$$
$$= k^R_{GPC} \cdot R_{t,N} + k^x_{GPC} \cdot \hat{x}_{t/t}$$

$$\tag{4.75}$$

where:

$$k_{GPC}^R = \begin{bmatrix} I & O & \cdots & O \end{bmatrix} \left(S_N^T Q_e S_N + Q_u \right)^{-1} S_N^T Q_e$$

$$k_{GPC}^x = -\begin{bmatrix} I & O & \cdots & O \end{bmatrix} \left(S_N^T Q_e S_N + Q_u \right)^{-1} S_N^T Q_e \Phi_N A$$

Substituting for $\hat{x}_{t/t} = \left(I - KD \right) \hat{x}_{t/t-1} + K y_t$ obtains:

$$u_t = k_{GPC}^R R_{t,N} + k_{GPC}^x \left(I - KD \right) \hat{x}_{t/t-1} + k_{GPC}^x K y_t \tag{4.76}$$

The state of the controller can be obtained by using the Kalman filter predictor equation described in Equation (4.18)

$$
\begin{aligned}
\hat{x}_{t+1/t} &= \left(A - AKD \right) \hat{x}_{t/t-1} + AK y_t + Bu_t \\
&= \left(\left(A - AKD \right) + Bk_{GPC}^x \left(I - KD \right) \right) \hat{x}_{t/t-1} + \left(AK + Bk_{GPC}^x K \right) y_t + Bk_{GPC}^R R_{t,N}
\end{aligned}
\tag{4.77}
$$

Therefore:

$$
\begin{aligned}
A_{co}^{GPC} &= A - AKD + Bk_{GPC}^x \left(I - KD \right) \\
B_{co}^{GPC} &= AK + Bk_{GPC}^x K \\
G_{co}^{GPC} &= Bk_{GPC}^R \\
C_{co}^{GPC} &= k_{GPC}^x \left(I - KD \right) \\
D_{co}^{GPC} &= k_{GPC}^x K \\
F_{co}^{GPC} &= k_{GPC}^R
\end{aligned}
\tag{4.78}
$$

LQGPC Controller
Taking into account Equation (4.66), the control action at time instant t can be extracted as the first n_u elements from the vector $U_{t,N}$:

$$
\begin{aligned}
U_{t,N} &= -\begin{bmatrix} I & O & \cdots & O \end{bmatrix} \left(Q_u + S_N^T Q_e S_N + \beta^T \tilde{H}^1 \beta \right)^{-1} \\
&\times \left[\left(S_N^T Q_e \Phi_N A + \beta^T \tilde{H}^1 A \right) \hat{x}_t + \left(\beta^T \tilde{H}^2 \Theta_N - S_N^T Q_e \right) R_{t,N} \right] = k_{LQGPC}^R R_{t,N} + k_{LQGPC}^x \hat{x}_{t/t}
\end{aligned}
\tag{4.79}
$$

where:

$$k_{LQGPC}^R = -\begin{bmatrix} I & O & \cdots & O \end{bmatrix} \left(Q_u + S_N^T Q_e S_N + \beta^T \tilde{H}^1 \beta \right)^{-1} \left(\beta^T \tilde{H}^2 \Theta_N - S_N^T Q_e \right)$$

$$k_{LQGPC}^x = -\begin{bmatrix} I & O & \cdots & O \end{bmatrix} \left(Q_u + S_N^T Q_e S_N + \beta^T \tilde{H}^1 \beta \right)^{-1} \left(S_N^T Q_e \Phi_N A + \beta^T \tilde{H}^1 A \right)$$

The rest of the derivation will be identical to that for the GPC controller, leading to:

$$A_{co}^{LQGPC} = A - AKD + Bk_{LQGPC}^{x}(I - KD)$$

$$B_{co}^{LQGPC} = AK + Bk_{LQGPC}^{x} \cdot K$$

$$G_{co}^{LQGPC} = Bk_{LQGPC}^{R}$$

$$C_{co}^{LQGPC} = k_{LQGPC}^{x}(I - KD) \tag{4.80}$$

$$D_{co}^{LQGPC} = k_{LQGPC}^{x} K$$

$$F_{co}^{LQGPC} = k_{LQGPC}^{R}$$

Time Evaluation of the Mean Value and the Variance
Given a system as described by either (4.1), or (4.7), or (4.9) and (4.10) then for a given multivariable controller as defined in Equations (4.73) it is possible to derive simplified system equations to evaluate the stochastic characteristics of the closed loop system. From Equation (4.76) the system parameters in the extended state equation can be defined as follows:

$$A_{sys} = \begin{bmatrix} A + BD_{co}D & BC_{co} \\ B_{co}D & A_{co} \end{bmatrix}, \quad B_{sys} = \begin{bmatrix} BF_{co} \\ G_{co} \end{bmatrix}, \quad G_{sys} = \begin{bmatrix} BD_{co} & BG \\ B_{co} & O \end{bmatrix}$$

$$\chi_{t+1} = \begin{bmatrix} x_{t+1} \\ x_{t+1}^{co} \end{bmatrix}, \quad \gamma_{t+1} = \begin{bmatrix} w_t \\ v_t \end{bmatrix}, \quad P_{sys} = [D_{co}D \quad C_{co}], \quad D_{sys} = [D \quad O] \tag{4.81}$$

Then taking the expectation of Equation (4.74) the mean of system states can be expressed as:

$$E[\chi_{t+1}] = \overline{\chi}_{t+1} = \begin{bmatrix} \overline{x}_{t+1} \\ \overline{x}_{t+1}^{co} \end{bmatrix} = A_{sys} \overline{\chi}_t + B_{sys} \tilde{R}_{t,N} \tag{4.82}$$

Define the sum of squares of the system states as $\underline{\chi}_{t+1} = trace\{E[\chi_{t+1}\chi_{t+1}^{T}]\}$, then using the analyses of the stochastic properties of the GPC/LQGPC algorithm as considered by Ordys [1993] and Blachuta and Ordys [1987], and by using Equation (4.82), deduce:

$$\underline{\chi}_{t+1} = trace\begin{cases} A_{sys}E[\chi_t\chi_t^{T}]A_{sys}^{T} + B_{sys}\tilde{R}_{t,N}\overline{\chi}_t^{T}A_{sys}^{T} \\ + A_{sys}\overline{\chi}_t \tilde{R}_{t,N}^{T}B_{sys}^{T} + B_{sys}\tilde{R}_{t,N}\tilde{R}_{t,N}^{T}B_{sys}^{T} + G_{sys}E[\gamma_t\gamma_t^{T}]G_{sys}^{T} \end{cases} \tag{4.83}$$

$$\underline{\chi}_{t+1} = trace\{A_{sys}\underline{\chi}_t A_{sys}^{T} + B_{sys}\tilde{R}_{t,N}\overline{\chi}_t^{T}A_{sys}^{T} + A_{sys}\overline{\chi}_t \tilde{R}_{t,N}^{T}B_{sys}^{T} + B_{sys}\tilde{R}_{t,N}\tilde{R}_{t,N}^{T}B_{sys}^{T} + G_{sys}\underline{\gamma}_t G_{sys}^{T}\} \tag{4.84}$$

Let the state covariance be defined as:

$$\tilde{\chi}_{t+1} = E[(\chi_{t+1} - \overline{\chi}_{t+1})(\chi_{t+1} - \overline{\chi}_{t+1})^{T}],$$

$$\tilde{\chi}_{t+1} = A_{sys}E[(\chi_t - \overline{\chi}_t)(\chi_t - \overline{\chi}_t)^{T}]A_{sys}^{T} + G_{sys}E[\gamma_t\gamma_t^{T}]G_{sys}^{T} \tag{4.85}$$

$$\tilde{\underline{\chi}}_{t+1} = A_{sys}\tilde{\underline{\chi}}_t A_{sys}^T + G_{sys}\underline{\gamma}_t G_{sys}^T \qquad (4.86)$$

The mean and the variances of the output (y_t) and input (u_t) signals can be obtained from Equations (4.74) and (4.84) as:

$$E[y_t] = \bar{y}_t = D_{sys}\bar{\chi}_t \qquad (4.87)$$

$$E[u_t] = \bar{u}_t = P_{sys}\bar{\chi}_t + f_{co}r_t \qquad (4.88)$$

and

$$\tilde{\underline{y}}_{(t)} = E\left[(y_t - \bar{y}_t)(y_t - \bar{y}_t)^T\right] = D_{sys}E\left[(\chi_t - \bar{\chi}_t)(\chi_t - \bar{\chi}_t)^T\right]D_{sys}^T + E\left[w_t w_t^T\right] \qquad (4.89)$$

$$\tilde{\underline{y}}_{(t)} = D_{sys}\underline{\chi}_t D_{sys}^T + \underline{w}_t \qquad (4.90)$$

$$\tilde{\underline{u}}_{(t)} = E\left[(u_t - \bar{u}_t)(u_t - \bar{u}_t)^T\right] = P_{sys}E\left[(\chi_t - \bar{\chi}_t)(\chi_t - \bar{\chi}_t)^T\right]P_{sys}^T + D_{co}E\left[w_t w_t^T\right]D_{co}^T \qquad (4.91)$$

$$\tilde{\underline{u}}_{(t)} = P_{sys}\underline{\chi}_t P_{sys}^T + D_{co}\underline{w}_t D_{co}^T \qquad (4.92)$$

4.2.4 Multivariable Benchmark Definition

The GPC cost function is a special case of the LQGPC cost function as can be seen from the definitions in Equations (4.38) and (4.48). Hence in deriving a predictive benchmark index it is more general to use the LQGPC cost function. The predictive performance cost function defined in Equation (4.48), can be re-written as:

$$J = E\left[\lim_{T_f \to \infty}\frac{1}{T_f+1}\sum_{t=0}^{T_f}\left[(Y_{t,N} - R_{t,N})^T Q_e (Y_{t,N} - R_{t,N}) + U_{t,N}^T Q_u U_{t,N}\right]\right]$$

$$J = \lim_{T_f \to \infty}\frac{1}{T_f+1}\sum_{t=0}^{T_f}E_{wms}(t) + U_{wms}(t) \qquad (4.93)$$

where $E_{wms}(t)$ and $U_{wms}(t)$ are defined as the weighted mean squares of the process output error and input respectively. Hence,

$$E_{wms}(t) = E\left[(Y_{t,N} - R_{t,N})^T Q_e (Y_{t,N} - R_{t,N})\right] \text{ and } U_{wms}(t) = E\left[U_{t,N}^T Q_u U_{t,N}\right] \qquad (4.94)$$

Ideally, predictive benchmarking measures the energy difference between the predicted trajectory of the sub-optimal control and the predicted optimal trajectory of the LQGPC or GPC control. The predicted trajectory of the sub-optimal controller is not available, only the actual trajectory. Instead, the amount of work required by the ideal predictive controller (LQGPC/GPC) at each time instance (t) to go from some state of the process under sub-optimal control to the desired terminal state in the interval ($t+1,t+N$) can be measured. The more sub-optimal the controller, the greater the amount of work required. This benchmark measure is then the actual amount of wasted energy in the system in the interval $(t+1, t+N)$.

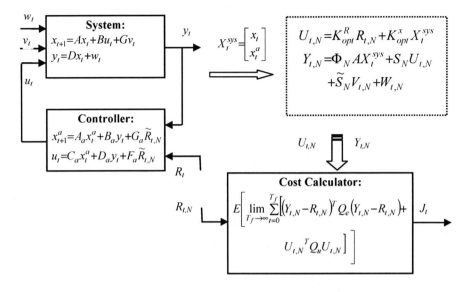

Figure 4.2. Benchmark calculation flow diagram

In Figure 4.2, $K_{opt}^i = K_{LQGPC}^i$ or $K_{opt}^i = K_{GPC}^i$, depending on the benchmarking method required. As indicated in the benchmark calculation flow diagram in Figure 4.2, to calculate the benchmark cost, at each time instance, the state of the system under the existing controller X_t^{sys} is measured and then used to compute the predicted inputs and outputs $U_{t,N}$ and $Y_{t,N}$ as follows:

$$U_{t,N} = k_{opt}^R R_{t,N} + k_{opt}^x X_t^{sys} \qquad (4.95)$$

$$= N R_{t,N} + M X_t^{sys} \qquad (4.96)$$

where $N = k_{opt}^R$, $M = k_{opt}^x$ and

$$Y_{t,N} = \Phi_N A X_t^{sys} + S_N \left(k_{opt}^R R_{t,N} + k_{opt}^x X_t^{sys} \right) + \tilde{S}_N V_{t,N} + W_{t,N}$$

$$= Q x_t + Z R_{t,N} + \tilde{S}_N V_{t,N} + W_{t,N} \qquad (4.97)$$

where $Q = \left[\Phi_N A + S_N k_{opt}^x \right]$ and $Z = \left[S_N k_{opt}^R \right]$

Equation (4.97) is then used to compute $E_{wms}(t)$ and $U_{wms}(t)$ as in Equation (4.94). Redefining the weighted mean squares of the process output error (E_{wms}) and process input (U_{wms}) as,

$$E_{wms} = trace\left\{ Q_e^{\frac{1}{2}} \mathrm{E}\left[\left(Y_{t,N} - R_{t,N}\right)\left(Y_{t,N} - R_{t,N}\right)^T\right] Q_e^{\frac{T}{2}} \right\} \qquad (4.98)$$

$$U_{wms} = trace\left\{ Q_u^{\frac{1}{2}} \mathrm{E}\left[U_{t,N} U_{t,N}^{\ T}\right] Q_u^{\frac{T}{2}} \right\} \qquad (4.99)$$

then

$$\mathrm{E}\left[\left(Y_{t,N} - R_{t,N}\right)\left(Y_{t,N} - R_{t,N}\right)^T\right] = \mathrm{E}\begin{bmatrix} \left(Q\chi_t + ZR_{t,N} + \tilde{S}_N V_{t,N} + W_{t,N} - R_{t,N}\right) \\ \times\left(\chi_t^T Q^T + R_{t,N}^T Z^T + V_{t,N}^T \tilde{S}_N^{\ T} + W_{t,N}^T - R_{t,N}^T\right) \end{bmatrix}$$

$$= Q\underline{\chi}_t Q^T + ZR_{t,N}\overline{\chi}_t^T Q^T - R_{t,N}\overline{\chi}_t^T Q^T + Q\overline{\chi}_t R_{t,N}^T Z^T + ZR_{t,N} R_{t,N}^T Z^T$$
$$- R_{t,N} R_{t,N}^T Z^T + \tilde{S}_N \underline{V}_{t,N} \tilde{S}_N^{\ T} + \underline{W}_{t,N} - Q\overline{\chi}_t R_{t,N}^T - ZR_{t,N} R_{t,N}^T + R_{t,N} R_{t,N}^T \qquad (4.100)$$

$$E_{wms} = trace\left\{ Q_e^{\frac{1}{2}}\left[Q\underline{\chi}_t Q^T + ZR_{t,N}\overline{\chi}_t^T Q^T - R_{t,N}\overline{\chi}_t^T Q^T + Q\overline{\chi}_t R_{t,N}^T Z^T + ZR_{t,N} R_{t,N}^T Z^T \right. \right.$$
$$\left.\left. - R_{t,N} R_{t,N}^T Z^T + \tilde{S}_N \underline{V}_{t,N} \tilde{S}_N^{\ T} + \underline{W}_{t,N} - Q\overline{\chi}_t R_{t,N}^T - ZR_{t,N} R_{t,N}^T + R_{t,N} R_{t,N}^T \right] Q_e^{\frac{T}{2}} \right\} \qquad (4.101)$$

where $\chi_t = \begin{bmatrix} x_t^{sys} \\ x_t^a \end{bmatrix}$ and $\mathrm{E}\left[U_{t,N} U_{t,N}^{\ T}\right] = \mathrm{E}\left[\left(M\chi_t + NR_{t,N}\right)\left(\chi_t^T M^T + R_{t,N}^T N^T\right)\right]$

$$= M\underline{\chi}_t M^T + N\tilde{R}_{t,N}\overline{\chi}_t^T M^T + M\overline{\chi}_t \tilde{R}_{t,N}^T N^T + N\tilde{R}_{t,N} \tilde{R}_{t,N}^T N^T \qquad (4.102)$$

$$U_{wms} = trace\left\{ Q_e^{\frac{1}{2}}\left\{ M\underline{\chi}_t M^T + N\tilde{R}_{t,N}\overline{\chi}_t^T M^T + M\overline{\chi}_t \tilde{R}_{t,N}^T N^T + N\tilde{R}_{t,N} \tilde{R}_{t,N}^T N^T \right\} Q_e^{\frac{T}{2}} \right\} \qquad (4.103)$$

The predictive performance cost for any controller can then be expressed as:

$$J_t = trace\left[Q_e^{\frac{1}{2}}\left\{ \begin{matrix} Q\underline{\chi}_t Q^T + ZR_{t,N}\overline{\chi}_t^T Q^T - R_{t,N}\overline{\chi}_t^T Q^T + Q\overline{\chi}_t R_{t,N}^T Z^T + ZR_{t,N} R_{t,N}^T Z^T \\ -R_{t,N} R_{t,N}^T Z^T + \tilde{S}_N \underline{V}_{t,N} \tilde{S}_N^{\ T} + \underline{W}_{t,N} - Q\overline{\chi}_t R_{t,N}^T - ZR_{t,N} R_{t,N}^T + R_{t,N} R_{t,N}^T \end{matrix} \right\} Q_e^{\frac{T}{2}} \right]$$
$$+ trace\left[Q_e^{\frac{1}{2}}\left\{ M\underline{\chi}_t M^T + N\tilde{R}_{t,N}\overline{\chi}_t^T M^T + M\overline{\chi}_t \tilde{R}_{t,N}^T N^T + N\tilde{R}_{t,N} \tilde{R}_{t,N}^T N^T \right\} Q_e^{\frac{T}{2}} \right]$$

$$(4.104)$$

where the parameters Q, Z, M, N are functions of the gains (k_{opt}^x and k_{opt}^R) of the optimal controller (GPC or LQGPC), and $\bar{\chi}_t$ and χ_t are evaluated using Equations (4.83) and (4.84) and hence depend on the actual controller used. Therefore, the predictive benchmark index for a given controller c_x is comparing the value of J_i from Equation (4.104) for the optimal and the actual controller, and can be defined as:

$$\eta = \frac{J_{opt}}{J_{c_x}} \tag{4.105}$$

The predictive performance cost function in Equation (4.93), can be modified to define a dynamic performance index over a finite period of time T_f, for a given reference trajectory $R = r(t)$, $t \in [0, T_f]$:

$$J\left(r(t)\right) = \frac{1}{T_f + 1} \sum_{t=0}^{T_f} E_{wms}\left(r(t)\right) + U_{wms}\left(r(t)\right) \tag{4.106}$$

$$\eta_t = \frac{J_{opt}\left(r(t)\right)}{J_{c_x}\left(r(t)\right)} \tag{4.107}$$

Using Equations (4.93) to (4.109) an iterative algorithm for obtaining the predictive performance index over a given reference trajectory (R) can be expressed as follows.

Procedure 4.1 MIMO LQGPC Predictive Benchmark

Step 1:
Define plant parameters A, B, D, G and nominal controller parameters A_{co}, B_{co}, D_{co}, G_{co}

Step 2:
Define the prediction horizon (N) and benchmarking parameters, i.e cost function weighting matrices Q_e, and Q_u. For benchmarking Q_e can be a diagonal matrix, with the elements of the diagonal indicating the relative importance of minimising each process output.

Step 3:
Derive optimal controller (either GPC or LQGPC) using Equations (4.39) to (4.68)

Step 4:
Derive optimal controller parameters: i.e for LQGPC
A_{co}^{LQGPC}, B_{co}^{LQGPC}, C_{co}^{LQGPC}, D_{co}^{LQGPC}, f_{co}^{LQGPC}, g_{co}^{LQGPC}, using Equation (4.80)

Step 5:
Arrange system matrices:

$$A_{sys} = \begin{bmatrix} A + BD_{co}D & BC_{co} \\ B_{co}D & A_{co} \end{bmatrix}, \ B_{sys} = \begin{bmatrix} BF_{co} \\ G_{co} \end{bmatrix}, \ G_{sys} = \begin{bmatrix} BD_{co} & BG \\ B_{co} & O \end{bmatrix}$$

$$\chi_{t+1} = \begin{bmatrix} x_{t+1} \\ x^{co}_{t+1} \end{bmatrix}, \ \gamma_{t+1} = \begin{bmatrix} w_t \\ v_t \end{bmatrix}, \ P_{sys} = [D_{co}D \ \ C_{co}], \ D_{sys} = [D \ \ O]$$

for the nominal controller and optimal controller respectively using Equation (4.81)

Step 6:
Initialise stochastic parameters:

- State mean vector $\bar{\chi}_0$, covariance matrix ($\tilde{\chi}_0$) and mean square matrix ($\underline{\chi}_0$)

- Output and state noise covariance matrices $\underline{w}_t$ and $\underline{v}_t$

Step 7:
Build predictive equation noise covariance matrices

$$G = \begin{bmatrix} DB & 0 & \vdots & 0 \\ DAB & DB & \vdots & \\ ... & ... & \vdots & ... \\ DA^{N_2}B & DA^{N_2-1}B & \vdots & DB \end{bmatrix}, \text{ and } V = \begin{bmatrix} \underline{v} & 0 & \vdots & 0 \\ 0 & \underline{v} & \vdots & \\ ... & ... & \vdots & ... \\ 0 & 0 & \vdots & \underline{v} \end{bmatrix}$$

Step 8:
Initialise output references r_t

Step 9:
Build reference vectors $R_{t,N}$

Step 10:
For the optimal and nominal controllers repeat benchmark procedure steps [11] to [15] for $t = 1 \ to \ T$

Step 11:
Calculate $\bar{\chi}_t$, $\tilde{\chi}_t$ and mean square matrix $\underline{\chi}_t$ using Equations (4.82) to (4.86)

Step 12:
Calculate output covariance ($\tilde{y}_t$) and input covariance ($\tilde{u}_t$) using Equations (4.89) to (4.92)

Step 13:
Calculate weighted mean square of process error (E_{wms}) and input (U_{wms}) using Equations (4.101) and (4.103) and then calculate $J_i = E_{wms} + U_{wms}$, where i is ether *optimal* or *Nominal*

Step 14:
Calculate the performance index $\eta_t = \dfrac{J_{opt}(t)}{J_{Nominal}(t)}$

Step 15:
Plot $\eta_{t:T_f}$

Procedure ends

Example 4.2.1 Applying LQGPC Benchmarking

In this example the MIMO LQGPC benchmark is used to compare the performance of two different controllers to determine which of the two meets the desired level of performance. The plant used in this example is a model of a pilot-scale binary distillation column used for methanol-water separation.

$$\begin{bmatrix} y_1(s) \\ y_2(s) \end{bmatrix} = \begin{bmatrix} \dfrac{12.8e^{-s}}{16.7s+1} & \dfrac{-18.9e^{-3s}}{21s+1} \\ \dfrac{6.6e^{-7s}}{10.9s+1} & \dfrac{-19.4e^{-3s}}{14.4s+1} \end{bmatrix} \begin{bmatrix} u_1(s) \\ u_2(s) \end{bmatrix} + \begin{bmatrix} \dfrac{3.7e^{-8.1s}}{14.9s+1} \\ \dfrac{4.9e^{-3.4s}}{13.2s+1} \end{bmatrix} d(s).$$

where output $y_1(s)$ is the overhead mole fraction of methanol, $y_2(s)$ is the bottom mole fraction of methanol, $u_1(s)$ is the overhead reflux flow rate, $u_2(s)$ is the bottom steam flow rate.

The original model has been utilised in several controller performance studies [Maurath et al., 1988; Rahul and Cooper, 1997]. Two different controller approaches are considered for the plant, the first being a simple multiloop (filtered) PID and the second, an adaptation of the multiloop PID control to multivariable control with the inclusion of decouplers. The transfer functions for the two designs are:

Multiloop (filtered) PID:

$$\begin{bmatrix} u_1(s) \\ u_2(s) \end{bmatrix} = \begin{bmatrix} \dfrac{0.7746s^2+0.3755s+0.1279}{s^2+2s} & 0 \\ 0 & \dfrac{-0.1697s^2-0.0658s-0.01007}{s^2+2s} \end{bmatrix} \begin{bmatrix} e_1(s) \\ e_2(s) \end{bmatrix}$$

Multivariable controller:

$$\begin{bmatrix} u_1(s) \\ u_2(s) \end{bmatrix} = \begin{bmatrix} \dfrac{0.04213s^2+0.02042s+0.00695}{s^2+2s} & \dfrac{-0.00763s^2-0.00296s-0.000453}{s^2+2s} \\ \dfrac{0.00936s^2+0.00454s+0.00154}{s^2+2s} & \dfrac{-0.00936s^2-0.00357s-0.000547}{s^2+2s} \end{bmatrix} \begin{bmatrix} e_1(s) \\ e_2(s) \end{bmatrix}$$

The desired control objective for the process was the minimization of the output variance and the reference value was assumed constant. The benchmark parameters were chosen to reflect this objective with:

prediction horizon: $N = 2$, control horizon: $N_u = 1$ and cost function weighting
matrices $Q_e = \begin{bmatrix} 1 & 0 \\ 0 & 1 \end{bmatrix}$, and $Q_u = \begin{bmatrix} 0.5 & 0 \\ 0 & 0.5 \end{bmatrix}$.

For use in the benchmarking algorithm the transfer functions for the plant and the
controllers were converted into state space form. The LQGPC controller gains
were computed as:

$$K_{LQGPC}^x = \begin{bmatrix} -31 & 0 & 1.1 & 0 & 40.2 & 0 & 5.6 & 0 & -215.9 & 118.6 & 1.4 & -0.1 \\ 8.7 & 0 & 21.6 & 0 & -9.9 & 0 & 54.8 & 0 & 118.6 & -346.8 & 23.3 & 0.4 \end{bmatrix} \times 10^{-3}$$

$$K_{LQGPC}^R = \begin{bmatrix} 8 & 5 & 669 & 200 \\ -8 & -12 & -65 & -626 \end{bmatrix} \times 10^{-4}$$

Using the controller gains, the controller parameters (A_{co}^{LQGPC}, B_{co}^{LQGPC},
C_{co}^{LQGPC}, D_{co}^{LQGPC}, f_{co}^{LQGPC}, g_{co}^{LQGPC}) were obtained using Equation (4.80).
Equations (4.81) to (4.92) were iterated over 1000 samples to calculate the mean,
variance and mean square of the extended state, output ($\tilde{y}_t$) and input ($\tilde{u}_t$). From
these the integrated MIMO and individual loop performance indices were then
computed using Equations (4.101) and (4.103).

The result of the benchmarking exercise as shown in Table 4.1, indicates that the
multivariable design with decouplers offers no additional benefits over the
multiloop design in the steady state. The interesting aspect of the multiloop
controller is its performance in Loop II, with an Individual Loop SISO
Performance Index greater than 1. This suggest better performance than LQGPC
control in the loop, however when the performance in Loop I (58% of the
performance of a LQGPC controller) is also considered, the results suggest that the
controller is too tightly tuned, i.e. the effect of a tight control in Loop I is a
deterioration in the performance in Loop II. Comparing the MIMO and SISO
performance indicators for both the multiloop and multivariable controllers
recorded in Table 4.1, it can be observed that if interaction exist between control
loops, then optimal performance in each individual loop does not translate into the
optimal global system performance. This is a key reason why multivariable
benchmarking methods are needed.

Table 4.1. Pilot Scale Column benchmarking result

Controller	Loop	Global MIMO Performance Index	Individual Loop (SISO) Performance Index	Output Variance	Input Variance
Multiloop	I	0.54	0.58	1.21	1.1
	II		1.2	1.02	0.003
Multivariable	I	0.55	0.52	2	0.045
	II		0.36	2.8	0.0007

4.3 Introduction to Restricted Structure Controller Benchmarking

The benchmarking techniques considered so far have assessed the performance of existing controllers against that of optimal full-order controllers. The order of most optimal controllers is related to the order of the plant as well as the cost weightings, and this is often high, relative to classical designs such as PID controllers. If a benchmark figure of merit is computed for a controller of much higher order than can actually be used on the plant, the achievable control performance is likely to be overestimated. In other words, if the actual control performance falls short of the benchmark, it is not clear whether this is due to poor controller tuning, or to the fact that matching the benchmark with the low-order controller structure is simply not feasible.

Recently, a method of restricted structure optimal control benchmarking has been introduced [Grimble, 2000], in which the controller structure may be pre-specified. Thus, for example, if a PID structure is required, the algorithm computes the best PID coefficients to minimize an objective function, where the dynamic weightings are chosen to reflect the desired economic and performance requirements. The benchmark in this case is more valuable, since an engineer is able to tune the actual system on the plant to achieve the benchmark cost values. Restricted structure optimal control benchmarking therefore provides figures of merit which an engineer on the plant may in fact achieve. Moreover, the low-order controller based on the Generalized Minimum Variance or LQG benchmarks will have good low frequency gain, good roll off at high frequencies and will also limit the variances, which are related to economic performance.

To obtain the restricted structure benchmarking solution, an optimal control problem is defined and a theorem is utilized from [Grimble 2000], which reveals that a simple cost minimization problem can be established and controller parameters directly optimized to achieve the low-order optimal solution. The actual optimization involves a transformation into the frequency domain and numerical optimization of the integral cost term. The results are in a convenient form where control activity and tracking accuracy can be assessed, and at the same time full order solutions compared with the restricted structure solutions.

4.3.1 Restricted-structure Optimal Control

The first step in the benchmarking procedure is to compute the optimal full-order controller. This result provides the simplified criterion for the subsequent numerical optimization, and can also be used to assess the existing controller against the full-order as well as reduced-order benchmark. The requirement is the knowledge of the plant model linearized around the operating point. It is also necessary to define the objective function which could be of the GMV or LQG type. In this section, the restricted-structure algorithm for the case of LQG control is presented.

The system shown in Figure 4.3 is assumed to be linear, continuous-time and single-input, single-output. The external zero-mean unity-variance white-noise sources $\xi(s), \zeta(s)$ are assumed to be mutually independent and drive colouring filters which represent the reference $W_r(s)$ and disturbance $W_d(s)$ subsystems. The system equations become:

$$y(s) = W(s)u(s) + d(s)$$ - system output   (4.108)

$$e(s) = r(s) - y(s)$$ - control error   (4.109)

$$r(s) = W_r(s)\zeta(s)$$ - reference generator   (4.110)

$$d(s) = W_d(s)\xi(s)$$ - disturbance model   (4.111)

$$u(s) = C_0(s)e(s)$$ - control signal   (4.112)

The system transfer functions are all assumed to be functions of the Laplace-transform variable in the complex-frequency domain, although it is straightforward to provide results for the discrete-time case [Grimble, 2002a]. For notational simplicity the arguments in the plant $W(s)$ and the other models are often omitted.

The system transfer functions are defined as:

$$W(s) = \frac{B(s)}{A(s)}$$ - plant transfer function   (4.113)

$$W_d(s) = \frac{C_d(s)}{A(s)}$$ - reference transfer function   (4.114)

$$W_r(s) = \frac{E_r(s)}{A(s)}$$ - disturbance transfer function   (4.115)

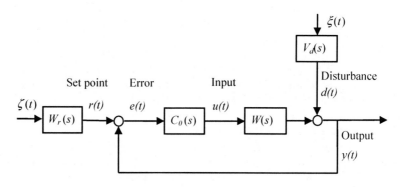

Figure 4.3. Stochastic linear continuous-time system description

The various polynomials are not necessarily coprime but the plant transfer function is assumed to be free of unstable hidden modes. The spectrum for the signal $r(t) - d(t)$ is denoted by $\Phi_{ff}(s)$ and a generalised spectral factor Y_f may be defined from this spectrum, using

$$Y_f Y_f^* = \Phi_{ff} = \Phi_{rr} + \Phi_{dd}$$ (4.116)

In polynomial form $Y_f = A^{-1}D_f$. The disturbance model is assumed to be such that D_f is strictly Hurwitz and satisfies

$$D_f D_f^* = E_r E_r^* + C_d C_d^* \tag{4.117}$$

These results enable the signal $f = r - d$ to be written in the equivalent form: $f(t) = Y_f \varepsilon(t) = A^{-1}D_f \varepsilon(t)$, where $\varepsilon(s)$ represents the transform of a zero mean white noise signal of unity variance.

Optimal LQG Controller
The results collected in this section come from [Grimble, 2000]. The optimal controller is defined to minimise the following LQG cost function:

$$J = \frac{1}{2\pi j}\oint_D \{Q_c(s)\Phi_{ee}(s) + R_c(s)\Phi_{uu}(s)\}ds \tag{4.118}$$

where the dynamic weighting elements $Q_c = \dfrac{Q_n}{A_q^* A_q}$ (positive-semi-definite) and

$R_c = \dfrac{R_n}{A_r^* A_r}$ (positive definite) act on the spectra of the error and control signals.

The optimal controller of restricted structure, to minimise the cost function (4.118), can be found by minimising the simplified criterion:

$$J_0 = \frac{1}{2\pi j}\oint_D T_0 T_0^* ds = \frac{1}{2\pi}\int_{-\infty}^{+\infty} T_0(j\omega)T_0(-j\omega)d\omega \tag{4.119}$$

where,

$$T_0 = \frac{H_0 A_q C_{0n} - G_0 A_r C_{0d}}{A_q A_r (AC_{0d} + BC_{0n})} \tag{4.120}$$

and G_0 and H_0 are the solutions of a set of two coupled Diophantine equations:

$$\begin{aligned}
D_c^* G_0 + F_0 AA_q &= B^* A_r^* Q_n D_f \\
D_c^* H_0 - F_0 BA_r &= A^* A_q^* R_n D_f
\end{aligned} \tag{4.121}$$

with F_0 of minimum degree.
These two equations are equivalent to the following implied equation:

$$AA_q H_0 + BA_r G_0 = D_c D_f \tag{4.122}$$

The polynomial D_c in the above equations is the stable solution of the following spectral factorisation problem:

$$D_c^* D_c = B^* A_r^* Q_n A_r B + A^* A_q^* R_n A_q A \tag{4.123}$$

Parametric Optimization Problem

To show how the parametric optimization is set up, assume that the existing controller is a PID controller with a filtered derivative term:

$$C_0 = k_0 + \frac{k_1}{s} + \frac{k_2 s}{1 + \frac{s}{\theta}} \tag{4.124}$$

The restricted-structure controller design problem then amounts to finding such values of k_0, k_1 and k_2 that minimise (4.119). Since the parameters appear both in the numerator and denominator of (4.120), this is a nonlinear optimization problem and hence has to be solved iteratively. The assumption is made here that a stabilising control law exists for the assumed controller structure.

To set up the optimization problem, rewrite (4.120) as

$$T_0 = \frac{C_{0n} L_1 - C_{0d} L_2}{C_{0n} L_3 + C_{0d} L_4} = C_{0n} L_{n1} - C_{0d} L_{n2} \tag{4.125}$$

where

$$L_1 = H_0 A_q \,, \; L_2 = G_0 A_r \,, \; L_3 = A_q A_r B \,, \; L_4 = A_q A_r A$$

$$L_{n1}(s) = \frac{L_1}{C_{0n} L_3 + C_{0d} L_4} \quad L_{n2}(s) = \frac{L_2}{C_{0n} L_3 + C_{0d} L_4} \tag{4.126}$$

and the controller numerator and denominator in L_{n1} and L_{n2} are assumed known (carried over from the previous iteration).

Evaluating the complex function $T_0(j\omega)$, the real and imaginary part of T_0 can be written as $T_0 = T_0^r + j T_0^i$. By separating the optimization parameters, the following expression is obtained:

$$T_0 = k_0 \left[(\alpha_0' L_{n1}' - \alpha_0' L_{n1}') + j(\alpha_0' L_{n1}' - \alpha_0' L_{n1}') \right] + k_1 \left[(\alpha_1' L_{n1}' - \alpha_1' L_{n1}') + j(\alpha_1' L_{n1}' - \alpha_1' L_{n1}') \right]$$
$$+ k_2 \left[(\alpha_2' L_{n1}' - \alpha_2' L_{n1}') + j(\alpha_2' L_{n1}' - \alpha_2' L_{n1}') \right] - \left[(C_{0d}' L_{n2}' - C_{0d}' L_{n2}') - j(C_{0d}' L_{n2}' - C_{0d}' L_{n2}') \right] \tag{4.127}$$

or

$$T_0 = k_0 \left[f_0^r + j \cdot f_0^i \right] + k_1 \left[f_1^r + j \cdot f_1^i \right] + k_2 \left[f_2^r + j \cdot f_2^i \right] - \left[g^r + j \cdot g^i \right] \tag{4.128}$$

where the definitions of $f_i(j\omega)$ and $g(j\omega)$ terms are obvious from (4.127).

For numerical optimization, the cost function (4.119) needs to be approximated with a finite summation:

$$J_0 \cong J_{00} = \frac{1}{2\pi} \sum_{l=1}^{N} T_0(j\omega_l) T_0(-j\omega_l) \Delta\omega \tag{4.129}$$

where N is the number of frequency points. Evaluating the terms in square brackets in (4.127) at each frequency point, separately for the real and imaginary parts, the above equation can be represented in the following matrix form:

$$\begin{bmatrix} T_0^r \\ T_0^i \end{bmatrix} = \begin{bmatrix} f_0^r(\omega_1) & f_1^r(\omega_1) & f_2^r(\omega_1) \\ \vdots & \vdots & \vdots \\ f_0^r(\omega_N) & f_1^r(\omega_N) & f_2^r(\omega_N) \\ f_0^i(\omega_1) & f_1^i(\omega_1) & f_2^i(\omega_1) \\ \vdots & \vdots & \vdots \\ f_0^i(\omega_N) & f_1^i(\omega_N) & f_2^i(\omega_N) \end{bmatrix} \begin{bmatrix} k_0 \\ k_1 \\ k_2 \end{bmatrix} - \begin{bmatrix} g^r(\omega_1) \\ \vdots \\ g^r(\omega_N) \\ g^i(\omega_1) \\ \vdots \\ g^i(\omega_N) \end{bmatrix} = Fx - g \qquad (4.130)$$

where $x = [k_0, k_1, k_2]^T$.

The optimisation can be performed by minimising the sum of squares of J_{00} at each frequency point as discussed in [Yukimoto et al., 1998]. The minimisation cost term becomes:

$$J_{00} = (Fx - g)^T (Fx - g) \qquad (4.131)$$

from which the optimal parameters for the current iteration can be obtained:

$$[k_0, k_1, k_2]^T = (F^T F)^{-1} F^T g \qquad (4.132)$$

The procedure requires the vector of initial optimization parameters to be provided, and is repeated until a specified stopping criterion is satisfied.

4.3.2 Restricted Structure Controller Benchmarking

Once the optimal RS controller has been determined, its performance can be compared with that of the actual controller. The value of the cost function for both the optimal RS controller J_0^{RS} and the actual controller J_0^{Act} can be calculated substituting the controller transfer functions into (4.119). The algorithm presented in [Åström, 1979] can be used, or alternatively a numerical integration can be performed.

Finally, the controller performance index can be calculated as

$$\kappa = \frac{J_0^{RS}}{J_0^{Act}} \qquad (4.133)$$

The performance index κ lies in the range $[0, 1]$, where '0' corresponds to very poor control and '1' to the optimal RS control. In the second case it is not possible to improve the controller tuning (in terms of the specified cost criterion). The point is that, unlike in the MV and GMV benchmarking, the '1' can actually be achieved by the existing controller. Moreover, the optimal tuning parameters are readily available as a 'by-product' of the benchmarking algorithm. The obvious drawback, on the other hand, is the need for model information.

The procedure for RS-LQG benchmarking will now be presented, assuming the filtered PID as the existing controller structure. It is assumed that the system model has been identified or is otherwise available.

Procedure 4.2 RS-LQG benchmarking

Step 1:   Compute the spectral factors (4.117) and (4.123), and solve the coupled Diophantine Equations (4.121)

Step 2:   Select the vector of initial optimization parameters $x_0 = [k_0^0, k_1^0, k_2^0]^T$, the frequency range, derivative filter time constant θ, and the stopping criterion

Step 3:   Define:

$$\alpha_0 = (1+\frac{s}{\theta})s, \ \alpha_1 = (1+\frac{s}{\theta}), \ \alpha_2 = s^2, \ C_{0d}(s) = (1+\frac{s}{\theta})s \tag{4.134}$$

Step 4:   Set $x = x_0$

Step 5:   Define

$$C_{0n}(s) = \alpha_0(s)k_0 + \alpha_1(s)k_1 + \alpha_2(s)k_2 \tag{4.135}$$

Step 6:   Calculate the transfers defined in (4.126), (4.127), (4.134) and (4.135) for all chosen frequency points.

Step 7:   Collect the above values in the matrix-vector form:

$$F = \begin{bmatrix} f_0^r(\omega_1) & f_1^r(\omega_1) & f_2^r(\omega_1) \\ \vdots & \vdots & \vdots \\ f_0^r(\omega_N) & f_1^r(\omega_N) & f_2^r(\omega_N) \\ f_0^i(\omega_1) & f_1^i(\omega_1) & f_2^i(\omega_1) \\ \vdots & \vdots & \vdots \\ f_0^i(\omega_N) & f_1^i(\omega_N) & f_2^i(\omega_N) \end{bmatrix}, \ g = \begin{bmatrix} g^r(\omega_1) \\ \vdots \\ g^r(\omega_N) \\ g^r(\omega_1) \\ \vdots \\ g^r(\omega_N) \end{bmatrix}$$

Step 8:   Solve the least-squares problem for the controller parameters:

$$x = [k_0, k_1, k_2]^T = (F^T F)^{-1} F^T g$$

Step 9:   If stopping criterion not satisfied, go to **Step 5**.

Step 10: Compute the optimal PID controller

Step 11: Compute the benchmark values for both the existing and the optimal PID.

Step 12: Compute the controller performance index (4.133)

Procedure End

An alternative formulation of the controller performance index is to define

$$\kappa_2 = \frac{J_{min}}{J_{min} + J_0} \tag{4.136}$$

where J_{min} is the minimum achievable cost, corresponding to the optimal LQG controller. This formulation involves the additional calculation of J_{min} but allows both values to be compared with the absolute optimum.

4.4 MIMO Restricted Structure Benchmarking

Restricted-structure controller design and benchmarking of multivariable systems are based on similar ideas as those presented in the previous section for the scalar case, with the exception that it is also possible to assess the structure of the controller as well as its tuning. In the case of an $r \times m$ multivariable plant, the controller transfer-function matrix is of size $m \times r$ and all elements of this matrix can have a specified structure. Of course, it is possible to further restrict this structure for some specified elements by forcing one, two, or even all of the parameters k_i to be zero – that last case simply means that no feedback exists between the specified output-input pair. By computing the benchmark figures corresponding to various configurations it is possible to determine the potential benefits resulting from the introduction of additional loops to the system. If this turns out to be greater than the cost of installation, wiring, maintenance etc., the optimal tuning parameters are then available. On the other hand, the procedure may certainly be useful in optimal tuning of existing controllers.

The results presented in this section follow the procedure for the RS-LQG benchmarking, which has been used in [Grimble, 2001b] and [Greenwood and Johnson, 2003]. In contrast with the results of Section 4.3, the discrete-time system description will be used. The notation that involves the tensor product of matrices comes from [Maciejowski, 1989].

4.4.1 Multivariable RS-LQG Optimal Control

By analogy with the scalar RS-LQG controller, the part of the cost function to be minimized by the restricted-structure controller is defined as

$$J_0 = \frac{1}{2\pi j} \oint_D trace\{T_0^* T_0\} \frac{dz}{z} \tag{4.137}$$

where

$$T_o = (GD_2^{-1}H_q^{-1}C_{0d} - HD_3^{-1}H_r^{-1}C_{0n})(AC_{0d} + BC_{0n})^{-1}D_f \tag{4.138}$$

and the polynomial matrices G, H, D_2, D_3, and D_f are defined by the following equations:

$$D_f D_f^* = E_r E_r^* + C_d C_d^* \tag{4.139}$$

$$A_q^{-1}A^{-1}BA_r = B_1 A_1^{-1} \tag{4.140}$$

$$D_f^{-1}AA_q = A_2 D_2^{-1} \tag{4.141}$$

$$D_f^{-1}BA_r = B_2 D_3^{-1} \tag{4.142}$$

$$D_c^* G_0 + F_0 A_2 A_q = B^* A_r^* Q_n D_2 \tag{4.143}$$

$$D_c^* H_0 - F_0 B_2 A_r = A^* A_q^* R_n D_3 \tag{4.144}$$

$$D_c^* D_c = B^* A_r^* Q_n A_r B + A^* A_q^* R_n A_q A \tag{4.145}$$

The details of the derivation of the optimal controller are omitted but the interested reader is referred to [Grimble, 1987].

The optimal full-order controller sets (4.137) to zero – however, when the controller structure is restricted, the minimum value of (4.137) will generally be nonzero. As already mentioned, the controllers are assumed to have a filtered PID structure and the multivariable controller can thus be represented in matrix form as

$$C_0 = K_0 + K_1 \frac{1}{1-z^{-1}} + K_2 \frac{(1-z^{-1})}{1-\tau_d z^{-1}} \tag{4.146}$$

where $K_i = \begin{bmatrix} k_{11}^i & k_{12}^i & \cdots & k_{1r}^i \\ k_{11}^i & k_{22}^i & \cdots & \vdots \\ \vdots & \ddots & \ddots & \vdots \\ \vdots & \ddots & \ddots & \vdots \\ k_{m1}^i & \cdots & \cdots & k_{mr}^i \end{bmatrix}$ $i = 0,1,2$

The controller can also be represented in the equivalent right matrix fraction form as $C_0 = C_{0n} C_{0d}^{-1}$, where

$$C_{0d}(z^{-1}) = (1-z^{-1})(1-\tau_d z^{-1}) \cdot I_m \tag{4.147}$$

and

$$C_{0n}(z^{-1}) = \alpha_0(z^{-1})K_0 + \alpha_1(z^{-1})K_1 + \alpha_2(z^{-1})K_2 \tag{4.148}$$

with

$$\alpha_0(z^{-1}) = (1-z^{-1})(1-\tau_d z^{-1}), \; \alpha_1(z^{-1}) = 1-\tau_d z^{-1}, \; \alpha_2(z^{-1}) = (1-z^{-1})^2$$

A parametric optimization algorithm is used to minimize the cost function (4.137) with respect to the restricted-structure controller parameters; then (4.146) gives the formula for the optimal restricted-structure controller. Backsubstituting into the cost function and comparing with the value obtained for the existing controller results in the controller performance index.

A useful way of checking if the algorithm works correctly is to calculate the values of the cost function for controllers of increasingly restricted structure (i.e. for which more and more parameters, or controller elements, are forced to be zero) – obviously, the corresponding cost values should increase.

There now follows a solution to the optimization problem.

4.4.2 Solution of the Parametric Optimization Problem

The algorithm is a direct generalization from the SISO case and involves representing the integral (4.137) in the frequency form:

$$\underset{C_0}{Min} \int_0^{2\pi} trace\{T_o(e^{j\omega})T_o(e^{-j\omega})\}d\omega \tag{4.149}$$

where all the polynomial matrices are functions of the complex frequency argument $e^{j\omega}$.

This is a nonlinear optimization problem because the controller parameters appear both in the "numerator" and "denominator" matrices in expression (4.138), and therefore an iterative method of solution is utilized. The basic idea is the same as in the SISO case: assuming the "denominator" matrix $(AC_{0d} + BC_{0n})$ is known (calculated in the previous iteration), a minimization is performed with respect to the parameters that appear linearly in the "numerator" matrix. This step is repeated a number of times, resulting in the successive approximation algorithm. The stopping criterion depends on the achieved convergence or on the number of performed iterations. The basic steps of the algorithm are presented in Figure 4.4.

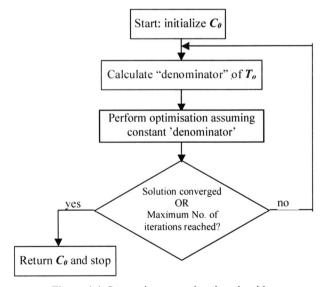

Figure 4.4. Successive approximation algorithm

The actual optimization algorithm (the shaded frame in Figure 4.4) will now be described. This problem is very similar to Edmunds' algorithm described e.g. in [Maciejowski, 1989] and in what follows the notation used there is adopted.

Begin by rewriting (4.138) as

$$T_o = Y - A_o C_{0n} B_o \qquad (4.150)$$

where:

$$Y = GD_2^{-1} H_q^{-1} C_{0d} (AC_{0d} + BC_{0n})^{-1} D_f \qquad (4.151)$$

$$A_o = HD_3^{-1} H_r^{-1} \qquad (4.152)$$

$$B_o = (AC_{0d} + BC_{0n})^{-1} D_f \qquad (4.153)$$

Since the least squares algorithm is to be applied, it is necessary to collect all the controller parameters in one column vector. For that purpose define the columns of matrices T_o, Y and C_{on} as:

$$T_0 = [t_1 ... t_r], \ Y = [y_1 ... y_r], \ C_{0n} = [n_1 ... n_r]$$

and introduce the notation $\otimes$ for the Kronecker or tensor product of two matrices: if P has p rows and q columns, and Q has r rows and s columns, then $P \otimes Q$ is a $pr \times qs$ matrix defined as:

$$P \otimes Q \triangleq \begin{bmatrix} p_{11}Q & p_{12}Q & \cdots & p_{1q}Q \\ p_{21}Q & p_{22}Q & \cdots & p_{2q}Q \\ \vdots & \vdots & & \vdots \\ p_{p1}Q & p_{p2}Q & \cdots & p_{pq}Q \end{bmatrix}$$

By stacking the columns on top of one another, obtain:

$$\begin{bmatrix} t_1 \\ t_2 \\ \vdots \\ t_r \end{bmatrix} = \begin{bmatrix} y_1 \\ y_2 \\ \vdots \\ y_r \end{bmatrix} - [B_o^T(z) \otimes A(z)] \begin{bmatrix} n_1 \\ n_2 \\ \vdots \\ n_r \end{bmatrix} \qquad (4.154)$$

Next, notice that the last vector on the RHS of (4.154) can be represented as

$$\begin{bmatrix} n_1 \\ n_2 \\ \vdots \\ n_r \end{bmatrix} = \begin{bmatrix} \alpha_0 k_{11}^0 + \alpha_1 k_{11}^1 + \alpha_2 k_{11}^2 \\ \alpha_0 k_{21}^0 + \alpha_1 k_{21}^1 + \alpha_2 k_{21}^2 \\ \vdots \\ \alpha_0 k_{mr}^0 + \alpha_1 k_{mr}^1 + \alpha_2 k_{mr}^2 \end{bmatrix} = \Sigma(z^{-1}) v \qquad (4.155)$$

where $\Sigma(z^{-1})$ is the matrix with mr rows and $3mr$ columns:

$$\Sigma(z^{-1}) = \begin{bmatrix} \alpha_0 & \alpha_1 & \alpha_2 & & & & 0 \\ & & \alpha_0 & \alpha_1 & \alpha_2 & & \\ & 0 & & & & \ddots & \\ & & & & \alpha_0 & \alpha_1 & \alpha_2 \end{bmatrix} \tag{4.156}$$

and v is a column vector containing all the controller parameters:

$$v = [k_{11}^0 \quad k_{11}^1 \quad \cdots \quad k_{mr}^2]^T$$

Finally, defining

$$\eta = [y_1^T \quad \cdots \quad y_r^T]^T, \ \varepsilon = [t_1^T \quad \cdots \quad t_r^T]^T$$

$$X = [B_0^T(z) \otimes A_0(z)] \cdot \Sigma(z^{-1})$$

gives

$$\eta = Xv + \varepsilon \tag{4.157}$$

which is in the standard form for the least squares calculation: η is a known vector, X is a known matrix, v is a vector of unknown parameters, and ε is a vector of "errors", the variance of which is to be minimized. This error does of course relate to the cost to be minimized with the restriction on the controller structure.

The next step is to calculate the integral (4.137). This is achieved by dividing the frequency range into a number of frequency points and approximating the integral with a finite summation:

$$\int_0^{2\pi} trace\{T_o(e^{j\omega})T_o(e^{-j\omega})\}d\omega \approx \sum_{i=1}^N \varepsilon^T(e^{j\omega_i}) \cdot \varepsilon(e^{-j\omega_i}) \tag{4.158}$$

The problem thus amounts to minimizing the sum of squares of (4.158) at each of the specified frequency points. Assembling the data from all these points, obtain:

$$\begin{bmatrix} \eta(e^{j\omega_1 T_s}) \\ \vdots \\ \eta(e^{j\omega_N T_s}) \end{bmatrix} = \begin{bmatrix} X(e^{j\omega_1 T_s}) \\ \vdots \\ X(e^{j\omega_N T_s}) \end{bmatrix} v + \begin{bmatrix} \varepsilon(e^{j\omega_1 T_s}) \\ \vdots \\ \varepsilon(e^{j\omega_N T_s}) \end{bmatrix} \tag{4.159}$$

The terms in (4.159) are split into their real and imaginary parts:

$$\eta = \eta_{\text{Re}} + j\eta_{\text{Im}}, \ X = X_{\text{Re}} + jX_{\text{Im}}, \ \varepsilon = \varepsilon_{\text{Re}} + j\varepsilon_{\text{Im}}$$

to create a new matrix equation:

$$
\underbrace{\begin{bmatrix} \eta_{Re}(\omega_1) \\ \vdots \\ \eta_{Re}(\omega_N) \\ \eta_{Im}(\omega_1) \\ \vdots \\ \eta_{Im}(\omega_N) \end{bmatrix}}_{\eta_1} = \underbrace{\begin{bmatrix} X_{Re}(\omega_1) \\ \vdots \\ X_{Re}(\omega_N) \\ X_{Im}(\omega_1) \\ \vdots \\ X_{Im}(\omega_N) \end{bmatrix}}_{X_1} v + \underbrace{\begin{bmatrix} \varepsilon_{Re}(\omega_1) \\ \vdots \\ \varepsilon_{Re}(\omega_N) \\ \varepsilon_{Im}(\omega_1) \\ \vdots \\ \varepsilon_{Im}(\omega_N) \end{bmatrix}}_{\varepsilon_1}
\tag{4.160}
$$

The final least-squares solution is then

$$
v^{opt} = (X_1^T X_1)^{-1} X_1^T \eta_1
\tag{4.161}
$$

Remarks:
The solution (4.161) assumes that the controller has a full structure with all the elements of the matrix being PID controllers. However, restricting the structure (forcing some of the parameters to be zero) can be achieved simply by omitting the corresponding columns in (4.159).

The alternative solution is to represent (4.137) as a quadratic form:

$$
J_o = v^T P(\omega)v + 2Q(\omega)v + C
\tag{4.162}
$$

where the matrices P and Q are frequency dependent and are approximately calculated for the specified frequency range. The solution follows from minimizing the gradient of (4.162) as

$$
v^{opt} = -[P(\omega)]^{-1} Q(\omega)
\tag{4.163}
$$

This approach has been used in [Greenwood and Johnson, 2003] to calculate the multivariable restricted-structure LQG controller.

4.4.3 Benchmark Cost Calculation and Controller Performance Index

The previous section described the algorithm for computing the optimal restricted-structure controller, which can be considered a useful result in itself. However, in order to benchmark the existing controller, there are a few steps left:

(a) Calculate the minimum value of J_o (corresponding to the optimal RS controller):

$$
J_0^{RS} = \frac{1}{2\pi j} \oint_{|z|=1} trace\{T_{RS}^* T_{RS}\} \frac{dz}{z}
\tag{4.164}
$$

(b) Calculate the value of J_o for the existing controller structure:

$$
J_0^{act} = \frac{1}{2\pi j} \oint_{|z|=1} trace\{T_{act}^* T_{act}\} \frac{dz}{z}
\tag{4.165}
$$

(c) Calculate the Controller Performance Index:

$$K = \frac{J_0^{RS}}{J_0^{act}} \qquad (4.166)$$

The Restricted-Structure LQG benchmarking algorithm can now be summarized. The required information for this algorithm is the knowledge of the full system model (polynomials matrices A, B, E and C_d).

Procedure 4.3 RS-LQG benchmarking: MIMO case

Step 1:   Compute the spectral factors (4.139) and (4.145), and solve the coupled Diophantine Equations (4.143) and (4.144).

Step 2:   Select optimization parameters: initial point, frequency range, stopping criterion

Step 3:   Set $K = K_0$

Step 4:   Calculate the transfers defined in (4.151), (4.152) and (4.153) for all chosen
frequency points.

Step 5:   Collect the above values in the matrices as in (4.160)

Step 6:   Solve the least-squares problem for the controller parameters as in (4.161)

Step 7:   Set $K = K_i$. If stopping criterion not satisfied, go to **Step 4**.

Step 8:   Compute the optimal PID controller

Step 9:   Compute the benchmark values for both the existing and the optimal PID.

Step 10: Calculate the controller performance index (4.166)

Procedure End

4.4.4 Numerical Example

The application of restricted-structure controller benchmarking to system structure assessment will now be illustrated.

The system description is as in Example 3.5. The low-order controllers are assumed to be of the filtered PID structure with the derivative filter "time constant" $\tau_d = 0.5$. In order to compare "like with like", the existing controller will also be of the above type. It will be a multi-loop PID controller tuned using Ziegler-Nichols rules separately for two loops. This controller has been assessed against the full-order GMV controller, the optimal full-structure PID, optimal diagonal PID and two optimal triangular PID controllers. The values of the restricted structure optimal cost term J_0^{RS}, defined in Equation (4.164), have been calculated for all these different configurations and compared with the value obtained for the existing controller, defined in Equation (4.165).Then, the CPIs (κ) have been obtained from Equation (4.166). The results are collected in Table 4.2.

Based on these results it is possible to tune the existing PID controllers "optimally" (in terms of the specified cost function) or predict how additional controllers would affect the performance of the system. This can be used as an indication of the profitability of the possible investment. In this example, it is clear that the upper triangular structure, i.e. introducing feedback between output 1 and input 2 (rather than between output 2 and input 1) would bring greater improvement. This simple example illustrates the potential of the technique in analyzing structure, pairing the input-output variables and as a tuning guidance.

Table 4.2. Structure assessment results

Controller	Multi-loop PID	RS full	RS diagonal #1	RS diagonal #2	RS upper triangular	RS lower triangular
J_0	2.803	0.015	0.086	0.502	0.036	0.066
κ	0.5%	100%	17%	3%	42%	23%

4.5 Weighting Selection for Benchmarking Algorithms

The selection of a cost function for benchmarking and performance assessment is always problematic. It is normally an iterative process and may require a simulation model being available. The form of the error weighting and the control weighting are probably defined beforehand but the actual size of the cost function weightings will depend upon the speed of response required from the system and other considerations.

In general, the more information that is available about the control system to be assessed, the easier it becomes to provide and interpret the weighting functions. In this section, the choice of weightings for the three following cases is discussed:

- no prior information available
- selection based on the existing stabilizing controller
- approximate process model available

The cost function weighting selection methods are presented for both GMV and LQG control and benchmarking. It may easily be shown that with reasonable choices of dynamic cost function weightings, a GMV controller gives similar responses to those of an LQG controller. Thus, a weighting selection method that works for GMV designs will also apply to LQG solutions [Grimble, 1981; 1988].

4.5.1 General Guidelines (No *a priori* Information)

In the rather rare case when no information is available about the control system, only very general guidelines for weighting choice can be given. Moreover, it is also in the nature of control design itself that the selection of cost weighting functions usually involves engineering judgement rather than precise rules. Hence it is difficult to state rules which ensure a given behaviour is obtained, since in most cases a number of criteria must be satisfied at the same time and trade-offs must be made. The following guidelines will however provide a basis for selecting and changing cost function weightings for GMV and LQG problems.

RULE 1   The weighting choice should be consistent with the existing controller structure – in practice this means that if the existing controller is of the PID type, the error weighting should include an integrator (see also RULE 2). The resulting optimal benchmark controller would then also contain integral action.

RULE 2   The common requirement is that the error weighting P_c should normally include an integral term

$$P_c = \frac{P_{cn}}{1 - z^{-1}}$$

P_{cn} may be constant or they may have the form $(1 - \alpha z^{-1})$ where $0 < \alpha < 1$ is a tuning parameter (the larger α, the sooner integral action is "turned off")

The general effect of introducing integral error weighting is to introduce high gain into the controller at low frequencies. This result is also valid for more general disturbance models. If, say, a system has dominantly sinusoidal disturbances of frequency ω_0, then the weighting can include a lightly damped second order system with natural frequency ω_0. In other words, the error weighting should have a high gain in the frequency range where the disturbances dominate or good tracking accuracy is required.

RULE 3   The control weighting F_c can be chosen as a constant or as a lead term:

$$F_c = \rho \text{ or } F_c = \rho(1 - \gamma z^{-1})$$

where ρ and γ can be considered as tuning parameters. In the case of the GMV design, ρ should normally be negative.

Introducing a high gain at high frequencies on the control weighting term ensures the controller rolls-off at high frequencies and does not amplify the measurement noise. Controller roll-off at high frequencies occurs naturally in LQG or H_2 designs due to the use of a measurement noise model.

If the system is to be made faster then the magnitude of the control weighting should be reduced. One method of getting into the vicinity of a good solution is to try small control weightings and then a much larger control weighting and interpolate between the two to obtain the type of response required. For example, if the small control weighting gives a one-second response and the large control weighting gives a 50-second response then something in between should give an intermediate value for the dominant time constant. Such a procedure does of course require iteration and a simulation model, and on some systems too low control weightings might lead to very harsh actuator movements. Nevertheless, cost-weighting selection can be achieved by such an iterative process.

4.5.2 Selection Based on the Existing Controller Structure

If a system is already controlled by a PID controller, or some other well defined classical control structure, then a starting choice of GMV cost function weighting is to choose the ratio of the error weighting divided by the control weighting equal to the aforementioned controller. There are several assumptions to make this result valid but it is a starting point for design. Moreover, it has realistic frequency response characteristics for weightings inherent in the approach. For example, a PID controller clearly has high gain at low frequency and if it includes a filter then it will have low gain at high frequencies. This is exactly the type of response needed for the ratio between the error and control weightings.

The GMV procedure is therefore to use the transfer of the existing controller to define the cost function weightings. An indirect benefit of this approach is that it is always difficult to sort out the type of scaling required for defining the cost weightings. Clearly, a system which has different physical parameters will require different cost function weightings, even though the underlying process is the same. By utilising the existing controller structure to define the cost weightings this scaling problem is avoided. Moreover, the type of transient response characteristics obtained for the unmodified optimal control solution will probably be of the same order as those for the classical design. This therefore provides a starting point for weighting selection [Grimble and Johnson, 1988, Grimble, 1994 and 2001a].

As a justification for the above discussion, it was shown in [Grimble, 2005] that the GMV controller may be implemented in a Smith Predictor structure. If the system is in the state space model form, the structure is as shown in Figure 4.5, which is intuitively reasonable and easy to explain. Note from the control signal u to the feedback signal p that the transfer is null when the model $z^{-k} W_k$ matches the plant model. It follows that the control action, due to reference signal r changes, is not due to feedback but involves the open-loop stable compensator and the *inner* feedback loop. This inner-loop has the ratio of the error to control weightings $F_{ck}^{-1} P_c$ acting like an inner-loop controller, with return difference operator: $(I - F_{ck}^{-1} P_c W_k)$. Thus, if the plant already has a *PID* controller that stabilises the delay free plant model, the weightings can be chosen equal to the *PID* controller. The choice of the weightings to be equal to a *PID* control law is only a starting point for design, since stability is easier to achieve. However, the control weighting will normally require an additional lead term (or alternatively a high frequency lag term may be added to the error P_c weighting). The high frequency characteristics of the optimal controller will then have a more realistic roll off. This may not be necessary if the *PID* solution already has a low pass filter for noise attenuation.

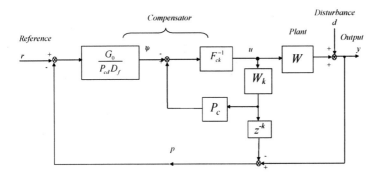

Figure 4.5. GMV controller in the Smith predictor form

The method described above provides a very fast way of generating the desired cost weighting functions. Once the existing controller structure is known then the required weightings follow almost immediately. It is true that some adjustment may be necessary after this initial selection, since it is generally the case that the magnitude of the control weighting function needs to be reduced to speed up the system. In this way, the initial design will normally be close to the existing classical controller but the design can be much improved by reducing the value of the control weighting term. Since the initial design will probably give reasonable responses this procedure reduces the danger of any experiments on the plant.

4.5.3 Process Model Available

The full utilization of the dynamic weightings is only possible when full knowledge of the process model linearized around the working point is available. The model does not need to be very accurate – in fact, a simple first or second order approximation is often quite sufficient.

The following paragraphs are concerned with the selection of the LQG weightings in the continuous-time case, as in (4.118). However, as noted before, a similar procedure applies also to the selection of GMV weightings, with the additional discretisation step involved.

The general rules given in Section 4.5.1 suggest a possible parameterization of the dynamic cost weightings, of the form:

$$Q_c = H_q^* H_q \quad \text{and} \quad R_c = H_r^* H_r$$

where

$$H_q = 1 + \omega_q / s \quad \text{and} \quad H_r = \rho(1 + s/\omega_r) \tag{4.167}$$

Clearly $H_q = (s + \omega_q)/s$ represents integral action, which is cut off at the frequency ω_q. If $\omega_q \to 0$ the minimum variance (constant) term dominates. If ω_c is the desired unity gain crossover frequency for the system, ω_q can be chosen as

$\omega_q = \omega_c/10$ to initiate a design. The resulting H_q has unity gain in the mid to high frequencies and the control weighting must be chosen relative to this value.

The weighting H_r is a lead term and ω_r should be selected to roll-off the controller, where measurement noise dominates. A starting value for a benchmark design is $\omega_r = 10\omega_c$. The value of ρ is chosen to determine the speed of response of the system. The intersection point for the frequency response magnitude plots of $H_q(s)W(s)$ and $H_r(s)$ will be denoted by ω_0 and this frequency is often close to the unity gain crossover frequency for the system. Let ω_g represent the corner frequency for the dominant time constant in the plant model W. Then $\omega_0 \cong \omega_c$ can be chosen to be $\omega_0 = 3\omega_g$ for a process plant and $\omega_0 = 10\omega_g$ for a machine control system.

To determine ρ, the point at which the plots of $H_q(s)W(s)$ and $H_r(s)$ intersect is required. That is,

$$|H_q(j\omega_0)|.|W(j\omega_0)| = |H_r(j\omega_0)| \tag{4.168}$$

At a particular frequency ω_0 write:

$$H_q(j\omega_0)W(j\omega_0) = H_w^r + jH_w^i \tag{4.169}$$

and

$$H_r(j\omega_0) = \rho(H_r^r + jH_r^i) \tag{4.170}$$

Then ρ can be found as the point where

$$\rho^2((H_r^r)^2 + (H_r^i)^2) = (H_w^r)^2 + (H_w^i)^2 \tag{4.171}$$

To summarise the design choices:
Typically $3\omega_g < \omega_0 < 10\omega_g$ and $\omega_c \cong \omega_0$

$$\omega_q = \omega_0/10 \quad \text{and} \quad \omega_r = 10\omega_0 \tag{4.172}$$

$$\rho = \sqrt{\frac{(H_w^r)^2 + (H_w^i)^2}{(H_r^r)^2 + (H_r^i)^2}} \quad \text{evaluated at } \omega_0. \tag{4.173}$$

Only three parameters are needed to fully determine the simple weightings given in (4.167):

- Cut-off frequency ω_q [rad/s]
- Cut-off frequency ω_r [rad/s]
- Control weighting gain ρ

The first two can be determined given only the dominant process time constant. To find parameter ρ, however, the knowledge of the plant model W is required. If no such accurate model is available, then there are two possible options to consider:

- plant model W is approximated with first order dynamics using the dominant time constant and the gain.
- ρ is considered a tuning parameter that determines the relative importance of the variance of the error signal and control activity – hence, changing this parameter will modify the speed of response of the system.

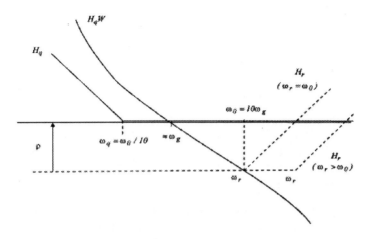

Figure 4.6. Error and Control Weighting Frequency Responses (two control weighting choices all with common intersection point at the frequency ω_0)

A procedure will now be given for the weighting selection, which assumes that a first-order lag approximation of the process model (i.e. the process gain and the dominant time constant) is available.

Procedure 4.3 LQG weighting selection based on the 1st order plant model

Step 1:	Determine the process gain K, dominant time constant τ, and choose the scaling factor β which is a number from 3 to 10 – higher numbers indicate faster desired response. For example, $\beta = 3$ may correspond to a process plant, and $\beta = 10$ to a machine control system.
Step 2:	Compute the plant corner frequency $\omega_g = 2\pi / \tau$ [rad/s]
Step 3:	Compute $\omega_c = \beta \omega_g$
Step 4:	Compute $\omega_q = \omega_c / 10$ and $\omega_r = 10\omega_c$
Step 5:	Compute $\rho = \dfrac{\omega_r}{K\omega_c} \sqrt{\dfrac{(\omega_c^2 + \omega_q^2)(\omega_c^2 \tau^2 + 1)}{(\omega_c^2 + \omega_r^2)}}$
Step 6:	Calculate the weightings as in (4.167)

Procedure End

4.6 Summary

The predictive benchmark provides a means of obtaining a metric that could judge the performance of predictive controllers as well as compare the performance of non-predictive controllers against the generalised predictive control standard. The LQGPC benchmarking index is an improvement on the MV benchmarking index because due to the formulation of the control problem, it does not have the same problem with non-minimum phase processes as the MV control law and just like the GMV index it introduces a penalty on the control action used. The predictive optimisation function includes a cost term on control action and can be readily modified to include constraint and provides the possibility of using the LQGPC/GPC performance index to benchmark systems with constraints.

Also the GPC benchmarking index does not rely on exact knowledge of the process time delay as does the MV algorithm and for the MIMO case avoids the need to obtain or manipulate interactor matrices. There are however some practical limitations with using the GPC benchmarking index, which assumes *a priori* knowledge of the system description and the calculation of the performance cost can be complex and computationally intensive.

The Chapter also introduced the concept of restricted structure performance assessment and using the RS-LQG benchmark index to compare the performance of PID controllers against the performance of an optimally tuned PID controller. Considering that the size of most optimal controllers is equal to the order of the plant and the cost weightings, then any benchmark using the MV, GMV or LQGPC controllers as a reference will be using controllers which are often of high order, relative to classical designs such as PID controller. If a benchmark figure of merit is computed for a controller of much higher order than can actually be used on the plant it is not a very effective measure. The restricted structure benchmarking index offers the ability of comparing "like for like" and also returns the control parameters required to achieve the desired performance.

The MV and GMV benchmarking indices all rely on output/input variance as being the key performance indicator (KPI) for processes. While this might be certainly true for continuous processes, where steady state performance is of more interest. For batch type processes however output/input variances are not necessarily the key KPI, since the dynamic performance of the process is often more important than its steady state performance. The model driven performance cost of the LQGPC and RS-LQG benchmarking index, offer the ability to consider the dynamic performance of controllers and hence are applicable to some batch type processes making them very versatile and adaptable.

5

Divided Wall Distillation Column Simulation Study

Damien Uduehi [1], Hao Xia [2], Marco Boll [3], Marcus Nohr[3], Andrzej Ordys [4] and Sandro Corsi [5]

[1] Strategy and Planning Coordinator (West Africa), BG Group, Reading, UK,
[2] Research Fellow, Department of Electronic and Electrical Engineering, University of Strathclyde, Glasgow, UK,
[3] BASF Aktiengesellschaft, Ludwigshafen, Germany,
[4] Professor of Automotive Engineering, Faculty of Engineering, Kingston University, London, UK,
[5] Manager, CESI SpA, B.U. Transmission & Distribution, Milan, Italy.

5.1 System Description

In this case study, the controller performance of a divided wall distillation column (Figure 5.1) is investigated. Considering a mixture of three components A, B and C, where A is the lightest boiling and C the heaviest boiling; in a conventional direct sequence arrangement, the separation of these components uses two cascade distillation columns: in the first column A would be separated from B and C, while in the second one, B would be separated from C. With the divided wall column technology, the two columns are compacted into one shell in which the middle part of the column is divided vertically by an internal wall. This technology generally leads to equipment cost reduction and significant energy savings.

From a thermodynamic point of view, a divided wall column can be substituted by a system of four columns (Figure 5.2), coupled by vapour and liquid flows. Each column in this structure represents one of the four compartments of the divided wall column (Figure 5.2). The part above the dividing wall corresponds to the upper column (C1) and the part below to the bottom column (C4). The feed side of the dividing wall corresponds to the pre-column (C2), and the reduction side to the main column (C3). This structure was also used for the simulation model.

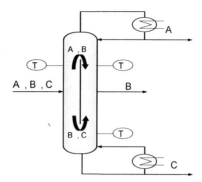

Figure 5.1. Divided wall column

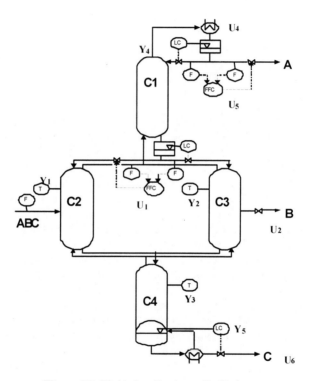

Figure 5.2. Divided wall column distillation process

The ultimate control objective of the divided wall column is the purity of all three components. As there are no online measurements of purity in the plant, three temperatures are controlled as substitutes: the temperatures in the pre-column C2 (Y_1), in the main column C3 (Y_2) and in the bottom column C4 (Y_3). Y_3 should be controlled in the interval between B and C boiling temperatures, whereas Y_1 and Y_2 should be maintained in the interval between A and B boiling temperatures. The temperatures Y_1 and Y_2 of the pre- and main columns are both influenced by the split ratio U_1 and the reflux ratio U_5: the reflux ratio U_5 has the same influence on both temperatures with slow dynamics, whereas the split ratio U_1 moves both temperatures rather fast and in different directions. Therefore, since the temperature Y_2 in the main column is more important for the product purity than the temperature Y_1 in the pre-column, the temperature in the main column Y_2 is controlled with the faster manipulated variable, the split ratio U_1. The temperature in the pre-column Y_1 is therefore controlled with the slower manipulated variable, the reflux ratio U_5. The temperature Y_3 in the bottom column is mainly dependent on the flow U_2 of component B, which has almost no influence on the other controlled variables. Two other variables are also under control: the level Y_5 in the bottom column and the pressure Y_4 inside the plant.

These input/output variables are summarized in Table 5.1 and the control loop pairings are given in Table 5.2.

Table 5.1. BASF process input/output variables

Inputs	Outputs
U_1 : Liquid split ratio between columns C2 and C3	Y_1 : Temperature in C2, TC2
U_2 : Flow of component B	Y_2 : Temperature in C3, TC3
U_3 : Heating energy for component C	Y_3 : Temperature in C4, TC4
U_4 : Cooling energy in the condenser	Y_4 : Pressure in C1, PC1
U_5 : Reflux ratio of component A	Y_5 : Level in the column sump, LC4
U_6 : Flow of component C	

Table 5.2. The control loop pairing

Loop number	Loop description	Measurement	Manipulated variable
I	Temperature control pre-column	Y_1	U_5
II	Temperature control main column	Y_2	U_1
III	Temperature control bottom column	Y_3	U_2
IV	Column pressure control	Y_4	U_4
V	Sump level control	Y_5	U_6

5.1.1 Divided Wall Column Simulator

The study was carried out on a plant model which was developed by BASF and validated against real plant data. The model has approximately 400 states, and a sample time of 60 seconds. The simulator uses the themodynamic substitute shown in Figure 5.2 with a total of 90 trays for the four columns. The model was developed to describe the divided wall distillation column process around its nominal condition, so no precautions were taken to handle abnormal thermodynamic situations like a vanishing liquid phase on one of the trays. This results in a limited range of operating conditions for the simulation model. In the simulator, besides the main variables introduced before, there are several other input variables as listed below. These variables are used as process disturbances and/or variables for modifying operating conditions:

- Feed flow
- Feed composition
- Temperature sensor noises
- Pressure sensor noise
- Heating energy for component C
- Energy lost to the environment

Table 5.3 shows the standard operating conditions of the BASF plant. The concentrations of components A, B and C in the feed stream are given as molar concentrations. Furthermore, to prevent the simulator from abnormality, the output temperatures must respect the constraints defined in Table 5.4.

Table 5.3. Standard operating conditions for BASF plant

Controlled Variable	Value	Manipulated Variable	Value	Other variables	Value
Temperature TC2	126 °C	Split ratio	1.242	Concentration A	0.136
Temperature TC3	126 °C	Flow of B	1437.81 g/h	Concentration B	0.717
Temperature TC4	143 °C	Cooling energy	3137 W	Concentration C	0.147
Pressure C1	90000 Pa	Reflux ratio	18.6	Feed Flow	2000 g/h
Level	0.5 (norm.)	Flow of C	291.47 g/h	Heating energy	2800 W
				Heat Loss	0 W

Table 5.4. Temperature constraints

For temperature TC4 = 145°C		
	Min.	Max
Temperature TC2	120°C	127°C
Temperature TC3	120°C	127°C
For temperature TC4 = 143°C		
Temperature TC2	120°C	132°C
Temperature TC3	120°C	132°C
For temperature TC2 = TC3 = 126°C		
Temperature TC4	141.5°C	150°C

Therefore, for a stable simulation the set points should be set at:

TC2= 126°C
TC3= 126°C
TC4= 143°C

The process has an integrating behaviour: this means that in open loop, the simulation could eventually run into an unstable situation. Hence the following precautions, recommended by BASF engineers, have been observed when running simulations:

- For process identification, open loop tests were performed but only small perturbations were used, as big steps in the inputs could trip the simulation.
- The temperature profile of the process is non-linear hence only variations of ±5°C around the set points should be used for step test. The only exception was the bottom column, which is, for stability, situated eccentrically in the almost linear part of the temperature profile: it is possible to get only −1.5°C but +7°C step tests.
- It is recommended to maintain the MVs (Manipulated Variables) under soft constraints as reported in Table 5.5.
- For concentration, values ±10% deviations are recommended.
- The heat loss is given as watt per Tray, so a value of 1W represents an overall loss of about 90W. It seems that 12W/Tray is the maximum loss which is possible, and for no more than a very short time.

Table 5.5. Ranges for manipulated variable

Manipulated Variable	Min	Max
Split ratio	0.5	2
Flow of component B	0	2500
Flow of component C	0	1000
Heating energy	2400	3500
Reflux ratio	5	25

5.2 SISO Benchmarking Profile

For the purpose of SISO benchmarking analysis, it is necessary to define the input/output pairing for each SISO control loop. In this study, only the currently applied pairings (see Table 5.2) between MVs (Manipulated Variables) and CVs (Controlled Variables) on the modelled BASF plant are considered. The heating energy is not used for SISO control.

BASF is especially interested in the steady state stochastic performance of the divided wall column. Therefore, each SISO loop will be analysed for different values of:

- feed flow rate
- feed composition (A, B and C)
- temperature sensor noises for Y_1, Y_2 and Y_3
- pressure sensor noise for Y_4
- energy lost to the environment

For SISO benchmarking, the loops will be benchmarked one at a time, and performance will be assessed with no consideration of how the performance of one loop affects the other. Three different benchmarking algorithms are used:

- MV: Minimum variance benchmarking index
- GMV: Generalised minimum variance benchmarking index
- RS-LQG: Restricted structure Linear Quadratic Gaussian benchmarking index.

5.3 MV and GMV Benchmarking

5.3.1 MV and GMV Benchmarking Test Set-up

To assess the performance of the PID controllers used on the divided wall column, MV and GMV benchmark test were conducted for 8 different operating conditions (Table 5.6). For each operating condition, a simulation was run and plant input/output/set-point data were collected. The data sets were then used to estimate the benchmark index with a tolerance of 0.01%.

Table 5.6. MV and GMV benchmarking tests

	Case 1	Case 2	Case 3	Case 4	Case 5
Temperature TC2	126 °C	126 °C	126 °C	126 °C	126 °C
Temperature TC3	126 °C	126 °C	126 °C	126 °C	126 °C
Temperature TC4	143 °C	143 °C	143 °C	143 °C	143 °C
Pressure C1	90000 Pa	90000 Pa	90000 Pa	90000 Pa	90000 Pa
Level Column Sump	0.5 cm	0.5 cm	0.5 cm	0.5 cm	0.5 cm
Concentration A	0.136	0.136	0.136	0.136	0.272
Concentration B	0.717	0.717	0.717	0.717	0.434
Concentration C	0.147	0.147	0.147	0.147	0.294
Loop Sensor Noise Variance	0.0001	0.001	0.001	0.0001	0.0001
Heat Loss	0	0	1	1	0
Feed Flow	2000 g	2000 g	2000 g	2000 g	2000 g

Table 5.6. Continued

	Case 6	Case 7	Case 8
Temperature TC2	126 °C	126 °C	126 °C
Temperature TC3	126 °C	126 °C	126 °C
Temperature TC4	143 °C	143 °C	143 °C
Pressure C1	90000 Pa	90000 Pa	90000 Pa
Level Column Sump	0.5 cm	0.5 cm	0.5 cm
Concentration A	0.272	0.136	0.272
Concentration B	0.434	0.717	0.434
Concentration C	0.294	0.147	0.294
Loop Sensor Noise Variance	0.0001	0.0001	0.0001
Heat Loss	.2	0	.2
Feed Flow	2000 g	1500 g	2200 g

To obtain good performance in the MV and GMV benchmark estimation, the general user parameters used for both algorithms were set as follows:

One benchmark value calculated per data set
Length of auto-regressive model: $m = 10$
Data set length = 1000
$T_s = 1$ min. (Note that the BASF model has a sample time of 60 seconds; so that 1 time instance in the model is equivalent to 60 seconds in the real plant).

5.3.2 Key Factors of MV and GMV Benchmarking

There are several key factors affecting the acceptability of the results of the benchmarking exercise:

Dead Times
To use the MV or GMV benchmark algorithm, knowledge of actual loop dead times is essential. In fact, the theory of MV and GMV benchmarking assumes that information on the process dead times is available. If the actual loop dead times are not used then the results returned by the algorithm would be biased. The sign of this bias would depend on whether the dead time used in the calculation of the benchmark index was greater or less than the actual loop dead time. Moreover, if the wrong loop dead times are used, the index value cannot be interpreted as MV/GMV benchmark but should be regarded as a sort of user defined performance index [Desborough and Harris, 1992]

Therefore, estimating the MV and GMV benchmark index for process loops without any knowledge of the loop delay is a real problem and it is not practical to use normal operating data for MV/GMV benchmarking without first having knowledge of the loop delay or trying to estimate it from the data. As suggested by both Thornhill *et al.* [1999] and [Desborough and Harris, 1992], the performance index can be treated as a function of dead time by plotting the value of the benchmark index over the range of the minimum and maximum estimated dead time. Then, analysing the patterns of these curves, it might be possible to obtain a reasonable guess of a suitable estimate of dead time for use in benchmarking. This has been performed for the MV benchmarking trial. The benchmark indices were computed for all loops over dead time ranging from 1 to 20 sample intervals.

For Loops I, II, III and IV the value of the benchmark index did not change significantly as the dead time increased. The value of changes in the benchmark index were less than the value of the standard deviation of the computed benchmark index, hence such change can be assumed to be stochastic and generated by the noise. The conclusion drawn was that it is highly probable that the dead times for these loops is either 1 or 2 sample intervals. Unfortunately, for Loop V no practical conclusion could be reached from such a test. However, it may be possible to overcome this obstacle with a physical analysis of process: as Loop V is a level control loop, it should be an integrator of global mass balance errors; so it can be assumed that for Loop V as well, dead time may be either 1 or 2 sample intervals.

Error and Control Weightings for GMV

The GMV benchmark algorithm needs a set of dynamic error and control weights to compute the performance index. These weights act as design parameters that specify the type of optimal controller required. Because of this it is difficult to compare the performance results returned for the same process control loop when two different set of weightings are used. Hence to efficiently use the GMV benchmark index as a performance assessment tool, the user is required to know and specify the optimal performance requirements for the control loop under assessment. In practice, knowledge of the performance specifications for process loops may be readily available from process control engineers or relevant documentation; these specifications are usually in terms of gain margins, phase margins, bandwidth, overshoot, rise time, *etc.* However, it must be clear that translating these specifications into frequency dependent dynamic error and control weighting functions may not be straightforward, and also that there might be sets of weightings for which no stable control solution is possible.

To obtain adequate weights, a more detailed knowledge of the plant dynamics is required. The weight selection procedure [Grimble, 2000b], summarized in Chapter 4, provides guidelines for the selection of weights once an estimate of the desired dominant closed-loop time constant (T) of the process is obtained. Let P_c and F_c denote the desired weights, then:

$$P_c = 1 + \frac{\omega_p T\left(1 + z^{-1}\right)}{2\left(1 - z^{-1}\right)} \tag{5.1}$$

$$F_c = -\rho(1 + \frac{2(1 - z^{-1})}{\omega_f T(1 + z^{-1})})$$

(5.2)

with $\omega_p = \omega_c / 10$, $\omega_c < \omega_f < 10\omega_c$ and $\omega_c \cong \omega_0$

Where ω_c denotes the desired unity-gain crossover frequency, ω_g represents corner frequency, for the dominant plant time constant and ω_0 is the frequency point at which intersection of P_c and F_c occurs, i.e. ω_0 will determine the point at which the control weighting has more influence than the error weighting term. Frequency ω_0 should typically be ten times greater than ω_g for a machine control system and less for a process plant.

Two methods are possible to assess the time constants for use in weighting selection:

- They are fixed by process control engineers as being the time constants that give the desired closed loop response pattern
- They are determined from the open loop time constants: then each desired closed loop time constant is taken j times lower than the open loop time constant, with j in the interval from 3 to 10.

The first method has been adopted for this study; the open loop time constants calculated for the five process loops are shown in **Table 5.7** as well as the computed error and control weightings.

Table 5.7. Time constants and error and control weightings for GMV benchmarking performance of BASF plant

	T	P_c	F_c	ρ
Loop I	100 sec.	$\dfrac{1.002 - 0.9985z^{-1}}{1 - z^{-1}}$	$\dfrac{0.07667 - 0.05667z^{-1}}{1 + z^{-1}}$	0.01
Loop II	80 sec.	$\dfrac{1.002 - 0.9981z^{-1}}{1 - z^{-1}}$	$\dfrac{0.0633 - 0.04333z^{-1}}{1 + z^{-1}}$	0.01
Loop III	15 sec.	$\dfrac{1.01 - 0.99z^{-1}}{1 - z^{-1}}$	$\dfrac{0.02}{1 + z^{-1}}$	0.01
Loop IV	400 sec.	$\dfrac{1 - z^{-1}}{1 - z^{-1}}$	$\dfrac{0.2767 - 0.2567z^{-1}}{1 + z^{-1}}$	0.01
Loop V	500 sec.	$\dfrac{1 - z^{-1}}{1 - z^{-1}}$	$\dfrac{0.3433 - 0.3233z^{-1}}{1 + z^{-1}}$	0.01

Operating Conditions
If the system is assumed to be linear and time invariant, the benchmark algorithm
which is based on linear system theory is invariant to operating points. However it
is understood that many processes contain some elements of nonlinearity, even if
they are assumed to be linear around their operating point. Hence the benchmark
index produced by the benchmark algorithm will only be applicable in regions
around the operating point at which it was computed. If the process operating point
is changed, a new set of benchmarks must be computed at the new operating point.

5.3.3 Benchmarking Results and Interpretation

For the MV and GMV benchmarking test, the operating point of BASF plant was
changed eight times corresponding to the eight cases of Table 5.6. The loop time
delay was estimated from an open loop step test and at each operating point,
process operating data was collected for each loop and benchmarked with both MV
and GMV algorithms. For the GMV test, it was assumed that the control
performance objectives could be approximated by using the control and error
weightings computed in Table 5.7 with the Grimble [2000b] guidelines and a
scalar term (ρ) in the control weighting equal to 0.01.

For both MV and GMV benchmarking, the test results are shown in Tables 5.8
and 5.9. These results showed a variation in the performance of the controllers as
the operating conditions were changed. This trend is more significant in some
control loops than in other, and also seems to be more characteristic for GMV
benchmarks. After an analysis of benchmark index sensitivity to parameters and
operating conditions as presented in Sections 0, a decision was made to assess the
performance of the existing loop controllers at the nominal operating conditions
(Case 1). The results of that assessment are discussed in Sections 0 and 0.

MV and GMV Benchmark Sensitivity to Operating Condition
For the MV benchmarking tests (Table 5.8), Loops III and V have both an index of
above 0.97 in all operating conditions. This suggests that compared to the MV
controller, these two loop controllers are optimal in all the above conditions. The
performance of the process controllers for Loops I and IV seem to be sensitive to
the operating conditions. In Loop I, for two variations of the operating point the
performance of the loop controller was degraded considerably from an average of
0.97 to 0.8. For Loop IV, there were significant variations from the average loop
performance of 0.15 to 0.07 and 0.3. Loop II shows only one case of performance
variation to changes in the operating conditions, with a degradation of performance
from an average of 0.97 to 0.39 (Case 4).

For the GMV benchmark (Table 5.9), only the performance of the controller in
Loop II can be regarded as being optimal in all operating conditions, with a
benchmark index of above 0.92 in all cases. The performance of the controllers in
Loops I, III, IV and V show significant variations in the value of the benchmark
index as the operating point of the process was changed. Unlike in the MV test, the
values of the performance index returned for Loop I seems to change with each
operating condition varying from a low of 0.27 (Case 7) to a high of 0.87 (Case 3).

This suggests that the characteristics of the process in loop I are significantly different in each of these operating regions.

Although the values of the benchmark index change significantly with the operating point for Loops III, IV and V, the performances recorded in all cases are so poor that such changes do not make any difference to the interpretation of the benchmark index.

The sensitivity of index to operating conditions indicates inherent non-linearities in the process, with the result that the optimal controller in one operating conditions may not be optimal when the operating conditions are changed. This suggests that:

- The benchmark index gives an idea of controller performance around the current operating conditions. For this reason, the benchmark analysis should be applied in operating conditions representative of normal plant conditions.

- If the normal plant functioning requires frequent or quite frequent changes of operating conditions, it is important to repeat the benchmark analysis at different characteristic operating conditions in order to have a realistic assessment of controller performance.

Table 5.8. MV benchmark table

Loop No.	Case	1	2	3	4	5	6	7	8
I	Index	1	1	.99	.80	.99	.99	.98	.99
	Stdev	.0145	.0145	.0145	.075	.016	.017	.02	.017
II	Index	1	.99	.99	.39	.97	.96	.99	.94
	Stdev	.013	.013	.013	.092	.025	.03	.016	.035
III	Index	.96	.97	.99	.99	1	1	.98	1
	Stdev	.018	.018	.018	.022	.012	.013	.015	.013
IV	Index	.15	.15	.2	.07	.15	.16	.3	.15
	Stdev	.015	.015	.035	.023	.02	.02	.037	.02
V	Index	1	1	.99	.99	.98	1	1	1
	Stdev	.0124	.0124	.0124	.016	.032	.012	.0129	.012

Table 5.9. GMV benchmark table

Loop No	Case	1	2	3	4	5	6	7	8
I	Index	0.41	0.86	0.87	0.431	0.76	0.76	0.27	0.80
	Stdev	0.10	0.049	0.044	0.10	0.077	0.076	0.085	0.067
II	Index	0.97	0.97	0.98	0.97	0.94	0.94	0.98	0.92
	Stdev	0.016	0.016	0.017	0.017	0.029	0.029	0.015	0.033
III	Index	0.004	0.045	0.052	0.005	0.014	0.014	0.009	0.012
	Stdev	0.002	0.018	0.02	0.002	0.006	0.006	0.004	0.005
IV	Index	0.055	0.055	0.07	0.051	0.038	0.039	0.092	0.032
	Stdev	0.009	0.009	0.011	0.011	0.005	0.005	0.012	0.005
V	Index $\times10^{-5}$	1.16	1.65	1.35	0.967	0.669	0.669	1.18	0.613
	Stdev $\times10^{-6}$	4.9	6.9	5.86	4.22	2.93	2.93	5.14	2.69

Existing Controller Performance

For the MV test it can be observed from Table 5.8 and Table 5.9 that loop controllers I, II, III and V are all operating at minimum variance performance levels with benchmark indices of 1, 1, 0.96 and 1 respectively. Loop IV however is very poorly tuned with a benchmark index of 0.15. For the GMV test it can be observed that Loops III, IV and V are all very badly tuned with benchmark indices of 0.004, 0.055 and 1.16×10^{-6}. Loop I is badly tuned having a benchmark index of 0.41, while Loop II is optimally tuned with respect to the GMV controller, with a benchmark index of 0.97. The need to retune the controller will depend on the requirements of the process operators. The pertinent question is what level of output variance is required and how much controller action is acceptable in achieving it.

In the case of Loop II, for nominal operating conditions no retune is required since it passes both the MV and GMV benchmark. Loops 1, 3 and 5 are all operating at minimum variance control level, but operate very poorly when compared to corresponding GMV controllers. This implies that the process output variance is as low as possible but that the control activity used in achieving this is quite high. If the process operator accepts that the variability in the control action can be high, i.e. if the objective is to have the minimum variance of the controlled variables, then they can be satisfied by the current controller tunings. However, if the operator considers the variability in the control action to be too high and is satisfied with the error and control weighting selections for the GMV benchmarking, then the controllers of Loops I, III and V should be re-tuned. For Loop IV, in either case MV or GMV, the loop is performing badly and if the operator is not satisfied with current loop performance levels, then the controller could be re-tuned to try to improve performance.

Controller Retuning Opportunity
For loops where performance improvement is required, the MV and GMV benchmarks only indicate the current level of performance and scope for improvement. However because the structure and order of the existing process loop controller is not taken into account, there is no way of knowing whether a controller of the pre-existing structure can actually achieve the same level of performance as MV or GMV controllers. Furthermore, there is also no indication of what performance levels can actually be achieved by the existing controller. Thus if, after being re-tuned and benchmarked, the loops always return a low performance score, it will still be unclear whether the lack of performance is due to sub-optimal tuning or limitations in the controller structure. Moreover the MV and GMV benchmark algorithms do not provide the controller parameters required to achieve the optimum performance.

5.4 RS-LQG Benchmarking

5.4.1 RS-LQG Benchmarking Test Set-up

For the RS-LQG performance assessment of the PID controllers used on the divided wall column, the benchmark tests were performed around the standard operating condition. Two benchmark tests were conducted; one test assumes a zero reference condition, which implies a benchmark of the steady state performance of the process, while the other test assumes that the process was driven by a changing reference in presence of a disturbance. This benchmark condition corresponds to trying to benchmark the dynamic performance of the process controllers.

According to recommendations of BASF, perturbations of 1°C were used as temperature set point step changes, whereas for level and pressure loops steps of 0.2 (norm) and 500 Pa were performed on set points.

In all scenarios the RS-LQG controller was chosen to be a PI controller, as all the existing controllers in the process were PI or P only. The operating conditions for the two benchmark tests are reported in Table 5.10.

5.4.2 Key Factors of RS-LQG Benchmarking

Transfer Function Model Description
To compute the performance index, the RS-LQG benchmark algorithm needs the polynomial transfer function description of the process, disturbance and reference. The process description should be a set of SISO transfer functions, one for each of MV/CV pairing corresponding to a control loop to benchmark. If a model of the process does not already exist then it would have to be identified by using an appropriate system identification method.

Table 5.10. Operating conditions and set points

	Steady state test	Dynamic test
Temperature TC2	126 °C	126 °C → 127 °C
Temperature TC3	126 °C	126 °C → 127 °C
Temperature TC4	143 °C	143 °C → 144 °C
Pressure C1	90000 Pa	90000 Pa → 90500 Pa
Level Sump	0.5 norm	0.5 norm → 0.7 norm
Concentration A	0.136	0.136
Concentration B	0.717	0.717
Concentration C	0.147	0.147
Loop Sensor Noise Variance	0.0001	0.0001
Heat Loss	0 W	0 W
Feed Flow	2000 g/min	2000 g/min

In general the best identification results may be obtained by identifying the process using an open loop test. However the process can still be identified adequately in the closed loop, if a dither signal is applied to the input of the process under feedback control. If only normal closed loop plant operating data are available, model identification is still possible, but in some cases the accuracy of the model obtained might be impaired. The first two methods should result in models that can be used for both dynamic and steady state controller design and benchmarking; the latter might only be suitable for steady state benchmarking and/or design.

If plant identification is required, then it is also necessary to identify a stochastic disturbance model for the noise. Generally this noise model can be described using the structure of a white noise driving colouring filters. Once the disturbance model is chosen, in most cases, the reference model can be assumed to be the same as the disturbance model; if however it is desired that the reference model contains some special characteristic, then it must be defined explicitly with a dedicated reference transfer function.

For the divided wall column simulation case study, SISO system identification was performed at nominal operating conditions. The process was put into manual at its nominal steady state operating point and a dither signal of appropriate variance and intensity applied to each input alternatively. The process input and output data were collected and a least squares identification method used to obtain the loop transfer functions. For each loop, the time delay was assumed to be 1 sample interval except for the third loop (Loop of bottom column temperature Y_3) for which the time delay was assumed to be 2 sample intervals. The identified transfer functions did not include any information on the interaction between loops.

The disturbance was assumed to be white noise driving a low pass filter with low intensity dc gains.

Higher order models were used to try to improve accuracy and to account for any mistake in time delay estimation. As already noted, this process identification, can be used for both dynamic and steady-state controller benchmarking. The identified models are listed in Table 5.11.

Table 5.11. Plant models for RS-LQG benchmarking

Loop	Plant $\left(\dfrac{B_0}{A_0}\right)$
I	$z^{-1}\dfrac{-0.02119 - 0.02668\,z^{-1} + 0.02611\,z^{-2}}{1 - 0.6959\,z^{-1} - 0.3301\,z^{-2} - 0.1108\,z^{-3} + 0.1358\,z^{-4}}$
II	$z^{-1}\dfrac{-3.493 - 0.4471\,z^{-1} + 0.7616\,z^{-2}}{1 - 0.7787\,z^{-1} - 0.08244\,z^{-2} + 0.1467\,z^{-3} - 0.06603\,z^{-4}}$
III	$z^{-2}\dfrac{0.003369 + 0.001255\,z^{-1} - 0.000486\,z^{-2} - 0.002708\,z^{-3}}{1 - 0.5871\,z^{-1} - 0.4113\,z^{-2} - 0.1559\,z^{-3} + 0.05964\,z^{-4} + 0.09504\,z^{-5}}$
IV	$z^{-1}\dfrac{-11630 + 4510\,z^{-1} + 5301\,z^{-2}}{1 - 0.6757\,z^{-1} - 0.4152\,z^{-2} + 0.0947\,z^{-3} + 0.0009064\,z^{-4}}$
V	$z^{-1}\dfrac{-8.541e\text{-}007 + 2.814e\text{-}007\,z^{-1} + 5.672e\text{-}007\,z^{-2}}{1 - 1.33\,z^{-1} - 0.3339\,z^{-2} + 0.6645\,z^{-3} - 0.0005495\,z^{-4}}$

Error and Control Weightings
Like the GMV algorithm, the RS-LQG benchmark algorithm requires a set of dynamic error and control weights to compute the performance index. These weights act as design parameters that specify the type of optimal controller required. Because of this it is difficult to compare the performance results returned for the same process control loop when two different sets of weightings are used. Hence to efficiently use the RS-LQG benchmark index as a performance assessment tool, the user is required to know and specify the optimal performance requirements for the control loop under assessment.

Unlike the GMV theory, the RS-LQG theory returns a stable control for all weighting choices. However, the choice of weightings must be consistent with the choice of controller. Once the consistency of the weightings with the controller requirements, for dynamic performance tests, has been verified, the guidelines of [Grimble 2002b] have been used to calculate the weights. For the five process loops, the desired closed loop time constants of Table 5.13 were used in calculating the error weightings; the control weighting was chosen only to be a scalar penalty term. The scalar control weighting should correspond to design or performance specifications. Although a nominal value of 10% (0.1) is usually chosen, it can be varied to reach a desired closed loop response speed.

Table 5.12. Noise and original PID models for RS-LQG benchmarking

Loop	Original PID	Noise$\left(\dfrac{C_d}{A_d}\right)$
I	$\dfrac{-0.728 + 0.7\,z^{-1}}{1 - z^{-1}}$	$\dfrac{.05}{1 - .3z^{-1}}$
II	$\dfrac{-0.011 + 0.01\,z^{-1}}{1 - z^{-1}}$	$\dfrac{.05}{1 - .4z^{-1}}$
III	$\dfrac{37.77 - 37.12\,z^{-1}}{1 - z^{-1}}$	$\dfrac{.05}{1 - .5z^{-1}}$
IV	$\dfrac{-2.004e\text{-}6 + 2e\text{-}6\,z^{-1}}{1 - z^{-1}}$	$\dfrac{.005}{1 - .4z^{-1}}$
V	-2000	$\dfrac{.005}{1 - .5z^{-1}}$

Table 5.13. Time constants used to compute error weightings for RS-LQG benchmarking performance of BASF plant

Loop I	30sec
Loop II	30sec
Loop III	30sec
Loop IV	100sec
Loop V	100sec

For the steady state performance test, the error and control weightings were chosen to be minimum variance weightings: the error weighting was set equal to unity and the control weighting set equal to zero. The resulting transfer functions of these weightings are shown in Table 5.14.

5.4.3 Loop Performance Analysis

Dynamic Performance Test
In assessing the performance of the process loop controllers, the loop transfer functions were identified and used in the RS-LQG algorithm along with the error/control weightings and disturbance and reference models. The RS-LQG

algorithm returned the set of benchmark indices and optimal PI parameters as documented in Table 5.15. It can be observed that controller performance for Loops I and II were 0.218 and 0.244 respectively: both controllers are performing at approximately 75% less than their optimal ability.

Table 5.14. Weighting selection for RS-LQG benchmarking

Loop	Dynamic Performance weighting selection		Static Performance weighting selection	
	Qc	Rc	Qc	Rc
I	$\dfrac{1.005 - 0.995\,z^{-1}}{1 - z^{-1}}$	0.2	1	0
II	$\dfrac{1.005 - 0.995\,z^{-1}}{1 - z^{-1}}$	0.2	1	0
III	$\dfrac{1.005 - 0.995\,z^{-1}}{1 - z^{-1}}$	0.02	1	0
IV	$\dfrac{1.002 - 0.9985\,z^{-1}}{1 - z^{-1}}$	0.1	1	0
V	$\dfrac{1.002 - 0.9985\,z^{-1}}{1 - z^{-1}}$	0.09	1	0

Table 5.15. RS-LQG controller and CPI obtained

No	Optimal (RS) PID	CPI	No	Optimal (RS) PID	CPI
I	$\dfrac{-6.162 + 6.108z^{-1}}{1 - z^{-1}}$	0.218	IV	$\dfrac{-9.433\times10^{-5} + 8.784\times10^{-6}\,z^{-1}}{1 - z^{-1}}$	0.006
II	$\dfrac{-0.2959 + 0.2905\,z^{-1}}{1 - z^{-1}}$	0.244	V	$\dfrac{-1.155\times10^{6} + 1.155\times10^{6}\,z^{-1}}{1 - z^{-1}}$	1.19×10^{-8}
III	$\dfrac{75.16 - 74.5\,z^{-1}}{1 - z^{-1}}$	0.832			

The controller performance for Loop III was 0.832, which indicates a 17% drop from its optimal performance. For Loops IV and V, the controller performances were 0.006 and 1.19×10^{-8} respectively: both controllers are performing at approximately 99.9% less than their optimal ability. So, excepting Loop III controller, it seems that, if the error and control weighting choices are consistent with the plant objectives, all the existing controllers are very poorly tuned.

To understand this result, the process was simulated with the existing loop controllers, and step changes were initiated with all the loops closed. The changes in loop set points were inserted in a staggered manner with intervals of 200 samples between the present loop and the preceding one. The process controllers were then re-tuned to the optimal settings returned by the RS-LQG algorithm and the simulations were repeated. The integral square error (ISE), the integral square control action (ISU) and the steady state variance were calculated for all loops and for the two controller settings. The results are documented in Table 5.16.

Table 5.16. RS-LQG controller dynamics test result

Loop	Controller (PID)	Steady State Variance	$ISE = \sum_{k=1}^{2200} \left(r_k - y_k \right)^2$	$ISU = \sum_{k=1}^{2200} \left(u_k \right)^2$
I	Original	0.0030	70.81	812172.83
	Retuned	0.0067	20.12	813286.39
II	Original	0.0027	100.58	3723.37
	Retuned	0.0037	10.92	3723.72
III	Original	0.0033	27.09	4866223201.63
	Retuned	0.0046	17.68	4866226956.09
IV	Original	0.86	60272821.82	23152.50
	Retuned	1.060	268881.43	23148.91
V	Original	3.364×10^{-5}	11.81	118360801.12
	Retuned	3.253×10^{-5}	0.117	157310194201.42

Because the ISE and ISU values are "real" plant data, they also contain components related to interactions between the different control loops so that they cannot be linked to the benchmark indices obtained from the RS-LQG algorithm; there is no direct relationship between these values and the RS-LQG benchmark. They serve only as a crude indication of performance of controllers, and verify the trend of the benchmark index.

Considering the ISE for all process loops, it can be concluded that the dynamic performance of the optimal (RS) controllers is better than that of the existing controllers. Much lower ISE indicates that the set points are effectively better tracked with the optimal RS controllers. However, as indicated by the ISU values, some of the loop controllers use more control action in achieving these low values of the ISE than the existing controllers.

It must be also observed that the RS-LQG algorithm does not take into account any constraints on input/output variables. This limitation can lead to results with poor physical sense; the Loop V re-tuned (RS) controller certainly improves level control but doing so, needs to have a negative flow for component C, that is not possible on real plant. Moreover performing such large changes on input variable, may invalidate the assumption that the process is described by a linear/linearised model. It means that if implemented on real plant, this controller will not act as the simulation results suggest, but will be limited by the physical constraints of the real plant.

Steady State Performance Test
For the steady state performance test of the process loop controllers, the reference model was assumed to be zero (i.e. no reference model) and the same transfer function models were used as in the dynamic case. The error and control weightings were chosen to be minimum variance weightings: the error weighting was equal to unity and the control weighting equal to zero. The RS-LQG algorithm returned the set of benchmark indices and optimal PI parameters documented in Table 5.17.

Table 5.17. RS-LQG controller and CPI obtained

Loop	Retuned (RS) PID	CPI	Loop	Retuned (RS) PID	CPI
I	P = -0.11025 and I $\cong$ 0	0.964	IV	P = -3.0865×10^{-5} and I $\cong$ 0	0.87
II	P = -0.08546 and I $\cong$ 0	0.899	V	P = -4.7801 and I $\cong$ 0	1.00
III	P = 0.091257 and I $\cong$ 0	0.957			

It can be observed that controller performance for Loops I, III and V were 0.96, 0.95 and 1 respectively. This would indicate that these controllers were performing at the limits of their optimal ability. The controller performance for Loops II and IV, were 0.89 and 0.87 respectively, which indicates that both controllers were performing rather well, approximately 11 % less than their optimal performance. In this case, because all the weightings were those of minimum variance, the optimal performance of all control loops corresponded to the minimum variance of the controlled variables.

Similar to the dynamic benchmarking study, the process was simulated with the existing loop controllers, and step changes were initiated with all loops closed. The changes in loop set points were inserted in a staggered manner with intervals of 200 samples interval between the present loop and the preceding one. The process controllers were then re-tuned to the optimal settings returned by the RS-LQG algorithm and the process was repeated. The integral square error (ISE), the integral square control action (ISU) and the steady state variance were again calculated for all loops and for the two controllers' settings. The results are documented in Table 5.18.

Table 5.18. RS-LQG controller static test result

Loop No.	Controller (PID)	Steady State Variance	Actual CPI	$ISE = \sum_{k=1}^{2200}\left(r_k - y_k\right)^2$	$ISU \dfrac{}{\sum_{k=1}^{2200}\left(u_k\right)^2}$
I	Original	0.0030	1.06	70.81	812172.83
	Retuned	0.0032		308.00	758718.44
II	Original	0.0027	0.96	100.58	3723.37
	Retuned	0.0024		20.29	3402.40
III	Original	0.0033	0.88	27.09	4866223201.63
	Retuned	0.0032		651.31	4548350478.05
IV	Original	0.86	0.004	60272821.82	23152.50
	Retuned	0.0035		3817893.84	21645.11
V	Original	3.413×10^{-5}	1	11.81	118360801.12
	Retuned	3.414×10^{-5}		12.42	106260375.74

As documented in Table 5.18 the existing controller for Loops I and III has a better ISE values than the re-tuned controllers (from the benchmark index both these controllers are performing at approximately 3% less than the optimal RS-LQG control performance). However the original controller uses more control action in achieving these low values of the ISE than the retuned controllers. This may be equivalent to an over aggressive opening and closing of the valve and may result in more utilization of energy and hence high operating cost. The ISE and ISU for Loop V is almost identical for both the existing and re-tuned controller, with the existing controller being marginally better (from the benchmark index the controller in Loop V is performing at its optimal ability). In Loops II and IV for which the existing controllers are performing at approximately 89% of the optimal RS-LQG control performance, the re-tuned controllers for those loops have a better ISE value. From the benchmark analysis, it should be said that even if the existing controllers were already performing well, the re-tuning with optimal parameters has further improved control performance. It may be however necessary to analyse in detail the optimal controller settings. Because of the choice of weightings (the error weight has no integral action at low frequencies) and also because of the choice of reference model, the RS-LQG algorithm identifies, for all the control loops, a proportional controller (only P term) as the best controller. This means that it would not be possible to have a zero steady-state error with set point changes.

Because the MV weighting was used in the benchmark algorithm, the steady state variance should be directly related to the benchmarking index. That is, if the steady state variance of the re-tuned controller is divided by the steady state variance of the existing controller for corresponding loop, the result obtained should be the same as the benchmark index. Using Table 5.18 and comparing the steady state variances for the two controllers it can be observed, that for Loops I, II, III and V, the division returns the values of 1.06, 0.96, 0.88 and 1, which are close to the theoretical values (0.964, 0.957,0.89 and 1) returned by the RS-LQG algorithm (Table 5.17). However for Loop IV, the value returned by this division is 0.004, which is far from the theoretical value (0.87) returned by the RS-LQG algorithm. Because, all the theoretical results obtained with RS-LQG algorithm assume that there are no interactions between the loops, the performance of each loop is calculated independently. Hence the value of the performance index returned by the benchmark algorithm can be interpreted as the performance of the loop controller if all other process loops were set to manual (i.e. open loop). However the reality is that interactions between loops exist and hence the predicted performances would not be the same as the actual performances. Moreover from simulation results it is evident that Loop IV is strongly influenced by interactions. Hence there are significant differences between the predicted and actual performance.

5.4.4 Opportunity for Controller Re-Tuning

One of the benefits of the RS-LQG benchmarking is that it does provide the optimal controller parameters that would enable the optimum performance reachable with the chosen error and control weightings to be achieved. However, the difficulty may be to know whether to use the controller optimum parameters calculated assuming steady state conditions or those calculated assuming dynamic conditions.

If the choice of weightings for steady-state and dynamic operating conditions are consistent with the plant objectives, the simulation and test results indicate that the existing controllers are quite well tuned for steady-state conditions but that the plant dynamic performance could be improved. So, it may be deduced that there is a need for two different controllers, one for transient stages and one for steady state operations since the controller that produces the lowest variance in steady state is different from that which produces the fastest response and lowest error during the transient stage.

However, the operator may not desire to complicate the structure of the controllers and keep only one controller for each loop. If they exist, the real optimal controllers will ensure good performance for both steady-state and dynamic conditions. The dynamic optimal controllers already guarantee high performance in dynamic conditions. If they also reach good performances in steady state, they can be considered as the appropriate choice. So the dynamic optimal controllers should be benchmarked in steady-state conditions. To benchmark the dynamic optimal controllers in the steady-state conditions, the RS-LQG algorithm was provided with the transfer functions of plant, noise, reference, error weighting and control weighting (i.e. GMV weightings) as described before, as well as the re-tuned (RS) PID controllers transfer functions of Table 5.15. The results are collected in Table 5.19.

Table 5.19. Steady-state benchmarking of dynamic optimal controllers

	CPI
Loop I	0.980
Loop II	0.686
Loop III	0.925
Loop IV	0.622
Loop V	0.091

Except for the case of control Loop V, the re-tuned controllers are still performing well as measured by steady-state criteria. So, for example, if the operator has to change the operating conditions around the nominal point frequently and if they are satisfied with the error variance and control actions achieved by these new controllers, then the plant performance may be enhanced when re-tuning the controllers with the dynamic optimum parameters. It must be still remembered that the plant performance improvement is intended as the improvement of each SISO control loop taken separately; the other control loops are shifted into manual functioning mode and their input variables are fixed at their nominal values.

5.5 MIMO Benchmarking Profile

SISO benchmarking algorithms do not take into account how the improvement in the performance of one loop could have an adverse effect and cause performance degradation in another loop. Hence there is a need for MIMO benchmark algorithms. MIMO benchmarking techniques can calculate both the total system variance as well as the individual variances of each loop when the overall system is in an optimal performance state. For the MIMO LQGPC/GPC performance assessment of the multi-loop PID controllers used on the divided wall column, the benchmark tests were conducted at the nominal operating condition. All the tests were conducted on the linearised state-space model obtained around the nominal operating point (see **Table 5.20**). The GPC performance index was used for the benchmarking tests. The index was chosen for computational reasons, as it was easier and faster to compute given the size of the matrix manipulations involved.

5.5.1 GPC Benchmarking Test Set-up

The benchmarking tests were grouped into two categories as follows:
1.   Steady State Performance Test
The steady state tests were set up to determine how well tuned the PID controllers were, to reject stochastic disturbances and to minimise the variances of the input and output signals as well as the steady state errors. The test conducted was analogous to minimum variance test conditions. All set points were held constant

at the nominal operating point over the benchmarking interval and no penalty was imposed on the variance of the inputs (control action).

2.   Transient Performance Test

The transient tests were set up to examine the tracking performance of the PID controllers and determine how well they were tuned to minimise rise times, settling times and overshoots. Assuming steady state conditions at the nominal operating points, the set points of all loops were then changed by a 5% deviation at exactly the same instance and the response of the PID controllers was benchmarked over the corresponding interval.

Table 5.20. Nominal Operating Conditions and Set Points

Variables	Operating Point	Setpoint Deviation
Temperature TC2	126 °C	6.3 °C
Temperature TC3	126 °C	6.3 °C
Temperature TC4	143 °C	7.15 °C
Pressure C1	90000 Pa	4500 Pa
Level Column Sump	0.5 cm	0.025 cm
Concentration A	0.136	
Concentration B	0.717	
Concentration C	0.147	
Loop Sensor Noise Variance	0.0001	
Heat Loss	0	
Feed Flow	2000 g	

5.5.2 Key Factors in GPC Benchmarking

State Space Model Description

The LQGPC/GPC benchmark algorithm needs the state space description of the process and disturbance models and knowledge of the reference trajectories to compute the performance index. If a model of the process does not already exist then it would have to be identified by using an appropriate system identification method. The difficulty and the number of tests required to identify a multivariable model increases with the product of the number of process inputs and outputs. For non-linear systems this number is even higher (usually doubled) since tests must be performed for both positive and negative increments of each variable.

Because of limited resources, it was not possible to conduct an extensive identification exercise on the divided wall column. An open loop MIMO system identification was performed. Opening all process loops and holding the inputs constant at their nominal operating point, 5% perturbations were applied in turn to each input and each time the values of the five corresponding outputs were measured. A least squares/regression method was used to fit the data to a discrete time transfer function model. The model was then augmented to include the transfer functions of the disturbance models. The same disturbance models as those in the SISO analysis were used. The resulting extended model was converted to a state-space form using MATLAB®. A measurement noise variable was then added to complete the state space model.

Apart from the difficulty involved in obtaining models for this large size MIMO system, the order and number of states in the model affects the ease of computation of the benchmark index. The divided wall column state space model identified by MATLAB® had 117 states. Given the number of predictions used in the test, it was required to make addition, multiplication and pseudo-inversion of matrices of the size 250 x 117. Hence to speed up the computations, the GPC benchmark, which requires fewer matrix manipulations, was used instead of the LQGPC benchmark

Error and Control Weightings and the Horizons
The LQGPC/GPC benchmark index also requires a set of constant error and control weights to compute the performance. The LQGPC/GPC controller does depend on the selection of the weights, they are mainly chosen to normalise or, more precisely, to scale to the same magnitude the square of each individual input and output variables as well as any cross product. This would improve benchmark effectiveness and help avoiding singularity in numerical optimisation. However, in addition, it is also possible to structure these weightings to reflect some priority given to individual input or output variables or a cross product between inputs or between outputs. This can be done by multiplying each individual term in the weighting matrices by a given scalar quantity which represents the chosen priority level. This procedure corresponds to the tuning of the LQGPC/GPC controller.

In most cases, LQGPC benchmarking should return a stable LQGPC controller, and therefore usable values of benchmark, for nearly all choices of weightings. This may not be the case for the GPC benchmark since

    i)   it does not have guaranteed LQG properties and therefore the GPC controller is more difficult to stabilize
    ii)   it is similar to the MV and GMV cost functions
    iii)   some weightings may not be applicable to non-minimum phase systems.

For the divided wall column, weightings were first chosen to scale the square of all inputs and outputs to the same magnitude. For the error weighting two sets of weightings were used in all the three benchmarking test categories listed previously. The first set of error weightings assumes equal priority for all process outputs after scaling, while the second set assumes that each of the three temperature loops had a 30% priority and that the pressure and level loops had a 5% priority each. The choice of priority resulted from information given by BASF, which had indicated that the temperature loops were the main loops of interest, while the level and pressure loops were regarded as utility streams. These priority levels were implemented after scaling by multiplying the weighting corresponding to the temperature loops by a scalar factor of 6. Then, as control action weightings are concerned (weighting matrix for inputs variables), a null scalar factor was used to zero the control weightings for the steady state performance tests whereas in all other tests, a factor of 0.2 corresponding to a 20% penalty was used for all the inputs. The resulting values of these weightings are shown in Table 5.21.

For the steady state performance tests, the error and control weightings were chosen to be minimum variance weightings in which case the error weighting was equal to unity and the control weighting equal to zero.

Table 5.21. GPC basic weighting design

Basic Error Weighting Matrix

$$\begin{bmatrix} 0.15 & 0 & 0 & 0 & 0 \\ 0 & 0.15 & 0 & 0 & 0 \\ 0 & 0 & 0.117 & 0 & 0 \\ 0 & 0 & 0 & 4.9\times10^{-8} & 0 \\ 0 & 0 & 0 & 0 & 1.6\times10^{-3} \end{bmatrix}$$

Basic Control Weighting Matrix

$$\begin{bmatrix} 0.0103 & 0 & 0 & 0 & 0 \\ 0 & 50.28 & 0 & 0 & 0 \\ 0 & 0 & 2.84\times10^{-6} & 0 & 0 \\ 0 & 0 & 0 & 2.52\times10^{3} & 0 \\ 0 & 0 & 0 & 0 & 8\times10^{-5} \end{bmatrix}$$

The output prediction horizon (N_2) was chosen to be equal to 50 while the control horizon (N_u) was chosen to be equal to 49. These values were chosen because empirical test on the linear state space model showed that they provide a good system response.

5.5.3 Loop Performance Analysis

For the performance analysis, two measures available in the MIMO benchmarking algorithm are utilized. Firstly, the global benchmark index, which is represented as one composite number (trended over time as in Figure 5.3), that indicates the combined performance of each loop in a multivariable process. This global index takes into account the interaction between loops and, therefore, is not a simple summation of individual loop performances. Secondly, the individual loop performance index, which compares the performance of the existing controller (multivariable or otherwise), on each single loop, against the performance of the appropriate inputs and outputs of the GPC control to obtain both a Loop Regulatory Index (LRI) and a Loop Control Effort Index (LCEI) as in Table 5.22 and Table 5.23

Using the loop performance index helps to provide an added layer of clarity as to how the performance of a loop within the multivariable system is affected by the performance of other loops and in identifying which loops need to be detuned or optimized. For analysis of the performance of the BASF controllers both features of the MIMO benchmark algorithm were used.

Steady State Performance Test

The minimum variance like performance test indicates that the multi-loop PID controller used to control the divided wall column was operating at minimum variance standard. The overall system performance benchmark is shown in Figure 5.3. This indicates that the combined minimum variance of the process outputs for the multi-loop control is optimal. These results are similar to those obtained for the corresponding SISO MV test case.

The results in Figure 5.3 also show that, for the stochastic steady state performance test, giving a 90% performance priority to the temperature loops does not have any impact on the overall performance.

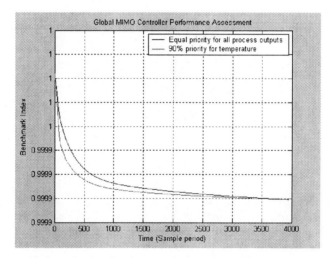

Figure 5.3. The global MV MIMO control performance

Table 5.22. Individual loop performance under MIMO MV weighting

	Loop No.	I	II	III	IV	V
Equal weighting	LRI	0.996	1	1	0.59	1
	LCEI	0.28	0.67	1	0.71	0.83
90% priority on temperrature	LRI	0.994	0.999	0.999	0.61	1
	LCEI	0.38	0.37	0.37	0.36	0.36

For the steady state test, there is no penalty on the input variables (control action). However, when comparing the results individually, although the variance of the process outputs under multi-loop control is close to that obtained with the optimal GPC controller, the multi-loop controller uses much more control action than the GPC controller. This results in large changes of control variables as indicated by the individual input benchmarks for Loops I, II, IV and V. Thus, the GPC controller reaches the same performance index as the multi-loop controllers but using rather less control action.

To further investigate the control effort needed during the steady state, another simulation test which penalises the control effort has also been carried out. With the control weighting [0.2 0.2 0.2 0.2 0.2 0.2], the LRI's recorded in **Table 5.23.** show that the regulation performance of the original multi-loop controller is almost identical to that of the optimal GPC controller. However, the LCEIs for Loops I,II,IV and V show that the control effort is much greater for the original controller than for the GPC controller.

Table 5.23. Individual loop performance with penalty on control action

	Loop No.	I	II	III	IV	V
Equal weighting	LRI	0.955	1	1.01	0.02	0.99
	LCEI	0.02	0.01	0.18	0.04	0.01
90% priority on temperrature	LRI	0.975	0.999	0.999	0.08	0.99
	LCEI	0.03	0.02	0.01	0.01	0.01

Transient Performance Test
For the transient performance test, the reference trajectory presents a simultaneous step change in set points at a time index of 100. As shown in Figure 5.4, the transient performance test indicates that the multi-loop PID controllers are operating at only 10 % of the GPC optimum. Since the prediction horizon is 50 samples and the step changes occur at a time instance of 100, this means that at the time instance 50, the GPC predictive controller would begin to adjust the input variables. The magnitude of the adjustment will of course depend on the process characteristics as well as on design parameters. The figure shows that the overall performance of the system begins to decline from 100% performance at the time instant 50, and drops to 10 % performance at the time instant 100 (when the step change occurs).

Once again the results of the MIMO analysis are in some respect similar to some of the information obtained from the SISO RS-LQG tests, which showed that the original controller should be tightly tuned for the dynamic changes. However, the results from the static performance test shows that the controller is achieving optimal MIMO MV performance, but with too much control action.

This clearly illustrates the conflicting requirements between good regulation and good tracking. However, as shown in Figure 5.5, the CPI will return to around 100% when the transient period is finished. With the system operated at the new steady state, the GPC CPI reflects the steady state performance.

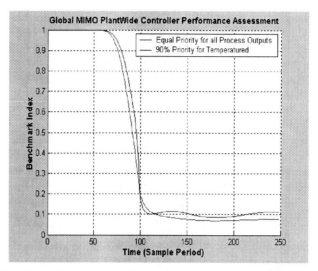

Figure 5.4. The global transient MIMO control performance with 250 samples

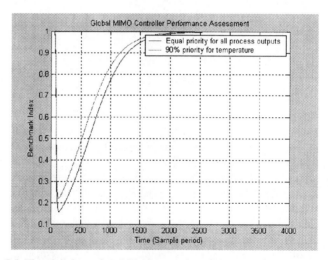

Figure 5.5. The global transient MIMO control performance with 4000 samples

5.6 Summary of Results and Conclusions

From the series of benchmarking tests run on the BASF divided wall simulator, the following conclusions about the current controller performance can be drawn:

- In terms of output regulation performance, both SISO MV and MIMO GPC controller benchmarking indicates that the current controller regulate the process output with minimum variance.

- However, both GMV and MIMO GPC benchmarks indicate that too much control energy may be used to obtain the minimum variance.
- In term of tracking dynamic changes, both RS-LQG and MIMO GPC indicate that the current controller is far from optimal. The performance level is around 10% of the optimal.
- This study clearly shows that different requirements are imposed by tracking goals and regulating goals.
- For the steady state controllers it would be essential to have a good disturbance rejection without upsetting the column through interaction between the loops. The controller setup for the dynamic case removed most of the interactions between the loops, so this may be the best setup as a compromise.
- The BASF divided wall column is mainly operating around a fixed operating condition, and this puts much more emphasis on the regulation performance. If the current control action is judged to be not too aggressive, then there is no need to change the current controller.

The results of the MIMO analysis are in some respects similar to some of the results obtained from the SISO RS-LQG tests; i.e. showed that the loop controllers IV and V were the most sluggish. The SISO analysis however suggested that the controller in Loop III was the best tuned. Clearly the MIMO results suggest, that while this might be the case for the SISO control objective, when the control objective is an integrated process optimization (i.e. measured against a MIMO criterion), the controllers in Loops I and II are the best tuned to meet the overall system objectives.

For the existing controller, the implication of the MIMO results are that all loop controllers in the process need to be tightly tuned and none need to be de-tuned, since none of the individual loop benchmark show the individual PID controllers to be out performing the optimal GPC performance specification for each loop.

Locating the Source of a Disturbance

Nina Thornhill

Professor of Control Systems, Department Electronic and Electrical Engineering, University College London, UK

6.1 Introduction and Motivation

The chapter discusses the state of the art in detection and diagnosis of plant-wide disturbances. Because a disturbance affects the control loops downstream of where it originates, being able to automate the localisation of the root cause is very important for effective performance assessment of controllers. The chapter gives a survey and an industrial case study. The industrial example uses oscillation and spectral methods for finding clusters of disturbed measurements and non-linear time series analysis to find the root cause.

The theme is the detection and diagnosis of distributed disturbances like the one in Figure 6.1. The key features of such an analysis are:

To use only historical process data, without taking the process off-line for special tests;

To detect distributed disturbances that affect many measurements in a unit, plant or site;

To determine which measurements are affected;

To characterize the disturbances, for instance with a spectral fingerprint;

To diagnose the root cause of each disturbance.

6.1.1 Distributed Disturbances

Disturbances to process operation are unwanted deviations in variables such as temperature, pressure and flow rates that move the process away from its optimum settings and thus degrade product quality or reduce output. The issue of the detection and diagnosis of distributed or plant-wide disturbances has been highlighted as a key issue facing the process industries [Qin, 1998; Desborough and Miller, 2002; Paulonis and Cox, 2003]. A disturbed plant generally operates less profitably than one running steadily because production and throughput may have to back away from their maximum settings to accommodate process variability so early detection and automated diagnosis will mean more profitable operation of large-scale chemicals and petroleum products manufacture [Martin *et al.*, 1991; Shunta, 1995].

A distributed disturbance is one which appears in several places in a process plant. It may upset a single unit such as a distillation column, it may be plant-wide if it affects a complete production process or even site-wide if utilities such as the steam system are involved. Automated methods are needed to detect a distributed disturbance. The challenge then is to trace the disturbance to its root cause because maintenance actions are more cost-effective if a root cause has been correctly identified.

The detection of distributed disturbances in process measurements requires their characterization with a signature that is distinctive of the disturbance. Examples of signatures are peaks in the power spectra or the pattern of zero crossings in the autocovariance functions. Detection of a distributed disturbance is accomplished through the recognition of measurements having similar signatures. For the diagnosis of the root cause of the disturbance, a signature that becomes stronger closer to the root cause is required.

6.1.2 Example

The concept is illustrated in Figure 6.1, courtesy of a BP refinery (Kwinana, W. Australia) which shows a unit-wide disturbance in a distillation column. The time trend of the analyzer shows the composition of the product leaving the top of column was varying in a undesirable way. The power spectrum of the measurement has a distinct spectral peak because of the oscillation. Similar oscillations and spectral peaks were present in the steam flow and temperature and so the upset can be classified as a distributed disturbance. The steam flow measurement showed non-linearity, as seen by the non-sinusoidal nature of the oscillation and the presence of a harmonic in the spectrum. As will be explained later, the appearance of non-linearity in a time trend is a diagnostic signature for the root cause of the disturbance; in this case it was a faulty steam flow sensor.

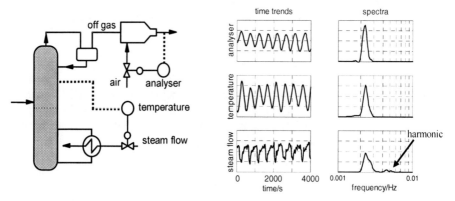

Figure 6.1. Illustration of a distributed disturbance in a distillation column

6.2 Survey of Fault Detection and Diagnosis

The focus in the chapter is on data-driven analysis with an emphasis on methods that incorporate signal processing, for example through the use power spectra, autocovariance functions, signal filtering and non-linear time series analysis. The survey in this section demonstrates that plant-wide disturbances can be detected and characterized by means of these techniques and also places the disturbance detection activity into a broader context.

A widely accepted taxonomy for fault detection and diagnosis is discussed first [Venkatasubramanian, 2001; Venkatasubramanian *et al.*, 2003*a,b,c*]. These papers describe a family tree of methods for analysis of process systems with an emphasis on fault detection and diagnosis. The main categories highlighted are:

Quantitative and qualitative process history based methods;
Quantitative and qualitative model-based methods.

The strengths and weaknesses have been examined systematically and comprehensively and a hierarchical structure has been imposed that has had great benefits, for instance in showing that seemingly unrelated approaches have a common mathematical basis. A recent text book [Chiang *et al.*, 2001] also presents broad coverage of knowledge-based, data-driven and analytical techniques.

The techniques introduced in this chapter can mainly be classified as *quantitative process history based* analysis. To place the chapter in context, all the main classifications within fault detection and diagnosis are outlined this section with a brief description of each and the citation of some key papers. Some additional practical issues are also addressed.

6.2.1 Model-based Methods

Quantitative Model-based Methods
Quantitative model based methods exploit a detailed model of a process for fault detection and isolation (FDI). The model is solved in real time subject to the same inputs as those of the real plant and the values of the outputs are compared to those of the plant and conclusions are drawn when any mismatch is observed.

Quantitative model-based methods have a history of at least 20 years, it is a mature area and there are text books on the subject as well as comprehensive reviews. Authors explaining the methods and setting out the foundations include Chow and Willsky [1984], Isermann [1984], Basseville [1988], Gertler and Singer [1990], and Frank [1990]. Frank *et al.* [2000] reviewed some recent applications. Text books by Patton [2000] and Gertler [1998] also cover model based and residuals methods.

The models may be first principles models based on physics and chemistry, empirically derived models such as time series or a hybrid of the two. FDI seeks for inconsistencies between expected and observed behaviour and in a multivariable system the geometrical properties of the residual can indicate the source of a fault as well as its occurrence. The residuals vector is tested for randomness; any deterministic or non-random behaviour indicates the occurrence of a fault while the pattern of non-randomness among the various elements in the residuals vector helps to isolate and diagnose the fault. The analytical processing carried out with the model and residuals is based upon transformations which project measurements into a *feature space* where any faults would show up more prominently and be decoupled from one another. In general the nature of the expected faults has to be known in advance to do the appropriate transformations.

Model-based FDI based on first principles models is hard to achieve in the process industries because accurate, well calibrated and well validated models are expensive to create and maintain for processes which are often not well understood. The reaction chemistry and the thermodynamic properties of the materials being processed are the main source of uncertainty. Time series and other identified models are more common (e.g. Li *et al.* [2003]).

The main industrial application at present for quantitative model-based methods is in condition monitoring of equipment and rotating machinery such as pumps and gas turbines but quantitative model-based methods are not yet very widely used in process applications.

Qualitative Model-based Methods
Qualitative model based methods include techniques that capture fundamental causal relationships of a process based on physics and chemistry in a non-numerical way using classification and ranking [Venkatasubramanian *et.al.*, 2003*b*]. Signed digraphs (SDG) are one of the main ways of representing causal qualitative knowledge. If a fault is observed then the model is used to find the root cause. Such causal models can be enhanced by the use of quantitative information concerning the magnitudes of influences, for instance that A influences B more than B influences A.

Abstraction hierarchies decompose a system in terms of its subsystems, again for the purposes of diagnostic reasoning. An abstraction hierarchy called Multilevel Flow Modelling [Petersen, 2000] gives a way to describe the flows and storage of mass and energy at various levels and the causal relationships between them. The diagnostic strategies for searching causal representations to find faults and fault propagation paths are reviewed in detail by Venkatasubramanian *et al.*, [2003*b*].

SDGs were extensively reviewed by Maurya *et al.*, [2003 *a,b*] who discussed graph-based approaches for safety analysis and fault diagnosis of chemical process systems. They highlighted the difficulties of accurate capture of the graph representation pointing out that it is a time consuming task and error-prone. They showed how to develop graph models systematically and how to eliminate spurious and redundant branches and showed how to construct signed digraph (SDG) models for the difficult causalities associated with feedback control loops. Fault detection and diagnosis using a SDG representation in the presence of feedback control has also been considered [Chen and Howell, 2001]. Early and influential papers which laid down the foundations were due to Kramer and Palowitch [1987], Petti, *et al.* [1990], Årzén [1994], Årzén *et al.* [1995], and Vedam and Venkatasubramanian [1999].

6.2.2 Process-history-based Methods

Process history based methods for fault diagnosis use historical process knowledge, both historical operating data and also qualitative historical knowledge such as operators' expertise [Venkatasubramanian *et al.*, 2003c]. Process history based methods aid rapid detection and diagnosis, have the ability to isolate as well as detect faults and are able to identify new and unseen faults and explain the reasons for decisions.

Qualitative Process-history-based Methods
Qualitative methods use process insights from operators or features described in natural language derived from past data. Examples include expert systems and qualitative process trend analysis.

The purpose of an expert system is to capture the knowledge of human experts and to perform the same sequence of actions as the human when the same conditions arise. Qualitative process trend analysis describes the shapes and patterns of the time trends of process and forms conclusions about underlying faults from the shapes. An example is the characteristic triangular shape sometimes seen in the controller output of a flow loop with a deadband [Rengaswamy and Venkatasubramanian, 1995]. Cheung and Stephanopoulos [1990] used a template-based representation to extract features from data sets while Bakshi and Stephanopoulos [1994] used wavelet packets and other templates for multiscale representation of process data. The work provides a string of symbols that describe the process trend. Learned and Willsky [1995] have also used wavelets for feature extraction and were able to classify transient signals.

A natural application domain for qualitative trend analysis is in batch processes to give a time domain description of a batch profile. For instance in a fermentation it might be very significant if the lag phase at the start of the fermentation lasts longer than usual. An application for batch processes called DB-miner Stephanopoulos *et al.* [1997] used multi-scale extraction of patterns, short term trends and local features in fermentation processes.

Quantitative Process-history-based Methods
Process history based methods use information captured from previous running of the process. The quantitative methods are numerical and data-driven. The main distinction drawn in Venkatasubramanian *et al.* [2003c] is between principal component analysis (PCA) as a statistical classification method and neural networks as non-statistical classifiers.

The methods and literature of PCA and related multivariate statistical approaches are discussed in detail shortly. The main outcome of a PCA analysis is a subspace with orthogonal axes which are the basis functions from which the original measurements may be reconstructed. As with the model-based residual methods, fault states are more distinct in the reduced space than in the original measurements [Gertler *et al.*, 1999].

Neural network classifiers are reviewed in depth in Venkatasubramanian *et al.* [2003c] which explains the methods and numerous applications. The attraction of neural networks is their flexibility in describing non-linear relationships, because neural network models are created from a non-linear combination of basis functions [Poggio and Girosi, 1990]. Bakshi and Utojo [1999] have placed all the quantitative process history methods under a common framework of projection onto basis functions, where the basis functions and type of projection depend on the type of modelling method selected.

An issue in the analysis of process data using neural networks has been to include time histories, as discussed in Zhang [2001]. Hybrid technologies have also been successful in which a multivariate statistical modelling method called PLS (projection onto latent structures) is combined with neural networks to accommodate non-linear relationships in input-output data [Qin and McAvoy, 1992]. Other successful hybrid approaches include a system in which a hybrid system using the process knowledge of the operators and a neural network improved the performance in making predictions and in parameter estimation [Thompson and Kramer, 1994]. Process knowledge can also be derived from a first principles process model rather than from operator experiences [Ignova et al., 1996] while Glassey et al. [1997] discussed combining neural network modelling with an expert system for bioprocess supervision.

Neural networks provide powerful input-output models for the purpose of real-time predictions during a process operation but they are non-linear black-box models and it is not normally easy to drill down into their structure to uncover insights into the behaviour of the plant. They can more easily be used to classify faults in a process if that fault has been encountered and analyzed before than to reveal the locations and nature of new and unseen process faults.

6.2.3 Supporting Methods for Fault Detection and Diagnosis

Gross Error Detection
Pre-processing of data can enhance fault detection, and anomalies detected in pre-processing may also give valuable insights for diagnosis. Elimination of gross errors and some filtering are usually to be recommended [Pearson, 2001]. Famili et al. [1997] discussed the topics of data visualization and multivariate methods as well as filtering and gross error (out-of-range data) removal. Their case studies showed the extent to which data pre-processing needs to be done manually. Practical guidelines about pre-processing to eliminate outliers can be found in Turner et al. [1996].

Gross error removal may be based just on the properties of the data set or it may use some process knowledge, for example that an apparent change between one value and the next is not physically feasible. When statistical tests are applied it is usually to test whether an individual measurement is further than, say, three standard deviations from the mean. Papers in the academic literature on this topic include those due to: Iordache et al. [1985], Jongenelen et al. [1988], Narashiman and Mah [1989], Rollins and Davis [1993], Narashiman and Mah [1989], and Pearson [2001]. Verneuil and Madron [1992] used process insights such as mass balances as well as statistical testing while Tham and Parr [1994] suggested ways in which bad data values can be replaced with estimated values. Tong and Crowe [1996] described the use of multivariate statistics in gross error detection.

The data can often be pre-processed by filtering which is beneficial for the removal of non-stationary trends such as diurnal variations. Filtering to remove non-stationary behaviour was used in the worked example at the end of this chapter.

Data Reconciliation
Gross error removal is often combined with data reconciliation in which the gross error is replaced by a more realistic value that meets process constraints. In some cases, the whole data set may be adjusted to better fit a mass or energy balance. The data reconciliation literature is huge and has been reviewed by Crowe [1996]. Albers [1994] gave a practical example of an application to a heat exchanger problem. A selection of academic authors who combine gross error detection with reconciliation is Crowe [1988], Holly *et al.* [1989], and Tjoa and Biegler [1991].

Gross error detection and data reconciliation tools are commercially available. They are used in on-line optimization where plant data are aligned with a process model in order to estimate unknown parameters before running the optimizer (e.g. in AspenPlus On-Line from AspenTech). They also have uses in fault detection, for instance the detection of leaks that prevent a mass-balance closure around a plant or unit. AspenAdvisor (AspenTech) is an example of a commercial tool for yield accounting and fiscal monitoring for accurate accounting of material entering and leaving a plant.

Data reconciliation is not as essential when dynamic features of the data are the focus of the work rather than the steady state alignment of data with a process model (e.g. for closure of a mass balance). Indeed, procedures for detection and diagnosis of process disturbances generally use mean-centred data where the steady-state value has been removed.

Data Compression
On-line data compression methods are implemented in the data historians of some commercial distributed control systems. The aim of data compression is to capture and store only those data points that represent the start and end of a trend [Hale and Sellars, 1981; Bristol, 1990; Mah *et al.*, 1995].

Data are reconstructed after compression by interpolation between stored data points. The reconstructed data trends therefore consist of piecewise linear segments and are not the same as the originals. The impact of compressed data in data-driven analysis of dynamic process data for control loop performance assessment and other applications has been investigated [Watson *et al.*, 1998; Thornhill *et al.*, 2004] where it was found that compressed data gave misleading results. With the cheaper costs of storage media, vendors are now placing less emphasis upon data compression. Honeywell's Loop Scout controller performance tool calls for uncompressed data, as does AspenTech's Performance Watch product [Aspentech, 2001].

Change Detection
Methods for detection of changes have been surveyed by Basseville [1998]. These include signal processing methods whose purpose is to rapidly identify jumps in

parameters in a dynamic time series model (e.g. Tugnait [1982], Basseville and Benveniste [1983]).

The topic of steady-state checking is also of interest for change detection, because a changing state implies the lack of a steady state. One approach converts the measurements into a qualitative description (i.e. "steady" or "not-steady") using statistical hypothesis testing. A null hypothesis is posed, for example that the measurements represent a steady state, and the measurements are tested to see if they have a distribution that can be accounted for by the null hypothesis [Cao and Rhinehart, 1995], or to determine if a change of steady state has occurred [Narashiman et al., 1988].

Non-parametric statistical tests have advantage that they do not depend on an assumption that the statistical distribution of the measurements is Gaussian. In a non-parametric test the null hypothesis concerns, for instance, the number of changes of direction of data points in a trend. The text book of Kendall and Ord [1990] describes a number of such statistical tests. A process application using the Wilcoxson test to detect the steady state of a single variable is reported in Koninckx [1988].

6.3 Disturbance Detection

6.3.1 Introduction

This section reviews specialized quantitative process history based methods for detection and diagnosis of distributed disturbances. This first sub-section places the methods in a broader context of related work in the area of chemical process control and gives some background.

The section then sets out the formulation of principal component analysis (PCA) as it is used for real-time multivariate statistical process control and then lays out an alternative formulation using the power spectra of process measurements that is appropriate for process auditing. Then, techniques of oscillation detection in chemical processes are examined and a recent method based on the concept of zero crossings is presented.

Spectral Analysis of Processes
It is necessary to find all the measurements or control loops in a plant having the same disturbance because the root cause will be among that group. Many plant-wide disturbances, especially those of an oscillatory nature, are not localized in time and thus the Fourier transform and power spectrum provide a natural means for their analysis. Pryor [1982] highlighted the usefulness of autocovariance functions and spectra for such a purpose.

Desborough and Harris [1992] used the power spectrum to determine whether poorly performing control loops had a long-term deviation from set point or oscillatory behaviour, and a spectral signature arising from a disturbance in data from a Shell refinery has also been detected using spectral methods [Tyler and Morari, 1996]. The presence of harmonics in a spectrum enables loop tuning faults to be distinguished from faults due to non-linearity such as valve friction leading to a limit cycle [Thornhill and Hägglund, 1997].

Early plant-wide assessments, done manually, were used by Harris *et al.* [1996] who reported plant-wide control loop assessment in a paper mill where they found that the spectral analysis of the univariate trends was useful. Ruel and Gerry [1998] found the non-linear root cause of an oscillatory plant-wide disturbance in a paper mill through an analysis of the harmonics of the power spectra of the measurements. The root cause was a circulating pump in a tank which periodically failed to pump as pulp consistency varied.

6.3.2 Principal Component Analysis

The methods of Principal Component Analysis (PCA) have been widely applied for the detection of correlated measurements such as would be expected in the presence of a plant-wide disturbance [Kresta *et al.*, 1991; Wise and Gallagher, 1996; Goulding *et al.*, 2000]). Descriptions of the methods of PCA may be found from many sources, for example in Chatfield and Collins [1980] and Wold *et al.*, [1987]. Kourti and MacGregor [1996] wrote an overview and commented on practical issues for industrial implementation.

Multivariate SPC
The idea in multivariate statistical process control (MSPC) is that a change in the correlations between variables can indicate a fault even if the variables themselves are all still within the limits of normal operation. The quantity plotted on a MSPC chart is a linear combination of several measured variables derived from projection of the current state of the plant onto a PCA model of normal in-control operation.

MSPC was placed in a broad context of quality and reliability engineering by Elsayed [2000]. Key issues addressed were change point detection, correlated measurements, multivariate systems and the integration of statistical control and process control. Goulding *et al.* [2000] reviewed multivariate statistical process control methods in the process industries while Yoon and MacGregor [2000] gave an in-depth comparison between MSPC methods and the model-based fault detection methods mentioned in section 6.2.1.

The concept of warning and alarm limits is the same in multivariate SPC charts as in univariate SPC charts. It involves finding the 95% and 99.9% contours of the multivariate distribution by estimating the statistical distribution of the quantity being plotted in the charts [Jackson and Mudholkar, 1979; Martin and Morris, 1996].

PCA Formulation for MSPC
PCA is derived from the singular value decomposition of the data matrix $\mathbf{X}$. In a MSPC application the columns of $\mathbf{X}$ are the time trends of m measurements each sampled at N time instants where in general $N > m$. The columns are usually mean centred and scaled to unit standard deviation. Each row of $\mathbf{X}$ is a *plant profile,* a snapshot at a particular time instant of the pattern of measurements across the plant. Of the numerical values in each row of $\mathbf{X}$, any positive values indicate

sensors reading above average at that time and negative values are sensors reading below average.

$$m\ measurements\ \rightarrow$$

$$\mathbf{X} = \begin{pmatrix} x_1(t_1) & \cdots & x_m(t_1) \\ .. & .. & .. \\ x_1(t_N) & \cdots & x_m(t_N) \end{pmatrix} \begin{matrix} N\ time \\ samples \\ \downarrow \end{matrix} \tag{6.1}$$

The singular value decomposition is $\mathbf{X} = \mathbf{UDV}^T$ where matrix $\mathbf{U}$ is N-by-m, $\mathbf{V}^T$ is m-by-m, and $\mathbf{D}$ is diagonal and its elements are the positive square roots of eigenvalues of the m-by-m matrix $\mathbf{X}^T\mathbf{X}$. The principal component decomposition is expressed as $\mathbf{X} = \mathbf{TW}^T$, where $\mathbf{T} = \mathbf{UD}$ and $\mathbf{W}^T = \mathbf{V}^T$.

A description of the majority of the variation in $\mathbf{X}$ can often be achieved by truncating the PCA description. The following is a three-PC model in which the variation of $\mathbf{X}$ that is not captured by the first three principal components appears in an error matrix $\mathbf{E}$:

$$\mathbf{X} = \begin{pmatrix} t_{1,1} \\ \cdots \\ t_{N,1} \end{pmatrix} \mathbf{w}^T_1 + \begin{pmatrix} t_{1,2} \\ \cdots \\ t_{N,2} \end{pmatrix} \mathbf{w}^T_2 + \begin{pmatrix} t_{1,3} \\ \cdots \\ t_{N,3} \end{pmatrix} \mathbf{w}^T_3 + \mathbf{E} \tag{6.2}$$

The $\mathbf{w}^T$ – vectors are orthonormal row vectors with m elements and are called *loadings*. They act as a set of orthogonal basis functions from which all the plant profiles (the rows of the $\mathbf{X}$ matrix) can be approximately reconstructed. The $\mathbf{t}$ – vectors are column vectors with N elements and are called *scores*. The scores indicate the amplitude of each normalized basis function in the reconstruction. For instance, with a $3-$PC model the approximate reconstruction of the plant profile in row 10 (denoted as $\mathbf{x}^T_{10}$) of the $\mathbf{X}$ matrix would be:

$$\mathbf{x}^T_{10} \equiv \begin{pmatrix} x_{10,1} & x_{10,2} & \cdots & x_{10,m} \end{pmatrix} = t_{10,1} \times \mathbf{w}^T_1 + t_{10,2} \times \mathbf{w}^T_2 + t_{10,3} \times \mathbf{w}^T_3 + \mathbf{e}^T_{10} \tag{6.3}$$

where $\mathbf{e}^T_{10}$ is the 10th row of matrix $\mathbf{E}$. The above formulation is the foundation for multivariate statistical process monitoring because the $\mathbf{w}^T$ – vectors capture typical plant profiles while the t – values indicate which plant profile is dominant at a given time. An MSPC application inspects the t – values and the magnitude of the corresponding row of the error matrix for anomalies.

PCA Formulation for Plant Auditing
In an alternative formulation the data matrix is transposed so that the rows are the time trends of the measurements:

$$N \text{ time samples} \quad \rightarrow$$

$$\mathbf{X} = \begin{pmatrix} x_1(t_1) & \cdots & x_1(t_N) \\ .. & .. & .. \\ x_m(t_1) & \cdots & x_m(t_N) \end{pmatrix} \begin{matrix} m \\ \text{measurements} \\ \downarrow \end{matrix} \qquad (6.4)$$

The $\mathbf{w}^T -$ vectors now have N elements and resemble time trends. In this formulation PCA is a function approximation method in which each time trend in a row of $\mathbf{X}$ is built up from a linear combination of the $\mathbf{w}^T -$ vectors which are a set of orthogonal basis functions for the observed time trends. Similarities in the scores may then be used to detect clusters of similar time trends.

The auditing approach has potential for plant-wide disturbance detection because all the measurements affected by a disturbance would have similar time trends. It is known, however, that time shifts in plant data caused by delays and lags in the process dynamics need special attention. The method of time shifted PCA can be used in which the rows of the data matrix representing measurements with time delays are either time shifted, or the matrix is augmented with one or more time shifted replicates of the vector [Ku *et al.*, 1995; Lakshminarayanan *et al.*, 1997]. Wise and Gallagher [1996] give a summary. Another approach to alignment of data trends used dynamic time warping [Gollmer and Posten, 1996; Kassidas *et al.*, 1998].

Time shifted PCA has been extended beyond the requirements of fault recognition. Time shifting can identify dynamic relationships in the form of a time series model or a dynamic model using multivariate modelling of input-output data [Wise and Ricker, 1992; Ku *et al.*, 1995; Lakshminarayanan *et al.*, 1997; Chen *et al.*, 1998]. Multiple time-shifted copies of data trends are included in the data matrix and a PCA analysis determines which of the time trends are linearly related. In this manner relationships in the form of a time series model may be established, for instance:

$$y_1[i] = a y_1[i-3] + b x_1[i-1] \qquad (6.5)$$

where y_1 is a measurement variable in the plant and x_1 is an input variable or some other plant measurement that is causally related to y_1. The data matrix becomes very large when an arbitrary number of lagged time trends augment the data matrix. Shi and MacGregor [2000] reviewed techniques for multivariate dynamic process modelling and gave a comparison of numerical methods for reduction of the large lagged variables matrix.

Disadvantages of Time Shifting
Time shifting may be problematical if the true delay is not a whole number of sampling intervals. For instance, if the delay were 3.2 minutes and the sampling interval is 1 minute then the closest alignment achievable would be a shift of 3.0 minutes and the time trends would remain out of alignment by 0.2 min. Secondly, there may be no unique time shift that achieves alignment if oscillations at two unrelated frequencies originate in control loops A and B and propagate plant-wide. Propagation delays from A to a given measurement point are not the same as from B because the routes through the plant are different and therefore it is not normally possible to find a unique time shift that aligns both oscillations simultaneously.

In the latter case, and also if time delays are unknown, it is necessary to include multiple time-shifted time trends in the data matrix. The data matrix then becomes very large and will generate more principal components than necessary. That makes it hard to reliably detect clusters of related measurements. These examples show that time delays pose significant problems in the detection of plant-wide oscillations and other types of time-varying disturbances.

Spectral PCA of Dynamic Data
In spectral PCA, the rows of the data matrix, X, are the single-sided power spectra $P(f)$ of the time trends of the plant measurements. The $\mathbf{X}$ matrix shown below uses N frequency channels, although only $N/2+1$ (if N is even) are distinct since the upper half of the power spectrum above the Nyquist sampling frequency is a mirror image of the lower half.

$$N \; frequency\, channels \quad \rightarrow$$

$$\mathbf{X} = \begin{pmatrix} P_1(f_0) & \cdots & P_1(f_N) \\ .. & .. & .. \\ P_m(f_0) & \cdots & P_m(f_N) \end{pmatrix} \begin{array}{l} spectra\ of\ m \\ \\ measurements \\ \downarrow \end{array} \tag{6.6}$$

The arrangement of the $\mathbf{X}$ matrix with the power spectra in the rows means that the $\mathbf{w}^T$ – basis functions are spectrum-like vectors (i.e. they have a frequency axis). The formulation given is therefore a functional approximation in which the spectra in the rows of $\mathbf{X}$ are reconstructed from the basis functions. Again, this interpretation of PCA lends itself to plant auditing.

Multivariate spectral analysis has several advantages over analysis in the time domain. It gives an improved signal to noise ratio if the spectral content of the wanted signal is narrow-band compared to the noise. The power spectrum is invariant to time delays or phase shifts caused by process dynamics. It is also insensitive to missing values and outliers because the transforms of such effects are spread thinly across all frequencies in the spectrum.

Detection of Spectral Clusters
The outcome of spectral PCA is the detection of clusters of plant measurements having similar spectra, which are interpreted to mean they are influenced by a common disturbance. The spectral cluster is detected by an examination of the scores. For instance, a three-PC model for the spectra in $\mathbf{X}$ is:

$$
\mathbf{X} \approx \begin{pmatrix} t_{1,1} \\ \ldots \\ t_{N,1} \end{pmatrix} \mathbf{w}^T_1 + \begin{pmatrix} t_{1,2} \\ \ldots \\ t_{N,2} \end{pmatrix} \mathbf{w}^T_2 + \begin{pmatrix} t_{1,3} \\ \ldots \\ t_{N,3} \end{pmatrix} \mathbf{w}^T_3 \tag{6.7}
$$

The above expression shows that the spectrum in the each row of the data matrix $\mathbf{X}$ can be approximately reconstructed from a weighted linear combination of the spectrum-like basis functions $\mathbf{w}^T_1$, $\mathbf{w}^T_2$ and $\mathbf{w}^T_3$, and that the weightings for the ith row spectrum are the scores $t_{i,1}$, $t_{i,2}$ and $t_{i,3}$.

If two spectra are similar then their scores are similar. Clusters are detected when a suitable distance measure such as the Euclidean distance or the angle between the score vectors is small.

PCA Using Other Integral Transforms
The use of autocovariance functions in the data matrix has been shown to give results similar to those from spectral PCA [Tan *et al.*, 2002; Thornhill *et al.*, 2002]. Other reported uses of PCA with transforms of the data matrix have focused on multiscale wavelet transforms for on-line multivariate statistical process monitoring. Bakshi and Stephanopoulos [1994] and Bakshi [1998] used wavelets and other templates for multiscale representation of process data in combination with PCA for the modelling of data having features that changed over time and frequency. Other researchers have also had success with multiscale PCA using wavelets [Kosanovich and Piovoso, 1997; Luo *et al.*, 1999; Shao *et al.*, 1999]. Wavelet functions are localized, however, so the wavelet transform is not invariant to time delays.

6.3.3 Other Factorizations of the Data Matrix

Independent Component Analysis
Independent Component Analysis (ICA) is a linear transform of a data matrix that minimises statistical dependence between the basis vectors. Its benefit is that it provides basis functions with a better one-to-one relationship with the physical sources of signals, for instance application of ICA to the sounds of a cocktail party is supposed to identify the individual speakers in the room. Its foundation is explained in Comon [1994] and a recent special issue on blind source separation in the *International Journal of Adaptive Control and Signal Processing* [Davies, 2003] shows that ICA is starting to attract attention in control applications.

PCA ensures that the $\mathbf{w}^T -$ vectors are orthogonal whereas ICA makes the vectors independent. The data matrix is represented as:

$$N \text{ time samples} \quad \rightarrow$$

$$\mathbf{X} = \begin{pmatrix} x_1(t_1) & \cdots & x_1(t_N) \\ .. & .. & .. \\ x_m(t_1) & \cdots & x_m(t_N) \end{pmatrix} \begin{array}{l} m \\ measurements \\ \downarrow \end{array} \qquad (6.8)$$

The PCA decomposition is $\mathbf{X} = \mathbf{TW}^T + \mathbf{E}$ and the rows of $\mathbf{W}^T$ are orthogonal:

$$\sum_{\ell=1}^{N} w^T{}_i(\ell) w^T{}_j(\ell) = 0, \text{ for } i \neq j \qquad (6.9)$$

where $w^T{}_i(\ell)$ and $w^T{}_j(\ell)$ are the ℓth elements in row i and j respectively of $\mathbf{W}^T$.

The ICA decomposition, by contrast, is $\mathbf{X} = \mathbf{AS}^T + \mathbf{F}$ where the rows of $\mathbf{S}^T$ are independent and $\mathbf{F}$ is an error matrix representing the difference between the ICA reconstruction and the data matrix $\mathbf{X}$. Independence means that, if s_i is an element of row vector $\mathbf{s}^T{}_i$ and s_j is an element of row vector $\mathbf{s}^T{}_j$, then

$$\Pr(s_i, s_j) = \Pr(s_i)\Pr(s_j) \qquad (6.10)$$

where $\Pr(s_i)$ and $\Pr(s_j)$ are the probability density functions of s_i and s_j and $\Pr(s_i, s_j)$ is the joint probability density function. Hyvärinen and Oja [2000] showed that statistical independence can be approximated when the kurtoses (the fourth moments) of the distributions $\Pr(s_i)$ of the row vectors are maximised and gave an algorithm called FastICA for combining the elements in the $\mathbf{X}$ matrix in a way that generates an $\mathbf{S}^T$ matrix whose rows have probability density functions with maximised kurtosis. The definition of independence treats $\mathbf{s}^T{}_i$ and $\mathbf{s}^T{}_j$ as random vectors with values s_i and s_j. In reality there is just one data set (one realization of the random processes) and the probability density functions are estimated from the histograms of the values in each row while statistical moments such as kurtosis are calculated from the values as:

$$kurt(y) = \frac{\dfrac{1}{N}\sum_{k=1}^{N} y_k{}^4 - 3\left(\dfrac{1}{N}\sum_{k=1}^{N} y_k{}^2\right)^2}{\left(\dfrac{1}{N}\sum_{k=1}^{N} y_k{}^2\right)^2} \qquad (6.11)$$

where y_k are the elements in a vector such as a row of $\mathbf{S}^T$.

Li and Wang [2002] gave the first example of an application of ICA to the analysis of dynamical chemical process trends and dimension reduction. ICA has also been used instead of PCA in on-line multivariate statistical process monitoring applications [Kano *et al.*, 2003; Lee *et al.*, 2004*a*] and explored for dynamic data using the time lagging method [Lee *et al.*, 200*b*].

ICA can also be applied to the power spectra of process data [Xia and Howell, 2005; Xia *et al.*, 2005]. A demonstration of what can be achieved with ICA as a post-processing step following a spectral PCA analysis was presented, where ICA produced a very useful set of spectral-like basis functions with just one spectral peak in each independent component. The task of diagnosis of the sources of oscillations in a chemical plant was greatly aided by the enhanced correspondence between independent components and plant oscillations.

Non-negative Matrix Factorization (NMF)
NMF is a method related to PCA and ICA. It seeks for an alternative set of basis functions that enable a better one-to-one correspondence between physical sources and basis functions. A key paper in *Nature* by Lee and Seung [1999] addressed the issue of face recognition from two dimensional images. They said in their paper: "The basis images for PCA are *eigenfaces* some of which resemble distorted versions of whole faces. The NMF basis is radically different: its images are localized features that correspond better with intuitive notions of the parts of faces." The matrix decompositions are:

$$\mathbf{X} = \mathbf{TW}^T + \mathbf{E} \qquad \mathbf{X} = \mathbf{AS}^T + \mathbf{F} \qquad \mathbf{X} = \mathbf{UH}^T + \mathbf{G}$$
$$PCA \qquad\qquad ICA \qquad\qquad NMF \qquad (6.12)$$

where in PCA the rows of $\mathbf{W}^T$ are orthonormal. In ICA the independence property discussed earlier $\Pr(s_i, s_j) = \Pr(s_i)\Pr(s_j)$ holds, while in NMF the elements of both $\mathbf{U}$ and $\mathbf{H}^T$ are non-negative. Matrices $\mathbf{E}$, $\mathbf{F}$ and $\mathbf{G}$ are the errors in reconstruction when a reduced number of basis functions are used.

NMF is applicable only when $\mathbf{X}$ is non-negative, as would be true if the rows of $\mathbf{X}$ are power spectra. The relationship between ICA and NMF for non-negative $\mathbf{X}$ has been established by Plumbley [2002] who showed the ICA decomposition can be transformed to an NMF decomposition by a matrix rotation. No applications of NMF have yet been reported in the area of process fault detection and diagnosis, however work is under way at the University of Alberta [Tangirala and Shah, 2005].

6.3.4 Oscillation Detection

Oscillations are a common type of plant-wide disturbance and their regularity permits some specialized techniques for their detection. Several authors have addressed the detection of oscillatory measurements in process data. Hägglund [1995] detected zero crossings of the error signal in a control loop and calculated the integrated absolute error (*IAE*) between successive zero crossings. An oscillatory time trend has larger *IAE* values than a random one. Hägglund's paper described a real-time method and its incorporation into the ECA400 autotuning PID control module from Alfa Laval Automation. The control module generated an alarm when oscillations were detected.

The presence of an oscillation may also be detected from the regularity of the zero crossings of a mean-centred time trend [Thornhill and Hägglund, 1997; Forsman and Stattin, 1999]. Use of zero crossing detection has to address the problem of noise in the time trend, however, and it requires a method for elimination of spurious zero crossings in the vicinity of the true zero crossing point caused by fluctuating noise.

Use of Autocovariance Functions

The horizontal axis in an autocovariance function (*ACF*) plot is time-like, representing the lag ℓ between the time trend and a delayed version of the time trend. The vertical axis is the correlation between the time trend and the delayed trend. An estimator for calculation of the *ACF* from a data set is:

$$ACF\left(\ell\right) = \frac{1}{N-\ell-1} \sum_{i=\ell+1}^{N} \tilde{x}(i)\tilde{x}\left(i-\ell\right) \tag{6.13}$$

where $\tilde{x}$ is the mean centred data scaled to unit standard deviation.

The *ACF* of an oscillating signal is itself oscillatory with the same period as the oscillation in the time trend. For example, if the absolute value of the *ACF* at the first minimum exceeds a threshold then the possibility of an oscillation is inferred, and additional cycles of the oscillatory autocovariance function can also be utilized to distinguish a decaying oscillation from a sustained oscillation [Miao and Seborg, 1999].

The method that will be demonstrated in the industrial case study at the end of the chapter was originally reported by Thornhill *et al.* [2003] and is briefly presented here. It has the following features:

Detection of the zero crossings of the *ACF*;
Filtering to remove unwanted interferences so that multiple oscillation may be detected;
Determination of the period of oscillation and the groups of measurements participating in each oscillation.

Oscillations can be detected if the pattern of zero crossings of the autocovariance function is regular. The benefit of the use of autocovariances is that they are much less noisy than the time trend data as can be seen in the example in Figure 6.2. For instance, in the case of white noise all the noise appears in the zero lag channel of the *ACF*. Therefore noise does not disrupt the zero crossing of the autocovariance function in the same way as it disrupts zero crossings in the time domain, and the pattern of zero crossings reveals the presence of an oscillation more clearly than the zero crossings of the time trend.

Each cycle of oscillation has two zero crossings and hence the intervals between zero crossings are:

$$interval = \frac{1}{2}\left(\overline{T_p} \pm \Delta T_p\right) \qquad (6.14)$$

where $\overline{T_p}$ is the mean period and ΔT_p a random variation in the period. Thus $\overline{T_p}$ is twice the mean value of the intervals and the standard deviation of the period is $\sigma_{T_p} = 2 \times \sigma_{intervals}$. An oscillation is considered to be regular if the standard deviation of the period is less than one third of the mean value. A three-sigma statistic is used and values of $r > 1$ are taken to indicate a regular oscillation with a well defined period:

$$r = \frac{1}{3} \times \frac{\overline{T_p}}{\sigma_{T_p}} \qquad (6.15)$$

Measurements with More Than One Oscillation
The zero crossings of the *ACF* may not be regular if more than one oscillation is present. The time trend and *ACF* in the uppermost two panels of Figure 6.2 have two superimposed oscillations of different periods. The third panel in the figure marks the positions of the zero crossings but the intervals between them reflect neither oscillation accurately because the zero crossings of the fast and slow oscillations each destroy the regularity of the other's pattern.

The problem is solved by frequency domain filtering. A filtered *ACF* is calculated from the inverse Fourier transform of a two-sided power spectrum in which the power in the unwanted frequency channels is set to zero. The lower three panels in Figure 6.2 show the example data after filtering to remove the lower frequency oscillation at about 330 samples per cycle. By removal of the interfering oscillation, the zero crossings of the faster oscillation become regular and therefore suitable for use in oscillation detection.

6.4 Disturbance Diagnosis

The focus in the previous section was the detection of plant-wide disturbances. The aim was to find all the measurements or control loops in a plant having the same behaviour because the root cause will be among that group. This section now examines methods available for diagnosis of the root cause.

The root causes of plant-wide disturbances include:

Non-linearities such as saturation, dead band, or hysteresis in control valves, sensors or in the process itself that cause limit cycle oscillations;
Control loop interactions;
Structural disturbances;
External disturbances;

Poorly tuned controllers.

This chapter addresses the diagnosis of non-linear root causes, the first item in the list. The diagnosis of the other root causes is an open and active research area at present. The motivation for making non-linear root causes a priority is that non-linearities such as valve friction constitute an important root cause of plant-wide disturbances [Desborough and Miller, 2002; Bialkowski, 1992; Ender, 1993; Shinsky, 2000]. Many vendors also emphasize these root causes, for instance Ruel [2001] placed deadband, backlash and stiction at the top of the list of checks to be made prior to tuning a process control loop.

6.4.1 Diagnosis of Distributed Disturbances

It is a challenging task to determine which single loop among many is the root cause of a distributed disturbance. The influences of a faulty control loop may propagate both downstream and upstream from the root cause because of recycles or the effects of control actions. For instance, the level in an upstream tank may be disturbed if the control valve on the (downstream) outflow causes a limit cycle oscillation. This section reviews and discusses the problems and presents some promising ways forward, though there remain many unsolved problems.

Normal Operations Versus Special Tests
The strong preference in the process industries is to conduct the first phase of detection and diagnosis using normal operating data without the need for any special tests. The concept was emphasized by Harris [1989] in the context of control loop performance assessment.

Special on-line tests may, however, be used for confirmation once a suspected root cause has been identified. Such tests initiate actuator movements (e.g. changes in valve position) by means of manual changes to the controller output [Åström, 1991; Gerry and Ruel, 2001]. The control loop is disabled during the tests which means the flow through the valve is not maintained at an optimum value and therefore such special tests have the potential to disrupt production. Many control loops may be affected by a plant-wide disturbance and it is especially undesirable to apply the special tests one by one to them all in an attempt to find the one faulty valve. An aim of the initial diagnosis step is to use normal operating data and thus to minimize the need for special tests by directing attention onto the control loop most likely to be the root cause.

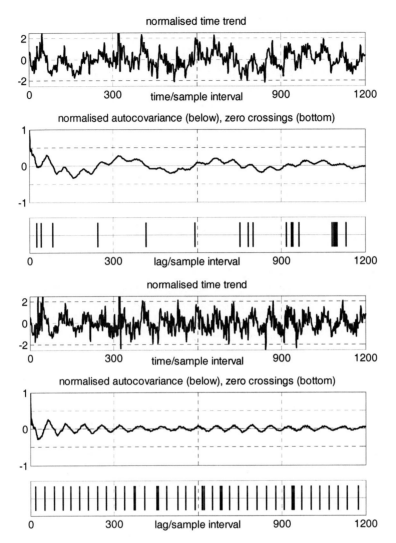

Figure 6.2. Upper three panels: Time trend, ACF and zero crossings with two simultaneous oscillations present. Lower three panels: after filtering to remove the low frequency oscillation.

Diagnosis of Individual Loops

Once a control loop has been identified as a likely root cause of a plant-wide disturbance a number of tests exist to test the hypothesis. Many diagnostic methods exist for individual control loops including:

Op–mv maps:
Even and odd covariance
Analysis of probability density functions
Higher order spectral methods

Oscillation shape analysis

Maps of process variable versus manipulated variable (i.e. flow through the valve versus valve position) can diagnose the nature of a valve fault and are used in several commercial packages. Other signatures such as odd and even cross correlations and an assessment of the probability density function of the measurement values have also been used [Horch, 1999; Horch, 2002]. Chapter 7 of this book written by Horch discusses several methods in depth.

If a candidate for the root cause has been identified and inspected then more invasive tests might be considered. For instance, the test of putting a control loop into manual mode is able to distinguish between an oscillation originating within the loop and one entering the loop as an external disturbance. Oscillations originating within the loop such as those due to too-tight tuning or a sticking valve will disappear if the loop is in manual mode. It is a rather invasive test because the loop may need to be in manual for several hours since plant-wide oscillations are often slow. Those due to sticking valves, for instance, often have periods of one to two hours.

Valve travel tests which put a control valve through several consecutive opening and closing strokes may be done while the loop is in manual. The results are definitive, nevertheless the tests have the potential to upset the process considerably. One of the benefits of data driven methods is that they can indicate which valve is the best candidate for testing and thus minimize disruption.

Control Loop Limit Cycles

A common source of oscillation is a faulty control valve. Control valves with non-linear behaviour such as a dead-band or excessive static friction cause hunting or limit cycling in the control loop.

Oscillating disturbances caused by non-linearity in the process itself can also propagate and upset downstream units. A well cited example relates to the stop-start nature of flow from a funnel feeding molten steel into a rolling mill [Graebe *et al.*, 1995], while a periodic hydrodynamic instability caused by foaming in a distillation column was described in Thornhill [2005]. Foam in the trays was periodically bubbling up and blocking the column causing a rise in differential pressure which eventually disrupted the foam and caused it to fall to the bottom of the column. The column then returned to normal for a spell until the foam built up again.

An automated means is needed to determine which among all the oscillating control loops is the root cause and which are secondary oscillations. Generally, non-linearity reduces as the disturbance propagates away from the source and the time trend with the highest non-linearity is thus the best candidate for the root cause. The reason why secondary oscillations have lower non-linearity is that as the signal propagates away from its source it passes through physical processes which give linear filtering and which generally add noise. Such a filter destroys the phase coherence that is characteristic of a time trend produced by a non-linear source and often reduces the magnitudes of the harmonics.

Finding Non-linear Root Causes
Diagnosis of a non-linear root cause can be accomplished using an unambiguous test for non-linearity. Non-linearity tests determine whether a time series could plausibly be the output of a linear system driven by Gaussian white noise, or whether its properties can only be explained as the output of a non-linearity. Early studies used the presence of prominent harmonics as an indicator of non-linearity, more recent advances make use of higher order statistics and surrogate data methods [Choudhury, 2004; Choudhury *et al.*, 2004; Thornhill, 2005].

The source of the limit cycle oscillation in Figure 6.1 was initially distinguished from the secondary oscillations in the distillation column because of the presence of the harmonic in the FC1 time trend [Thornhill *et al.*, 1998]. It is not always true, however, that the time trend with the largest harmonic content is the root cause because linear signals can also have harmonic content, as demonstrated by Choudhury *et al.* [2002]. In particular, the second and third harmonics of a non-sinusoidal oscillatory disturbance are sometimes amplified in the secondary disturbance yet the linearity is still increased. This can happen when a control loop compensates for higher harmonics in an external disturbance. In that case the harmonic content of the manipulated variable is higher than that of either the disturbance or the controlled variable, yet non-linearity tests show the manipulated variable to be less non-linear than either. The presence of harmonics usually suggests a non-linearity is present somewhere but for the reasons given above they may not reliably locate the source.

Data-driven Methods for Other Root Causes
Other root causes of distributed disturbances include:

Poorly tuned controllers.
Control loop interactions arising when two controllers have a shared mass and/or energy store (e.g. pressure and level controllers may compete for control of the contents of a reactor);
Structural disturbances caused by coordinated transfers of mass and/or energy between different process units, especially when a recycle is present;
External disturbances (e.g. due to site utilities such as steam).

The determination these other root causes from normal operating data remains an active area of research.

Xia and Howell [2003] used a comparison of the signal-to-noise ratios of controlled variable and controller output to give a loop status that determined if an individual loop was affected by long or short term transients or a slow or fast oscillatory disturbance. Fast oscillatory disturbances were attributed to poor tuning. Methods of advanced signal processing using causality analysis are also starting to be used. The aim is to map out the structure of the process by using the measured data. An elementary case is when the time trend of Tag A correlates with a time-lagged trend of Tag B. It can then be concluded that disturbances are propagating from A to B. Another approach is path analysis which can determine direction as well as correlation in a data set and distinguish between the case where A directly influences both B and C and the case where A influences B and B influences C, [Huang *et al.*, 2002; Johnson and Wichern, 1992].

Entropy measures are starting to be used to determine causality and directionality. The term *entropy* implies that the analysis is being done using the probability density functions (*pdf*) of the measurements. Chiang and Braatz [2003] looked for changes in the *pdf* or changes in the relationships between the *pdf*'s of two measurements. They concluded that the information was most useful when combined with a knowledge of the process. A transfer entropy method based on conditional probabilities has been applied to data from an industrial process showing changes the causal relationships in the process when a process fault occurred [Bauer *et al.*, 2004; Schreiber, 2000]. The ability to map out cause and effect paths in a plant from analysis of process data is potentially an attractive way forward for root cause diagnosis.

Use of a Qualitative Model
Stanfelj *et al.* [1993] provided a decision-making tree which included cross-correlation between a feed forward signal and the controlled variable of the loop under analysis. Likewise, Owen *et al* [1996] showed an application of control-loop performance monitoring in paper manufacturing which accounted for upset conditions of the whole mill and interactions between control loops. These cases needed a knowledge of the process flowsheet, in particular knowledge about which loops might disturb one another. Chiang and Braatz [2003] made similar comments. It is clear that the purely data-driven approach would be enhanced by the capture and integration of cause and effect information from a process schematic.

The trend in large petrochemical facilities which can justify the model development costs, for instance in ethylene plants, is increasingly towards model-based control and optimization in which each plant has a first-principles process model based on the physics and chemistry of the process. Greater opportunities will therefore exist in future to exploit the linkage of process models and data-driven analysis for cause-and-effect reasoning.

6.5 Industrial Case Study

6.5.1 The Process

Figure 6.3 plots a full data set from the refinery separation unit of Figure 6.1 showing the steam flow, analyser and temperature measurements (*pv*) and set points (*sp*) and also the controller outputs (*op*). Measurements from upstream and downstream pressure controllers PC1 and PC2 are also included. The sampling interval was 20 s. The data set includes some set point changes and evidence of non-stationary low frequency trends.

The data were filtered to remove low frequency trends with periods longer than 200 samples (67 minutes) by a frequency domain filter that removed the unwanted frequencies from the discrete Fourier transform [Thornhill *et al.*, 2003]. Then a subset of the data where there were no set point changes was selected (samples 581 to 1080) as shown in Figure 6.4 which is the data set used for analysis. Figure 6.5 shows the power spectra on a normalized frequency axis, where normalization is

achieved by scaling with the sampling frequency f_s such that a spectral peak on
the normalized frequency axis at $f/f_s = 0.05$ corresponds to an oscillation with a
period of $1/0.05 = 20$ samples per cycle. The controller errors and controller
outputs show the presence of unit-wide oscillation.

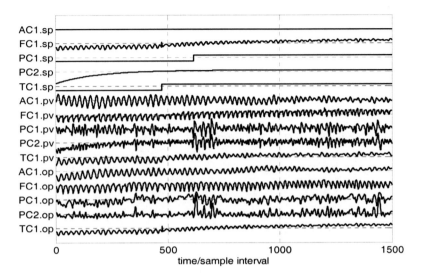

Figure 6.3. Data set for the separation column analysis

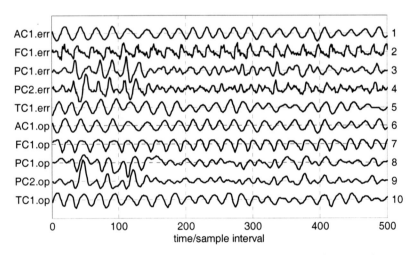

Figure 6.4. Samples 580 to 1080 with low frequency trends removed

It is known that there was a faulty steam sensor in the steam flow loop FC1. It
was an orifice plate flow meter but there was no weep-hole in the plate which had
the effect that condensate collected on the upstream side until it reached a critical

level, and the accumulated liquid would then periodically clear itself by siphoning through the orifice. The challenge for the analysis of this unit is to characterize the oscillatory disturbances and to verify that the methods correctly highlight the faulty steam flow loop as the root cause of the disturbance.

6.5.2 Spectral and Oscillation Analysis

Figure 6.6 shows autocovariance functions for the data in Figure 6.4 together with the zero crossings. Table 6.1 gives the results of the oscillation analysis which show there are two distinct oscillations present. One has a period of 21 samples per cycle (7 minutes) and the other about 18 samples per cycle (6 minutes). The reason why two groups of oscillations have been reported is that the tags with the highest oscillation indexes in each group have oscillation periods $\overline{T}_p$ that are different by more that the standard deviation of either (Tag 4 has 18.9 ± 1.5 and Tag 7 has 21.1 ± 1.1).

It is doubtful whether a visual inspection of the time trends and spectra in Figure 6.4 and Figure 6.5 would have revealed the presence of two separate oscillations because the spectral peaks look so similar, but in fact a close examination of the spectra in Figure 6.5 shows the spectral peaks of PC1 and PC2 are at a slightly higher frequency.

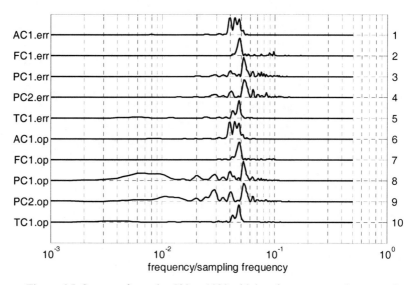

Figure 6.5. Spectra of samples 580 to 1080 with low frequency trends removed

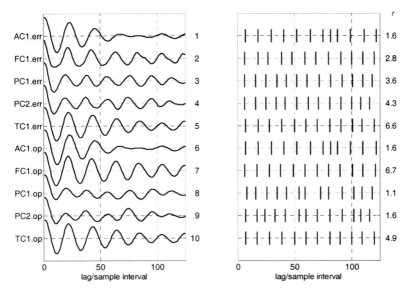

Figure 6.6. Autocovariance functions and zero crossing for the refinery data set. Tags with $r > 1$ are oscillating

Table 6.1. Oscillation analysis for the industrial case study

Analysis by Tag				Plant-wide analysis	
tag no	$\overline{T}_p$	σ_{T_p}	r	average period	tags
1	20.4	4.3	1.6	18.9	4 3 9 8
2	20.9	2.5	2.8	20.7	7 5 10 2 6 1
3	19.1	1.8	3.6		
4	18.9	1.5	4.3		
5	20.9	1.1	6.6		
6	20.4	4.3	1.6		
7	21.1	1.1	6.7		
8	18.7	5.5	1.1		
9	18.9	3.9	1.6		
10	20.7	1.4	4.9		

The results of spectral principal component analysis are shown in Figure 6.7. The analysis used 5 PCs to capture 99% of the variability in the data set. Figure 6.7 is a spectral classification tree that groups together tags having similar spectra and hence similar dynamic features. In the tree, each whole spectrum is represented as a spot on the horizontal axis. Spectra form a cluster if they are connected to each other by short vertical lines and are well separated from all other spectra. If the vertical lines between clusters are long then it means the spectra are very different in the two clusters. There are two main clusters in the data. Tags 3, 4, 8 and 9 have similar spectra, as do 1, 2, 5, 6, 9, and 10.

Tags 3, 4, 8 and 9 are the controller errors of PC1 and PC2 and their controller outputs. Spectral PCA shows that their spectra are distinctly different thus confirming the finding from oscillation analysis that their period of oscillation was different from the oscillation period in other tags.

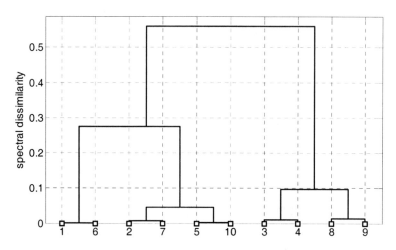

Figure 6.7. Spectral classification tree

Among the remaining tags, the tree shows 2, 7, 5 and 10 are the most similar. These are the controller errors and controller outputs from FC1 and TC1. The finding that they are similar is to be expected because the FC1 and TC1 loops are in a master-slave configuration. Tags 1 and 6 are the controller error and controller output of AC1. AC1 at the top of the column is physically well separated from FC1 and TC1, however, it shares similar dynamic behaviour because of the unit-wide oscillation.

Root Cause Analysis
The numerical values on the right hand side of Figure 6.8 show the results from non-linearity detection using the surrogate data analysis method of Thornhill [2005]. Other non-linearity assessment methods may also be used to give similar results, for instance the bicoherence test [Choudhury, 2004; Choudhury et al., 2004].

The group with the 21 samples per cycle (7 minute) oscillation period has non-linearity in the FC1 controller output, FC1 controller error and the TC1 controller output. The TC1 controller output is the set point of the FC1 control loop because of the cascade configuration. The non-linearity test therefore points unambiguously to the FC1 control loop as the source of the 7 minute oscillation and also shows that the root cause is a non-linearity. This is the correct result, the FC1 control loop was in a limit cycle because of its faulty steam flow sensor.

The non-linearity tests showed no non-linearity present in tags 3, 4, 8 and 9 associated with the 6 minute oscillation in PC1 and PC2. Therefore a root cause other than non-linearity has to be sought for those. Figure 6.3 shows that the set point change in PC1 at about sample 600 initiated a large oscillatory transient response in *both* these pressure controllers which suggests (a) the tuning might be rather tight and (b) that these two pressure loops are interacting with one another.

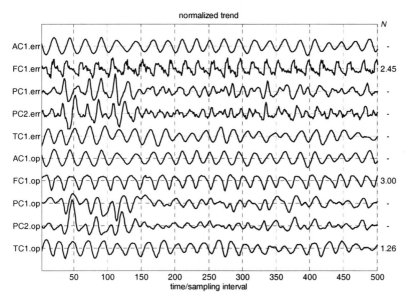

Figure 6.8. Non-linearity analysis

6.6 Chapter Summary

This Chapter has given an overview of the state of the art in detection and diagnosis of distributed disturbances to establish that, in the area of chemical process control, *quantitative process history based* methods are currently more widely used than model-based methods for fault detection and diagnosis. Some recent methods for plant-wide disturbance detection and diagnosis were covered in more detail. These methods used the power spectra and autocovariance functions and the methods of non-linear time series analysis.

The chapter included a worked example from a refinery separation unit. Oscillation detection based on autocovariance functions showed clearly that there were two distinct oscillations present at two very similar frequencies. The finding was confirmed by a spectral principal component analysis. Diagnosis using non-linear time series analysis successfully located the source of one of the two oscillations to the known root cause in a steam flow control loop. The same method showed the other oscillation was not due to a non-linearity, and a detailed visual examination of the data set suggested the most likely cause to be an interaction between two tightly tuned pressure controllers.

Benchmarking Control Loops with Oscillations and Stiction

Alexander Horch

ABB Corporate Research, Germany

7.1 Introduction

Control loop oscillations are the most prominent indications of deteriorated controller performance. In many cases, a predominant oscillation thwarts the original intention of feedback control. The feedback mechanism may increase the variability of the control error rather than keeping the actual process variable close to the desired target. Oscillations often indicate a more severe problem than irregular variability increase and hence require more advanced maintenance than simple controller re-tuning.

Nowadays, the detection of oscillations is a solved problem. Diagnosis of oscillations, however, still is an area of active research. Fortunately, during the last years, significant progress has been made in this direction, even though some questions remain unanswered. Nonlinearity analysis, as reported in this book (Chapter 6) has enabled important differential diagnoses mainly for oscillating loops and will be presented for stiction detection later on.

Still partly unsolved is the problem of diagnosing arising oscillations due to linear coupling between two – well-tuned, if seen individually – control loops.

This chapter first presents an overview of the main methods for oscillation detection in control loops, followed by a survey of the main reasons for such oscillations.

The most relevant tasks in control loop benchmarking or oscillation diagnosis are:

- Reliable detection of oscillatory behaviour in control loops
- Reliable indication of the most relevant root causes for such oscillations
  - Stiction
  - External disturbances
  - Tight tuning
  - Others
- Assessment and benchmarking of the identified root cause and remedy suggestion.

The requirements are set by the industrial large scale application of these methods:

- Model-free technology
- Use of minimum information about the control loop
- Works on control error data (SP-PV) and controller output (CO) only

Still most activity, both academic and industrial, is directed towards the detection of control valve stiction (= **st**atic fr**iction**). Several methods for stiction detection have been suggested during the last years. Therefore a significant part of this chapter is devoted to the different methods for stiction detection as one important cause of control loop oscillations.

7.2 Periodic Variations in Process Control Loops

Oscillations in process control loops are a very common problem. In Canadian studies it was found that up to 30% of all loops may be oscillating in typical pulp and paper processes, see [Bialkowski, 1992] and [Ender, 1993]. The fact that some loops oscillate is often known to the operators. However, since the cause is usually unknown, appropriate corrective measures may not be obvious. Therefore, oscillating loops tend to be neglected or put in manual mode. Unfortunately, none of these measures maintains the benefit of automatic control.

However, the automatic diagnosis of oscillations without performing any experiments is a difficult task and may often not even be possible for some causes. In this section the most important reasons for oscillation in process control loops are briefly discussed.

7.2.1 Important Reasons for Oscillations

Static friction (Stiction)
High static friction (stiction) in a control valve often causes oscillations. The mechanism is simple: Assume that a valve is stuck (stick-phase) in a certain position due to high static friction. The (integrating) controller then increases the setpoint to the valve until the static friction can be overcome. Then the valve breaks off and moves (note that the dynamic friction is smaller than the static friction) to a new position (slip-phase) where it sticks again. The new position is usually on the other side of the desired setpoint such that the same process starts again in the opposite direction. A very simple simulation model, based on the work by [Olsson, 1996] was presented in [Horch and Isaksson, 1998]. A recent very useful and intuitive simulation model has been proposed in [Choudhury *et al.*, 2005]. The characteristic (nonlinear) behaviour of friction in control valves in shown in Figure 7.1.

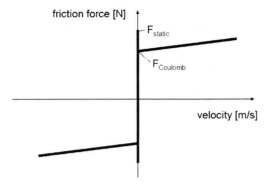

Figure 7.1. Characteristic friction plot

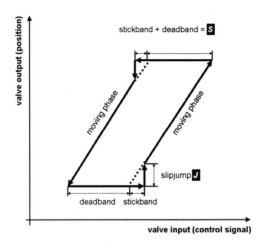

Figure 7.2. Input-output behaviour for sticking valves

The describing function for stiction nonlinearity was derived by [Choudhury *et al.*, 2005] and is given as

$$N(C) = -\frac{1}{\pi C}[A(C) - iB(C)], \tag{7.1}$$

where:

$$A(C) = \frac{C}{2}\sin 2\phi - 2C\cos\phi - C\left(\frac{\pi}{2} + \phi\right) + 2(S - J)\cos\phi \tag{7.2}$$

$$B(C) = -\frac{3C}{2} + \frac{C}{2}\cos 2\phi + 2C\sin\phi - 2(S - J)\sin\phi \text{ and} \tag{7.3}$$

$$\phi = \sin^{-1}\left(\frac{C - S}{C}\right). \tag{7.4}$$

The parameters S and J parameterize the stiction behaviour where S is the sum of stickband and deadband and J is the slipjump in the valve input-output behaviour, see also Figure 7.2.

The negative inverse of the describing function is shown in Figure 7.3.

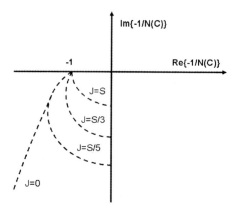

Figure 7.3. Negative inverse of the stiction describing function for different values of S and J

Typical stick-slip behaviour results in a rectangular process output and a triangular control signal, see Figure 7.4 for an industrial example. Occurrence of oscillations (not amplitude and frequency) due to stiction is usually tuning-independent as long as there is integral action. Oscillations due to stiction are usually not removed by re-tuning the controller. An *ad hoc* control strategy - which could be used until the next maintenance stop – could be the use of a dead-zone in the controller as suggested by [Ender, 1997]. The tuning of PI-controllers in loops with stiction was also discussed by [Piiponen, 1998].

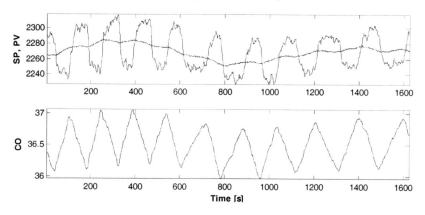

Figure 7.4. Typical behaviour of a control loop that exhibits stiction

Dead-zone

Dead-zone is the combined nonlinearity that represents the amount the input signal needs to be changed before the actuator actually moves. This kind of nonlinearity can, for example, occur in the control system software. Figure 7.5 shows the time-domain behaviour of a dead-zone element.

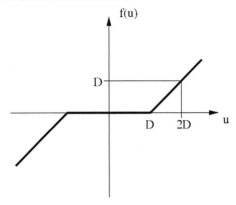

Figure 7.5. Time-domain characteristic of a dead-zone element

The describing function is given by

$$N(C) = \begin{cases} 1 - \dfrac{2}{\pi}\left(\arcsin\left(\dfrac{D}{C}\right) + \dfrac{D}{C}\sqrt{1 - \dfrac{D^2}{C^2}} \right) & , C \geq D. \\ 0 & , C < D \end{cases} \qquad (7.5)$$

See Figure 7.6 for a plot of $N(C)$.

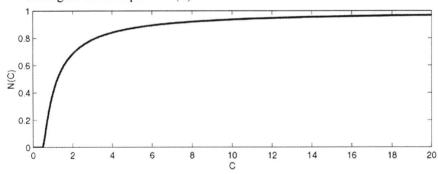

Figure 7.6. Describing function for a dead-zone nonlinearity with $D=0.5$

The negative inverse of this function always covers the negative real axis to the left of $-1 + j0$. A *stable* limit cycle can only be obtained if the open-loop without the nonlinearity itself crosses the negative real axis left of the critical point $-1 + j0$. A possible scenario to satisfy such a condition can, for example, be a PID-controlled plant which contains a first-order dynamics and an integrator (e.g. a level control loop). An industrial example of a level control loop where an oscillation is caused by a dead-zone in the control system is shown in Figure 7.7.

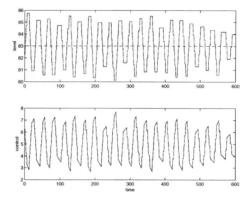

Figure 7.7. Example of a level control loop that oscillates due to dead-zone

Backlash
Backlash is present in every mechanical system where the driving element is not directly connected with the driven element. Figure 7.8 shows the time-domain behaviour for a backlash element.

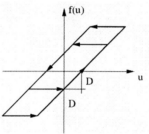

Figure 7.8. Time-domain characteristic of a backlash element

Backlash is a dynamic nonlinearity, i.e. its state depends on the input *and* past states. The describing function for the backlash is therefore a complex valued function. Assuming a unit slope in the input-output relation, the describing function for $C \geq D$ where D is the width of the dead-band is given as

$$\text{Re}\{N(C)\} = \frac{1}{\pi}\left(\frac{\pi}{2} + \arcsin\left(1 - \frac{2D}{C} \right) + 2\left(1 - \frac{2D}{C} \right)\sqrt{\frac{D}{C}\left(1 - \frac{D}{C} \right)} \right)$$

$$\text{Im}\{N(C)\} = -\frac{4D}{\pi C}\left(1 - \frac{D}{C} \right)$$

(7.6)

It is not possible - for general process models - to state simple conditions for a limit cycle to occur. For a limit cycle to occur the process must be higher than first order, which is always true for a first-order plus dead-time process controlled by a

PI-controller. An example of the negative inverse of the describing function is shown in Figure 7.9.

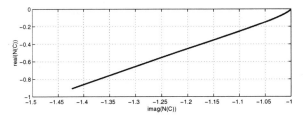

Figure 7.9. Plot of the negative inverse of the describing function for backlash

As can be seen, de-tuning of the controller will probably remove a backlash-induced oscillation in many cases.

Saturation
Real control signals always have constraints. Therefore, alarm limits are set for each control signal. If the controller is tuned such that the loop is unstable, there will be an oscillation due to the saturation nonlinearity. Figure 7.10 shows the time-domain behaviour for a saturation element.

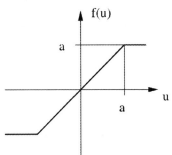

Figure 7.10. Time-domain characteristic of a saturation element

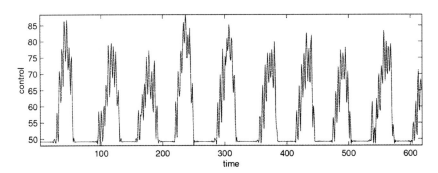

Figure 7.11. Data from a preheater level control loop. The oscillation is due to tight tuning in connection with saturation

The negative inverse of the describing function for a saturation nonlinearity is located on the negative real axis left of the critical point. For an oscillation due to saturation, the phase difference between control signal and the process output will therefore ideally be 180 degrees. An industrial example is shown in Figure 7.11.

Quantisation

Quantisation is always present in modern control systems since analog values have to be discretised and *vice versa*. Especially in older systems where the converters have a relatively low resolution this may become a problem.

The input-output characteristic of a quantiser with a quantisation level of q is given by $Q(x) = q \; round \; (x/q)$. This nonlinearity is plotted in Figure 7.12.

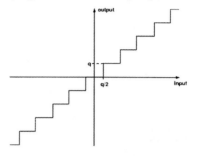

Figure 7.12. Time-domain characteristic for a quantiser

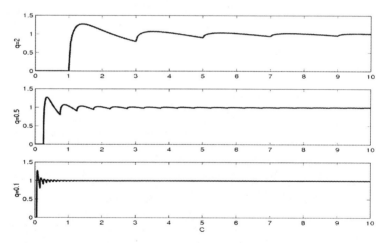

Figure 7.13. Describing function of a quantiser for some levels of quantization

The describing function for this nonlinearity can either be found by extrapolating results for the three-point and five-point nonlinearities or by considering the addition of signum-with-dead-zone functions. The describing function is

$$N(C) = \begin{cases} 0 & 0 \le C \le \dfrac{q}{2} \\ \dfrac{4q}{\pi C} \sum_{i=1}^{n} \sqrt{1 - \left(\dfrac{2i-1}{2C} q\right)^2} & \left(n - \dfrac{1}{2}\right)q \le C < \left(n + \dfrac{1}{2}\right)q \end{cases} \tag{7.7}$$

and is plotted in Figure 7.13. The variable n denotes the number of the step in the quantisation curve (Figure 7.12) which corresponds to the actual value of the amplitude C.

It is noteworthy that the maximum of $N(C)$ does not change with the quantisation level q. It is calculated as

$$\max_{C} N(C) = N\left(q / \sqrt{2}\right) = 4/\pi \tag{7.8}$$

The fact that the maximum of $N(C)$ is constant even for $q \to 0$ reminds of the Gibbs phenomenon, since the describing function is based on a truncated Fourier series. This has an interesting implication on the analysis using the describing function method. Since $N(C)$ is purely real, the negative inverse of $N(C)$ will be on the negative real axis only. It starts in -∞ when $0 \le C < q/2$. The maximal value is then given by

$$\max_{C} \frac{-1}{N(C)} = -\frac{\pi}{4} \approx -0.785. \tag{7.9}$$

Note that the function $-1/N(C)$ alternates around the critical point $-1+j0$ which can lead to ambiguities in the estimation of the oscillation amplitude.

It remains to discuss what happens for $q \to 0$, i.e. when the quantisation nonlinearity vanishes completely. As can be seen from Figure 7.13, if $q \to 0$, the function $N(C)$ will be squeezed together without changing its shape in the vertical direction. For the negative inverse, $-1/N(C)$, this means that the complete shape of the curve will not change with q but the amplitudes corresponding to particular points on the curve will change. Assume that there is an intersection between the Nyquist curve and the describing function for some finite q, then there will still be an intersection no matter how much q is decreased. The amplitude of the oscillation will, however, tend to zero.

For applying describing function analysis to a loop with quantization, assume that the controller $F(s)$ stabilises the linear part of the loop. Intersection of $-1/N(C)$ and the frequency function $L(j\omega) = F(j\omega)G(j\omega)$ can then occur in several ways. Firstly, if $\angle L(j\omega) < -180°$ for small frequencies. In that case, the intersection lies in the interval $[-\infty, -1]$. Secondly, if $L(j\omega)$ crosses the negative real axis in the interval $[-\infty, -\pi/4]$.

Industrial example. Consider a flow control loop which is oscillating, see left part of Figure 7.14. The reason for this oscillation was found to be an 8-bit resolution in the D/A-converter, yielding a quantisation level $q=0.5\%$ of the control signal. After a 12-bit converter was installed, the oscillation has decreased drastically, see right part of Figure 7.14. The explanation is that the quantisation level was now decreased to $q=0.06\%$. The theory then predicts that the amplitude should decrease to $0.06/0.5 = 0.12$ times the original amplitude.

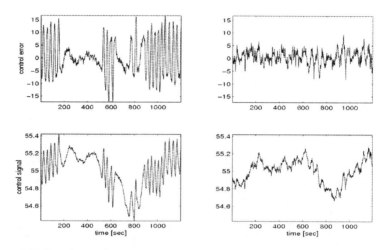

Figure 7.14. Data from a flow control loop. Left: Oscillation due to quantisation in the D/A-converter. Right: Same loop with higher resolution in D/A-converter. Top: control error, bottom: control signal.

External oscillating disturbances
Despite the use of single-loop controllers in the process industry, many basic control loops are coupled. Therefore, if one loop is oscillating, it is likely to influence other loops too. In many cases the oscillations are in a frequency range such that the controllers cannot remove them. Then an oscillation is present even though the controller is well tuned (it might have been tuned for some other control task). Industrial data for this kind of root cause is shown in Figure 7.15.

The experiment shows that the oscillation is not removed by putting FC105 in manual (at time 500) since the oscillation is induced by Loop QC193. The oscillation was generated in the QC loop which is verified by putting it in manual at time 1900.

External disturbances are a challenge for an automatic diagnosis algorithm. When having detected an oscillation, it is important to distinguish between internally and externally caused oscillation. Using nonlinearity analysis and other heuristics, this seems to be a solvable task for a number of typical cases [ABB, 2005].

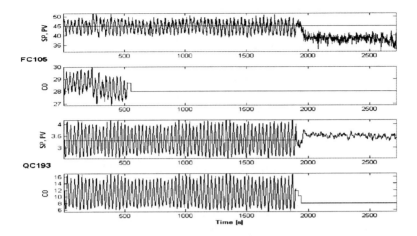

Figure 7.15. Example of an external oscillating disturbance in Loop FC105

Other reasons

There are of course other reasons for oscillations. It is for example possible that two well-tuned single loops start oscillating when they are coupled (e.g. due to nonlinearity or multivariable effects). However, using normal operating data and without any process knowledge it seems to be almost impossible to diagnose such a cause.

Yet another cause which has not been mentioned so far is an oscillating loop set point, see Figure 7.16. This problem has been analysed by [Ettaleb *et al.*, 1996] and will typically only appear in cascaded control loops.

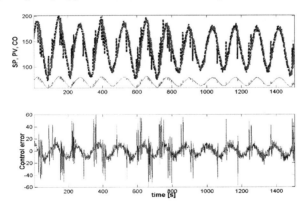

Figure 7.16. Oscillating setpoint (top) that yields a slightly oscillatory control error (bottom)

Of course, often there may be several causes at the same time responsible for a certain oscillation. This makes the diagnosis even more difficult.

7.3 Automatic Oscillation Detection

Obviously, the first step in oscillation diagnosis needs to be reliable oscillation detection. This task can be solved by several methods, most of which will be summarized below.

A discussion about the online detection of oscillation vs. running the detection method in offline mode now follows. Usually, for offline detection, only small amounts of data are required. Therefore the conclusion is that for practical applications it is not really critical if some method can only be applied to an offline data batch. This use in quasi 'online' mode enables normally a 'fast enough' oscillation detection in practice. Oscillations are a wide-spread problem but the detection speed is not critical in terms of minutes or hours. Many plants exhibit oscillations that have been going on for years, such that the operators have accepted them as being 'normal' for the process.

Oscillation detection can be done somewhat heuristically since there is no clear mathematical definition of an 'oscillation' that could be applied. It is therefore common sense to speak of oscillations as *periodic variations that are not completely hidden in noise,* and hence visible to the human eye. There are other applications, e.g. signal processing for communication applications where periodic signals with a very low signal to noise ratio are very common. For the process industry, such signals are usually not considered.

This raises philosophical questions about the size of oscillation amplitude. Is a regular oscillation with a very small amplitude important enough to be highlighted by the assessment system? This is an ambiguous question since small oscillations are usually not a large problem for the complete process, however, oscillations usually pinpoint an underlying problem. Therefore, it is suggested never to hide information about detected oscillations from the user. Of course, an expert assessment cannot be done automatically.

The oscillation detection methods and their references are listed next:

Method A [Forsman and Stattin, 1998]
Regularity of upper and lower IAEs and zero-crossings in the time-domain.

Method B [Hägglund, 1995]
Regularity of 'large-enough' IAE in the time-domain.

Method C [Miao and Seborg, 1999]
Damping factor approach of the auto-correlation function.

Method D [Thornhill *et al.*, 2003]
Regularity of zero-crossings of the auto-covariance function.

Method E [Salsbury and Singhal, 2005]
Identification of undamped poles of time series ARMA model.

Method F (without reference)
Detection of single peaks in the signal spectrum.

Method G [Matsuo *et al.*, 2003]
   Wavelet-based visual detection of multiple oscillations in process data.

There are, of course other approaches but this collection shall be sufficient to give an overview. Most of these methods will in most cases give a reliable detection of oscillation in the sense discussed above.
   Table 7.1 shows a comparison of the most important methods for oscillation detection.

Table 7.1. Comparison of methods for oscillation detection

	Domain	Mode	Load	Pros	Cons
A	Time	Offline	Low	- handles asymmetrical oscillations - simple & intuitive	- filtering required - regular oscillation needed
B	Time	Online	Low	- simple & intuitive - online applicable	- needs to specify ultimate frequency - noise sensitive
C	ACF	Offline	Medium	- detects short-term oscillations - automatic filter	- patented - hard if several frequencies
D	ACF	Offline	Medium	- automatic filter - intuitive	- hard if several frequencies
E	Model	Online	Medium	- noise modelled - sound method	- hard if several frequencies
F	Frequency	Offline	Medium	- classical method - finds main oscill.	- difficult to evaluate automatically
G	Time / Frequency	Offline	High	- handles multiple oscillations	- no automatic evaluation

Hägglund [1995] has proposed a method in the time-domain which considers the integrated absolute error (IAE) between all zero-crossings of the signal. If the (IAE) is *large enough*, a counter is increased. If this counter exceeds a certain number, an oscillation will be indicated. In order to quantify what *large enough* means, the ultimate frequency of the loop in question (as may be known from identification using relay-experiments) is used. Alternatively, one may use the integral time of the controller, assuming that the controller is reasonably tuned. This method is very appealing but has two disadvantages: Firstly, it is assumed that the loop oscillates at its ultimate frequency which may not be true, e.g. in the case of stiction. Secondly, the ultimate frequency is not always available and the integral time may be a bad indicator for the ultimate period. A strength of the method is that it quantifies the size of the oscillation.

A very similar idea was used in the detection method by Forsman and Stattin [1998]. There, the time instants of all zero-crossings and all IAEs are computed and compared pairwise. Then, all IAEs and zero-crossings which are pairwise approximately equal are counted. This number is divided by the total number of half-periods yielding an index between 0 (no oscillation) and 1 (perfect oscillation). A reasonable threshold was found to be 0.3-0.4. An advantage with this method is that an estimate of the frequency and oscillation amplitude are computed as a by-product. There are some tuning parameters which determine what *approximately equal* means but these do not usually have to be changed. This method relies on some pre-filtering in order to cope with signal noise.

The problem of signal noise is elegantly removed by considering the auto-correlation function ACF (or auto-covariance) of the control error. By estimating the ACF, a large extent of the signal noise is removed. This approach has been used by [Thornhill *et al.*, 2003] where a regularity factor of the zero-crossings of the ACF has been proposed. An oscillation is detected if the standard deviation of the estimated oscillation period from all periods in the available ACF is not more than e.g. 33% of the mean value. This approach is very sensible, but may not work well if more than one frequency is dominant in the signal. This can be the case when two frequencies that are rather different are present in the same signal, see Figure 7.17.

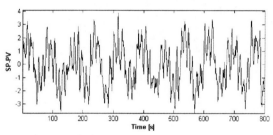

Figure 7.17. Example of an oscillating loop with two distinct frequencies. Which oscillation shall be detected?

Yet another method was presented by Miao and Seborg, [1999]. Here the auto-correlation function of the signal in question is estimated. Then a measure similar to the damping ratio is computed. If this measure exceeds a given threshold, an oscillation is indicated. This measure also detects oscillatory behaviour that is not yet persistent. The method is patented.

A more complex detection of oscillation is certainly frequency analysis which reveals both broad peaks, indicating a non-sinusoidal oscillation and multiple oscillation frequencies very easily. However, typically, an automatic detection is not trivial since, in practice, non-oscillatory signals can also have a maximum at some frequency in the spectrum.

The next natural step is to combine time and frequency domain and investigate data using wavelet analysis. This has been proposed by Matsuo *et al.*, [2003]. The wavelet plots are, however, difficult to be evaluated automatically and a better tool for human expert analysis. They have the ability to cover both temporary oscillations and multiple frequencies.

As an example consider the control error signal in Figure 7.18 (top plot) and its ACF (middle) and spectrum (bottom). All the methods above will be able to detect the (asymmetrical) oscillation of this flow control loop.

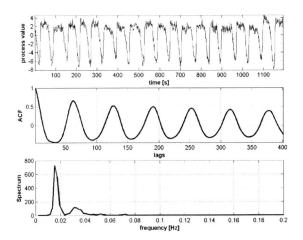

Figure 7.18. Example for detection of oscillation. Top: control error time trend. Middle: Auto-Correlation Function. Bottom: Signal Power Spectrum.

The next step after oscillation detection is oscillation diagnosis. The minimum-variance based controller performance index proposed by Harris, [1989] is a main ingredient in performance monitoring tools nowadays. The following section will therefore investigate the problem of applying this index to oscillating signals.

7.4 Application of Minimum-variance Based Index

This section describes the use of a minimum-variance based control performance index [Harris, 1989] when applied to an oscillating signal. The use of this index in such cases may give unrealiable results. As a motivating example consider the process variable in a flow control loop in a pulp mill, see Figure 7.19.

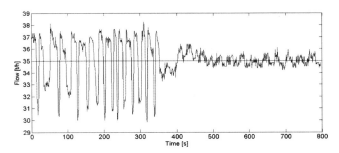

Figure 7.19. Measured data from an oscillating flow control loop

The signal is oscillating for the first 350 samples, after that the oscillation vanishes. The reason for the oscillations is likely to be a nonlinear phenomenon and a change in the noise structure is not expected within the short time period considered. It would therefore be expected that the performance index should be able to identify the change in the performance.

From identification experiments it is known that the process dead-time is two samples. Table 7.2 shows the estimated indices and variances for both parts of the data.

Table 7.2. Minimum-variance results for the data set in Figure 7.19

Data range	Actual variance	Minimum variance	Estimated noise variance	Harris Index [0...1]
[1:350]	5.87	1.92	0.75	0.36
[500:800]	0.14	0.087	0.081	0.54

The time-series model used is an AR model of order 25. Notice that the variance in the process output decreased by a factor of about 40. The performance index however, decreased by a factor less than two. It is also noteworthy that both the estimated minimum-variance terms and the noise variances differ considerably for both cases. On the other side, the output variances were obtained using the estimated time-series models and are in good agreement with the variances obtained directly from the data. The question then arises:

Why is the obvious change of performance not properly reflected in the index?

This question has been discussed in [Horch, 2000] where an attempt has been made to motivate the recommendation to take the minimum-variance based performance indices with "a pinch of salt" in the presence of oscillations. It was found that both the actual variance and the noise variance can be estimated reliably but that the minimum-variance expression in some cases is not a reliable indicator, especially when nonlinearities are involved.

It has also been noted in [Horch, 2000] that the actual noise variance should be computed via the estimated time-series model rather than directly from the data. Since the detection of oscillations is reliably possible nowadays, it is recommended not to rely on the minimum-variance based index in these cases. If an oscillation is known to be present, there is no real need to assess the loop performance via the Harris index.

For the Harris index numerous modifications and extensions have been proposed. An approach that tries to eliminate the problems described above and induced by oscillations has been proposed and patented by the Pulp and Paper Research Institute of Canada [Owen, 1997]. The idea is to evaluate the Harris index in the frequency domain and to exclude the frequencies where the oscillation is mainly located from the analysis. The heuristics behind this approach is that a controller that is so badly tuned that the loop actually oscillates also has a poor performance for other frequencies – and hence a poor index value after removing the oscillation contributions.

A similar approach could be developed in the time-domain [Horch, 2000; Chapter 12], e.g. by using a notch filter, however, such an approach does not seem to exist in practice so far.

7.5 Oscillation Diagnosis

The remaining part of this chapter will be devoted to the oscillation diagnosis with the focus on the automatic procedures, i.e. without active experiments or interaction with the process. The most important challenge about oscillation diagnosis is to distinguish internally from externally generated oscillations. This information will help the maintenance effort to be directed to the correct source of problem, i.e. which loop to address. Valve stiction seems to be the most predominant cause of internally generated oscillations. The other internal reasons – in connection with other static nonlinearities, such as saturation etc. – have been discussed above. The distinction between all these causes seems to be less important than being able to state if the cause is internal or external.

7.5.1 Automatic Detection of Stiction in Control Loops

The stiction generation mechanisms in final control elements have been described by several authors. Also, some mathematical models to simulate and understand the effects of sticking valves in automatic control have been proposed. See [Choudhury *et al.*, 2005], [Olsson, 1996], [Deibert, 1994] and the references therein. The main purpose of this section is to present, classify and evaluate most currently published methods for stiction detection on some benchmark data from the process industry. Before the benchmark data is introduced, the methods will be presented briefly. Due to space limitations, no detailed presentation can be given. The complete methodology including assumptions, details, implementation issues and examples are in most cases found in the published references given below. The focus here is to give an overview and to understand where the differences of the methods are. The following methods will be presented:

Method A [Horch, 1998]
Cross-correlation information between controller output and process variable

Method B[1] [Horch, 2001]
Signal distribution of derived and filtered control error signal

Method C [Stenman *et al.*, 2003]
Detection of abrupt changes in process variable

---

[1] Patent pending ("A method and a system for evaluation of static friction")

Method D[2] [Choudhury *et al.*, 2004]
Detection of nonlinearity and non-Gaussian properties based on higher-order statistics

Method E [Thornhill, 2005]
Detection of nonlinearity based on phase spectrum analysis

Method F [Singhal and Salsbury, 2005]
Detection of asymmetrically distributed areas in oscillation half-periods

Method G1 & G2 [Rengaswami *et al.*, 2001]
G1) Comparison of oscillating signals with pre-defined shapes.
G2) Identification of Hammerstein models *(no benchmark results available)*

Methods H1 & H2 [Kano *et al.*, 2004]
H1) Detect when change in CO is not followed by a change in PV
H2) Detect a parallelogram in the phase plot of PV vs. CO

Method H3 [Yamashita, 2004]
H3) Fit qualitative shapes to the phase plot of PV vs. CO.

Method I [Scali and Ulivari, 2005], [Rossi and Scali, 2004]
Comparison of signal fit to pre-defined shapes

A simple classification of methods can be given as shown below.

Time domain shape analysis	A, B, F, G1, H2, H3, I
Nonlinearity analysis	D, E
Fault detection & identification analysis	C, G2, H1

The most important properties of all methods together with the advantages and disadvantages are summarized in **Table 7.3**.

There are other methods to solve this task that have been proposed during the last decade, see e.g. [Deibert, 1994], [Ettaleb *et al.*, 1996], [Hägglund, 1995], [Horch and Isaksson, 1998], [Taha *et al.*, 1996] or [Wallen, 1997]. Unfortunately all the approaches mentioned require either detailed process knowledge, user-interaction or rather special process structures and will not be reviewed here.

The described methods have been applied to a number of benchmark data sets. The data is plotted in Figure 7.20. For all control loops, the top plot contains the set point (SP) and process variable (PV), followed by the controller output (CO) in the bottom plot.

The loops originate from different industries and processes. They are by no means typical in all cases and neither do they represent an exhaustive selection of all possible causes.

---

[2] Patent pending ('Methods for detection and quantification of valve stiction')

Table 7.3. Comparison of different methods for stiction detection

	Load	Pros	Cons
A	Low	- simple - intuitive	- strong assumptions - not for all loop types
B	Low	- simple - independent on signal shape	- dependent on good filtering - method patented
C	High	- theoretically sound idea - no regular oscillation needed	- complex computations - problems with noisy signals
D	High	- theoretically sound - can do more than human eye	- complex computations - not stiction specific
E	High	- theoretically sound - can do more than human eye	- complex computations - not stiction specific
F	Low	- simple - intuitive	- filtering required - method patented
G1	Medium	- intuitive - flexible	- noise sensitive - sensitive to shape deformation - high complexity (slow)
G2	High	- theoretically sound	- complex method - need excitation - parameter required
H1	Medium	- intuitive - simple	- threshold to be set for each loop
H2	Low	- classical approach	- noise sensitive - classification difficult
H3	Medium	- intuitive	- needs classical PV/CO pattern
I	Medium	- intuitive - flexible	- noise sensitive - sensitive to shape deformation - high complexity (slow)

However, here the purpose is to illustrate the methods rather than letting them compete against each other. All of them have strengths and drawbacks. Future research will most likely aim to learn from each method such that reliable stiction detection can be done as easily as oscillation detection can be done nowadays.

The benchmark control loops were identified as having the problems listed in Table 7.4.

The benchmark data was evaluated using the methods above. The rest of the chapter describes briefly each of the methods and shows the results of the evaluations.

Method A

The detection method is based on the cross-correlation between controller output (CO) and process variable (PV). In [Horch, 1999] it was shown that the cross-correlation information can correctly distinguish static friction (stiction) in control valves from other sources of oscillation. The decision rule is stated as follows:

If the cross-correlation function between controller output and process output is an odd function, the likely cause of the oscillation is stiction. If the cross-correlation function is even, then the likely cause is an external oscillation or too tight controller tuning.

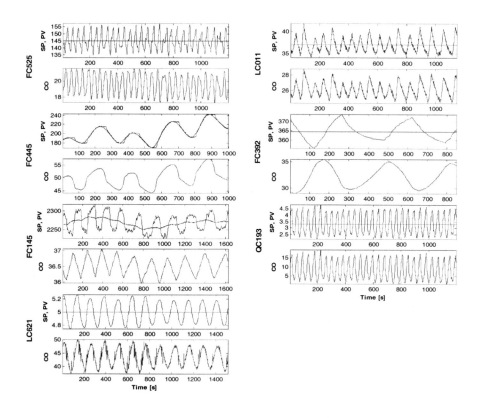

Figure 7.20. Benchmark data for stiction detection algorithms

Table 7.4. Stiction benchmark data. Description of data sets

Loop name	Type	Oscillation cause
FC525	Flow	Stiction
LC011	Level	Stiction
FC445	Flow	Stiction
FC392	Flow	Stiction
FC145	Flow	Stiction
QC193	Concentration	Valve dead-zone & tight tuning
LC621	Level	Tight tuning

The assumptions made here are very important:
    An oscillation has already been detected
    The process is not integrating
    The controller has (significant) integral action
    The loop does not handle compressible media

Note that the method may give wrong indications if any of these assumptions are violated.

In order to distinguish between odd and even correlation functions, it would be sufficient to consider the cross-correlation at lag zero. However, for a practically working algorithm it is better to make use of the cross-correlation function up to the first zero-crossing in each direction.

For a human it is rather simple to state whether or not a function is odd or even. For automatic diagnosis, different measures to distinguish between odd and even functions can be defined. As examples, consider the results for FC525 and QC193 in Figure 7.21.

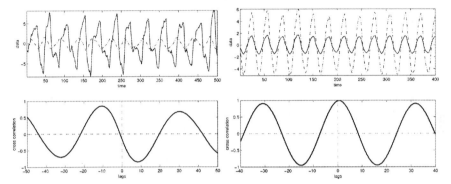

Figure 7.21. Detection of stiction, FC525, (left) and non-stiction, QC193, (right) with Method A. Top plots: Time series in [s] of CO and PV, bottom plots: CCF (normalized)

Method B

The main idea in [Horch, 2001] is to detect abrupt changes (which are typical in the case of stiction) in the process value. This is done by considering the probability distribution of the signal derivative. In the stiction case the distribution is close to Gaussian, otherwise it will have two peaks. For integrating processes (level control), the second derivative has to be used, for self-regularing processes, the first derivative is sufficient. In the following, the process is described for integrating loops only. The only difference for self-regulating loops is one differentiation less to be applied to the data.

Consider the second (filtered) derivative of the process value. Ideally, such a signal is a pulse train with additional white noise. In the non-stiction case, the (oscillating) loop output is usually close to sinusoid. Hence it has a probability distribution with two separate maxima. The detection algorithm tests which of both possible distributions is more likely to fit the data, see Figure 7.22.

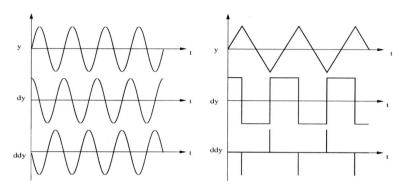

Figure 7.22. Basic idea of Method B. Left: Non-stiction case. The second derivative still is sinusoidal and exhibits a distribution with two separate maxima. Right: Stiction case. The second derivative is mainly centered around 0 except for some peaks, hence the distribution is close to Gaussian.

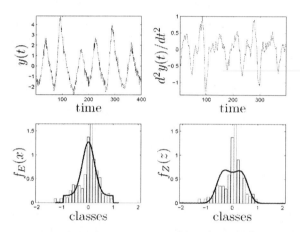

Figure 7.23. Stiction detection for Loop LC011 using Method B

The next step is to fit two different distribution functions to the sample histogram of the second (filtered) derivative of the measured control error. It is assumed that the loop in question is known to be oscillating, e.g. by using one of the oscillation detection methods described above.
The decision is then made as follows:

> *Consider the second (filtered) derivative of the loop output. Check whether the probability density function is approximately Gaussian (neglecting the pulse train) or the one of a sinusoid with additional Gaussian noise. A better fit to the first distribution indicates stiction, a better fit for the second one non-stiction.*

An example for the application of this method to loop LC011 is shown in Figure 7.23. The mean square error for the stiction case is 1.12 compared to 2.01 for the non-stiction case. Hence the presence of stiction is correctly concluded in this case.

Method C
The model-based detection method in [Stenman *et al.*, 2003] is inspired by ideas from the fields of change detection and multi-model mode estimation. This method does not require any oscillation being present in the data. The valve model used here is very simple:

$$x_t = \begin{cases} x_{t-1}, & if \, |u_t - x_t| < d \\ u_t, & otherwise \end{cases} \qquad (7.10)$$

The model for detecting stick-slip behaviour in the valve is $x_t = (1-\delta_t)x_{t-1} + \delta_t u_t$, where δ_t is a binary *mode parameter* which is 1 if a jump in the valve position x_t occurs and 0 otherwise. The process dynamics is assumed to be linear. The model is used as shown in Figure 7.24.

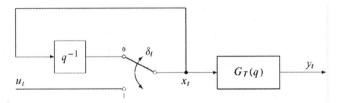

Figure 7.24. The assumed stiction model for Method C

The detection procedure then aims to identify a suitable mode sequence $\delta^N = (\delta_1,...,\delta_N)$. This step is called *segmentation*. The algorithm has some tuning parameters, most of which can be set to default values for practical application.

Method D
The main idea of [Choudhury *et al.*, 2004b] is to compute nonlinearity measures. Two measures are used, a nonlinearity index (NLI) and a non-Gaussianity index (NGI). Both indices are deduced from the signal bicoherence. Bicoherence is the same as the normalised bi-spectrum of a signal. A characteristic of a non-linear time series is the presence of phase coupling such that the phase of one frequency component in the spectrum is determined by the phases of other frequency components. This characteristic does not exist for linear signals (i.e. signals that are generated by linear systems). The test in [Thornhill, 2005] is based on the same property as discussed here.
The NLI and NGI are defined such that they analyse the non-zeroness and the constancy of the bicoherence plot. If both indices are greater than zero, the signal is described as nonlinear and non-Gaussian. The test is performed on the control error, as all stiction detection tests usually do.
The described test statistics does not directly refer to stiction. If one excludes the presence of process nonlinearity and non-linear disturbances, any detected nonlinearity is attributed to the final control element. The successful detection of nonlinearity for Loop FC525 is shown in Figure 7.25.

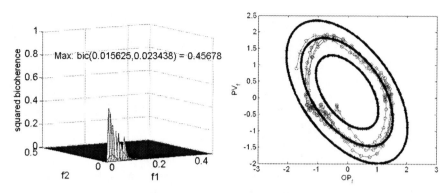

Figure 7.25. Squared bicoherence plot (left) and fitted ellipse for stiction quantization in the phase plot (right) for loop FC525

It should be noted that this method is one of the few to try to quantify the size of stiction which is very important for the assessment of its relevance [EnTech, 1998]. The other method that makes an attempt to quantify stiction is Method H2.

Method E
The description of this method is taken from [Thornhill, 2005]. The main feature is a comparison of the predictability of the time trend compared to that of its surrogates. See Figure 7.26 for illustration where the top plot shows the time trend of FC525. It has a clearly defined pattern and a good prediction of where the trend will go after reaching a given position.

The lower panel shows a surrogate (a signal with a randomized phase) of the time trend. By contrast to the original time trend the surrogate lacks structure even though it has the same power spectrum. The removal of phase coherence has upset the regular pattern of peaks.

This property can be used to define an index that quantifies the predictability of the signal when the phase is randomized and hence an index that quantifies nonlinearity.

Method F
The underlying idea of this method [Singhal and Salsbury, 2005] is very simple and appealing. An index is computed that characterizes if an oscillation half period is symmetrical, i.e. the maximum (minimum) is located in the middle of the half period. The more skew the areas left and right of the maximum (minimum) are, the more likely the oscillation is generated by a sticking valve. The index is simply the ratio of the areas left and right of the peak, see Figure 7.27.

Obviously, the index value depends heavily on two things: Signal to noise ratio and the *similarity* of the half periods. Both aspects have been improved by the authors but not published yet. Noise changes both zero crossings and the location of the peaks. The noise problem is handled by applying filtering to the area ratio R for each half period rather than the raw data.

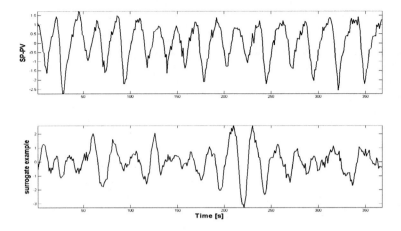

Figure 7.26. Time trend of LC011 (top) and surrogate example (bottom)

The index has been re-defined further such that not the ratio is reported but a probability measure in order to incorporate uncertainty in the value of R. Instead of checking if $R > 1$, the probability of $R > 1$ is considered and it is computed via a *t*-statistic.

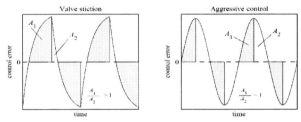

Figure 7.27. Definition of the stiction detection index $R = A1/A2$

Method G1

The method proposed in [Rengaswamy *et al.*, 2001] is based on a qualitative pattern recognition approach. The algorithms aims to distinguish square, triangular and saw-tooth like signal shapes in both controller output and process variable. The technique used to classify the signals is Dynamic Time Warping (DTW) and this is used on each oscillation cycle individually rather than on a complete data set at once. DTW is a classic technique from speech recognition used to compare signals with stored patterns. DTW takes into account that different parts of the signal under investigation may be shorter or longer than the reference and also enables the detection of similarities in such cases.

This approach allows for varying frequencies often encountered in industrial data. Table 7.5 shows the pattern shapes used for diagnosis.

Table 7.5. Stiction pattern shapes for Flow, Pressure, Temperature and Integrating Processes

Measurements	Fast Process(Flow)		Slow Process and (Pres. & Temp)	Integrating Process (Level)	Level with PI control
	Dominant (I) action	Dominant (P) action			
OP	Triangular (Sharp)	Rectangular	Triangular (Smooth)	Triangular (Sharp)	Triangular (Sharp)
PV	Square	Rectangular	Sinusoidal	Triangular (Sharp)	Parabolic

Method G2

The method proposed in [Rengaswamy et al., 2001] is based on a classical system identification approach. A quantitative Hammerstein model approach is used that estimates a stiction parameter based on the available data (CO and PV). The block diagram of a Hammerstein model is shown in Figure 7.28.

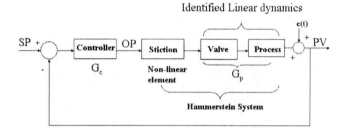

Figure 7.28. Block diagram of a Hammerstein model

Method H1

The method described in [Kano et al., 2004] is based on the observation that with stiction present, there are sections where the valve position does not change, even though the controller input changes. The algorithm proposed by the authors can be summarized as follows: The difference in the valve position (or the measured process value after the valve) is compared to a given threshold,

$$\Delta y(t) = y(t) - y(t-1). \tag{7.11}$$

All time intervals are identified where $|\Delta y(t)| < \varepsilon$. For each time interval where this condition is satisfied, the difference between minimum and maximum controller output, denoted $\tilde{u}$, and the difference between minimum and maximum valve position (or process variable), denoted $\tilde{y}$, are compared against given thresholds ε_u and ε_y respectively. In all cases where the position difference (process variable) is below the threshold and the control action difference is above the threshold, stiction is concluded. This comparison is done in each time step.

Then the ratio of the times where stiction was concluded is divided by the number of total time steps. This results in an index bounded by 0 and 1. The closer the index is to 1, the more likely stiction has occurred.

Method H2

This method was also proposed in [Kano *et al.*, 2004]. It is based on the relationship between controller output and the valve position (or process variable in the case where the valve position is not available). In the case of stiction, the observation is that the phase plot of the controller output versus valve position is a parallelogram.

The challenge is the automatic detection of the difference between a parallelogram and an ellipse which is the typical shape for more linear oscillations. The authors propose a function F such that

$$F(t) = \max\{\min(F(t-1) + \Delta u(t), F_{max}), 0\}. \qquad (7.12)$$

This function is evaluated in each time step. The maximal value of $F(t)$ and the initial value $F(0)$ are needed to quantify stiction and can be obtained from normal operating data by solving an optimization problem that maximises the correlation coefficient between $CO\text{-}F(t)$ and PV. Reliable results are obtained if the correlation coefficient is close to unity.

Method H3

The method proposed by [Yamashita, 2004] applies qualitative shape analysis to the phase plot of PV vs. CO rather than on the control error signal itself. Qualitative primitives are defined and a sequence of these primitives are fitted to data within a certain time period, see Figure 7.29.

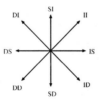

Figure 7.29. Basic shapes that describe valve movement in the PV - CO phase plane

The different indices describe singular movements: (**I**)ncreasing, (**S**)teady and (**D**)ecreasing. For each of the basic movements, the time periods of their presence are computed. An index is defined that relates times of sticking to the total time,

$$\rho_3 = (\tau_{IS\,II} + \tau_{IS\,SI} + \tau_{DS\,DD} + \tau_{DS\,SD})/(\tau_{total} - \tau_{SS}), \qquad (7.13)$$

where τ_{xy} denotes the time where the movement is in direction xy. Double subscripts denote a combination of patterns. The index hence denotes phase plots that are often found for sticky valves, see Figure 7.30.

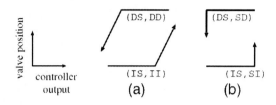

Figure 7.30. Qualitative shapes in the phase plot for stiction cases

Method I

A fitting of oscillations of the control error is performed by changing the three parameters of a first-order plus dead-time (FOPTD) process (K, θ, τ). This causes changes in the shape of the signal, allowing the best fit to data in order to minimize the prediction error (E_R).

The fitting of PV data is also performed by means of sinusoidal functions (parameters: A, ω, φ) and by means of triangular waves (also three parameters, related to slope, scaling factor and constant of the linear equation), with an error equal to E_S and E_T, respectively. Sinusoidal waves indicate the presence of periodic disturbances. Triangular waves are associated with the presence of stiction: the minimum error between relay and triangular waves is chosen to represent the approximation with "stiction" primitives

$$E_{RT} = \min(E_R, E_T).\qquad(7.14)$$

The fitting is performed on each half cycle of oscillation and the average error is computed for two case of stiction and sinusoids, respectively $E_{RT,m}$, $E_{S,m}$.

On this basis a stiction index is defined for all the examined half cycle of oscillations, as

$$S_I = \frac{E_{S,m} - E_{RT,m}}{E_{S,m} + E_{RT,m}}.\qquad(7.15)$$

The stiction index varies in the range $[-1,1]$, with positive values indicating stiction, while negative values indicate sinusoids. An uncertainty range, where no clear indication are given, has been assumed, for $|S_I| < 0.21$. Computation of the stiction index is carried out on positive (S_{I+}) and negative (S_{I-}) half cycles, in order to detect the presence of asymmetrical stiction.

Summary of results

The results of the application of all methods to the described benchmark data are summarized in **Table 7.6**.

Table 7.6. Results of the evaluation of the stiction benchmark data

	FC525	LC011	FC445	FC392	FC145	QC193	LC621
Type	Flow	Level	Flow	Flow	Flow	Conc.	Level
Stiction?	yes	yes	yes	yes	yes	no	no
A	yes	n.a.	yes	yes	yes	no	n.a.
B	n.a.	yes	n.a.	n.a.	n.a.	n.a.	no
C	yes	no	yes	yes	yes	no	no
D	yes	yes	yes	yes	yes	(yes)	(yes)
E	yes	yes	yes	no	yes	(no)	no
F	yes	yes	yes	yes	yes	yes	no
G1	yes	yes	no	no	yes	no	yes
G2							
H1	no	no	no	no	no	no	no
H2	no	no	no	no	no	yes	no
H3	yes	no	no	no	no	yes	no
I	yes	yes	?	yes	yes	yes	no

Key: For dark grey entries, no results were available. For light grey entries, the results differed from the correct reason. Parenthesis denote that the result is not incorrect but delivers different information. The term *n.a.* denotes cases where a certain method must not be used.

Conclusions from benchmark evaluations

From the above comparison, the following conclusions can be made:

The available methods rely on different assumptions.

All methods have their strengths and application areas.

There is not one method that can cover all cases reliably.

A special challenge is to decide when an oscillation is not due to stiction.

Shape-based methods are reliable for 'typical' cases mainly

Non-stationary data are a challenge

7.5.2 External Oscillations

Most industrial measurements are corrupted by disturbances. In many cases, oscillations that are generated by stiction (for example) will propagate through the plant by energy, signal or mass transport mechanisms. The distinction of an actuator problem from an external disturbance is in many cases not simple to perform. If the actuator problem can be reliably detected, external causes can usually be excluded. The case where both are present will typically be very difficult to handle and there is no 'correct' answer in such a case. A detection method will most likely identify the more prominent aspect of an oscillation.

If the final control element is not responsible for an oscillation, the question remains whether an oscillation was generated by the controller or is external. This problem has not been solved satisfactorily and solutions for this kind of problem seem to be difficult to find. The academic research activity in the area is still low.

From the linear control theory point of view, it can be shown in (noise-free) simulation and analytically that external (sinusoidal) disturbances and fast controller tuning (without nonlinearity involved) lead to identical PV and OP signals in the control loop [Horch, 2000]. This means that the automated discrimination between the two cases is left to the information that is carried by disturbances of other frequencies and / or noise properties.

There are several ways to solve this dilemma:

- Include knowledge about the process transfer function. If the transfer function is known, then it is possible to compute the ultimate oscillation frequency of the loop and compare to the current frequency. If they are different, it can be concluded that the oscillation is externally generated. This, however, depends heavily on the reliability of the available transfer function.
- An approach used in [Owen, 1997] could be applied where the oscillation is removed from the signal (either in the time or frequency domain) and the remaining signal is benchmarked against minimum-variance. See also the discussion of this reference in Section 7.4 of this chapter.
- It has been noted that a control loop at the stability boundary has a rather strong variation in the oscillation amplitude, see Figure 7.31. This property could be exploited to distinguish internal from external oscillations. It is, however, not a simple task.

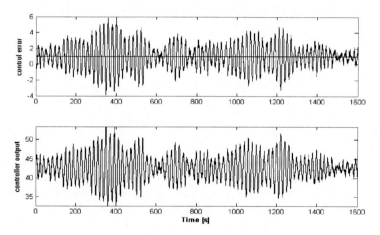

Figure 7.31. Example of a loop that is tuned at the stability boundary

7.6 Conclusions

The detection of oscillations in process control loops is a solved problem. The challenge is to reliably diagnose a detected oscillation without human interaction. For the important problem of stick-slip behaviour in control loops, most of the currently available methods have been reviewed and exemplified on some examples.

It became clear that assumptions and results need to be handled with care, especially when automatic diagnosis is targeted. All methods have their special strengths and weaknesses and it is necessary to know these when applying them to real-world data.

What remains to be done in the future?
There are still very interesting problems to solve, some of which have been mentioned above: the discriminatory diagnosis of problems that are related to linear control theory (e.g. tight tuning vs. external oscillation vs. linear coupling).

Secondly, even though there *are* many methods for stiction detection available, no exhaustive comparison has been made. As stated before, the overview in this chapter is for illustration and general understanding only. A thorough comparison would need to involve many different data sets.

Thirdly, control performance monitoring research has proposed a large number of measures and indices. The future challenge is to find a mapping from the plethora of indices to understandable diagnoses that help the plant technician to maintain acceptable plant performance.

7.7 Acknowledgments

The author acknowledges the friendly and very spontaneous enthusiasm and agreement of the authors of the reported stiction detection methods to test some benchmark data as a contribution of this chapter.

The help of the following people contributed considerably to this comparative overview: Claudio Rossi, Fabio Ulivari, Timothy Salsbury, Ashish Singhal, Nina Thornhill, Rahgunathan Rengaswamy, Manabu Kano, Hidekazu Kugemoto, Tore Hägglund, Fredrik Gustafsson, Anders Stenman, Sirish Shah, M.A.A. Shoukat Choudhury, Ranganathan Srinivasan, Yoshiyuki Yamashita.

8

Controller Benchmarking Algorithms: Some Technical Issues

Damien Uduehi [1], Andrzej Ordys [2], Hao Xia [3], Martin Bennauer [4], Gerta Zimmer [5] and Sandro Corsi [6]

[1] Strategy and Planning Coordinator (West Africa), BG Group, Reading, UK,
[2] Professor of Automotive Engineering, Faculty of Engineering, Kingston University, London, UK,
[3] Research Fellow, Department of Electronic and Electrical Engineering, University of Strathclyde, Glasgow, UK,
[4] Siemens AG Power Generation, Muelheim an der Ruhr, Germany,
[5] Siemens AG Power Generation, Muelheim an der Ruhr, Germany,
[6] Manager, CESI SpA, B.U. Transmission & Distribution, Milan, Italy.

8.1 Introduction

Earlier chapters of this book describe the derivation of control loop benchmarking theory. Different versions, appropriate for different industrial needs have been presented in Chapters 3 and 4. Based on that theory, it is possible to devise software algorithms to perform benchmarking on real systems, as presented in Chapter 5. In fact, such algorithms have been produced in MATLAB®/Simulink®, for research purposes. If one wants to use the benchmarking in an industrial environment, the algorithms should be extensively tested, to establish their boundaries of applicability. This Chapter discusses the issues related to practical implementation of benchmarking software in performance assessment of industrial processes and provides guidelines on using benchmarking techniques in particular process systems.

Sections 8.2 and 8.3 describe the tests performed on the various theoretical algorithms introduced in Chapters 3 and 4. For the purpose of the tests, difficult situations have been arranged with the difficulties caused either by the nature of the systems considered, the stochastic characteristics of the disturbances or the extreme settings of the benchmarking algorithms themselves. This provides useful information on the applicability and robustness of the algorithms. A separate issue is the implementation of the algorithms in a real-time industrial control environment, which has not been tested here.

To facilitate the tests, a group of reference plants and noise models that cover different extreme conditions were utilised, with PID controllers, designed - for each plant - using Ziegler Nichols tuning rules, implemented to serve as nominal controllers. During the validation tests, the user defined parameters (e.g. auto-regressive length, time-delay, design weightings) were varied to fully exercise the benchmarking algorithms. The plant, noise models and benchmark parameter set used for the validation exercise were chosen to test the boundaries of the benchmark theory, and not the coding of the benchmark algorithms in software. The validation is achieved through the comparison of the theoretical benchmark and the estimated benchmark. Because the simulation used in the validation contains signals which have stochastic properties, calculating variances using only one data ensemble introduces a limit on the accuracy of the results. Furthermore the solution for the theoretical benchmarks, sometimes involve complex and high order mathematical operations. The strict accuracy of the results depends on the ability of the tools used, in this case MATLAB®, to accurately perform these calculations. The theoretical benchmark computed from plant transfer functions should be comparable within a tolerance of 5% to the benchmark estimated from plant data-sets using the benchmark algorithms. When the results are within this tolerance, the MATLAB® benchmark algorithm is deemed be conforming to theory.

Finally, in Section 8.4, this chapter also presents the results of an industrial case study performed on the control system of the SIEMENS steam turbine. The challenges of this particular application are discussed and some conclusions drawn as to applicability and the need for the future extension of benchmarking algorithms.

8.2 Validating the SISO Benchmarking Algorithms

For the SISO validation testing for each plant/noise combination, the RS-LQG, MV and GMV controller were designed for different variations of the benchmark test parameters. Simulations were then conducted for each case for four controllers, namely: nominal PID, RS-LQG, GMV and MV, and process data comprising plant output, controller output and process error was collected and saved. Using collected data, the actual variances and benchmark cost were then calculated using the developed algorithms and compared against results calculated theoretically.

8.2.1 Plant Models

The plant and noise models are assumed to fit the following generalised description:

$$y_k = z^{-k} \frac{B_0(z^{-1})}{A_0(z^{-1})} u_k + \frac{C_1(z^{-1})}{A_1(z^{-1})} \xi_k \qquad (8.1)$$

where $B_0(z^{-1})$ and $A_0(z^{-1})$ are the polynomial numerator and denominator of the plant transfer function , while $C_1(z^{-1})$ and $A_1(z^{-1})$ are the stochastic noise numerator and denominator polynomials. ξ_k is assumed to be white noise with zero mean and unit variance. Plants 1 to 6 are designed to be variations of a first order system with time delay. The plant models are defined in **Table 8.1** as follows:

Table 8.1. Test system characteristics

No.	Time Delay	Poles	Zeros	Transfer Function
Plant 1	1	$z = 0.67$	none	$z^{-1} \dfrac{0.33}{1-0.67z^{-1}}$
Plant 2	8	$z = 0.67$	none	$z^{-8} \dfrac{0.33}{1-0.67z^{-1}}$
Plant 3	1	$z = -0.67$	none	$z^{-1} \dfrac{0.33}{1+0.67z^{-1}}$
Plant 4	1	$z = 1.67$	none	$z^{-1} \dfrac{0.33}{1-1.67z^{-1}}$
Plant 5	1	$z = 1.67$	$z=0.75$	$z^{-1} \dfrac{1.33-z^{-1}}{1-0.67z^{-1}}$
Plant 6	1	$z = -1.67$	$z=1.5$	$z^{-1} \dfrac{-1.33+2z^{-1}}{1+0.67z^{-1}}$
Plant 7	2	$z_1 = 0.408 + j0.5315$ $z_2 = 0.408 - j0.5315$	$z=0.6861$	$z^{-2} \dfrac{0.36+0.247z^{-1}}{1+0.816z^{-1}+0.449z^{-2}}$
Plant 8	1	$z_1 = 1$ $z_2 = 1$	none	$z^{-1} \dfrac{1}{1-2z^{-1}+z^{-2}}$

   Plant 1 represents a simple "base" case and it was expected that for well-defined user parameters of the benchmarking algorithms and for "normal" disturbances, the theoretical and simulation results should be very close for all the benchmarking algorithms. Plant 2, was designed to test the effect of time delays, with plant time delay greater than the plant rise time. Comparing the test results between plant 1 and 2 should highlight some of the effects of time delay on the algorithms and on the theory. The focus in Plant 3 was to test the benchmarking of an oscillatory 1st order plant; a situation that could arise when the sampling time used in the control system is too low (this would not normally arise in a process plant, but could arise in a digital servo controller). Plant 4 was an unstable 1st order plant, while plant 5 was designed to test the effect of minimum phase zeros. Plant 6 has a non-minimum phase zero. Because the benchmarking theory, based on the minimum variance control law, assumes that the plant is minimum phase hence the testing for this plant should highlight the effect of calculating an MV/GMV benchmarking index for a non-minimum phase plant. Plant 7, was an oscillatory underdamped 2nd order system and used to tests the effect of complex poles on the benchmarking algorithm. Finally Plant 8, was a simple double integrator used as an extreme case to test the effect of integrating poles on the benchmark algorithms.

8.2.2 Disturbance Models

From Equation (8.1) the disturbance part of the transfer function is given by:

$$v_k = \frac{C_1}{A_0}\xi_k \tag{8.2}$$

Four different transfer functions are considered for the disturbance channel of the model. Two are ARMA (Auto Regressive Moving Average) models, with different C polynomials. The third has ARIMA (Auto Regressive Integrated Moving Average) structure, while, for completeness, the fourth stochastic disturbance model is implemented as a pure white noise source. Therefore, comparing (8.1) and (8.2) obtains:

ARMA: $A_1 = A_0$, $C_1 = C_{10}$ or $C_1 = C_{11}$ $\tag{8.3}$

ARIMA: $A_1 = \Delta A_0$, $C_1 = C_{12}$ $\tag{8.4}$

White Noise: $A_1 = 1$ and $C_1 = 1$ $\tag{8.5}$

Pairing each undisturbed discrete time plant with the 4 stochastic disturbance models results in 32 stochastic benchmark test processes.

Note: For the unstable plant (Plant 4), it was assumed that for the ARMA case $A_1 = A_x$, and for the ARIMA case $A_1 = \Delta A_x$, where A_x is a stable polynomial obtained by spectral factorization of A_0.

8.2.3 Nominal PID Controller

For each of the different plant noise combinations, the performance of two different controllers were benchmarked. The first controller was a nominal controller, which would be of the PID class and the second controller was an optimal controller. Because the classical tuning of PID controllers does not depend on stochastic disturbance characteristics, one nominal PID was tuned using Ziegler- Nichols tuning rules, for each plant type. Therefore, eight nominal PID controllers were designed and benchmarked for the 32 plant/noise combinations. The nominal controllers for each plant are as follows:

1) $PI_{controller} = \dfrac{3.4 - 1.5z^{-1}}{1 - z^{-1}}$ $\tag{8.6}$

2) $PI_{controller} = \dfrac{0.08338 - 0.7555z^{-1}}{1 - z^{-1}}$ $\tag{8.7}$

3) $PI_{controller} = \dfrac{0.6652 - 0.4358z^{-1}}{1 - z^{-1}}$ $\tag{8.8}$

4) $PI_{controller} = \dfrac{9.519 - 4.759z^{-1}}{1 - z^{-1}}$ $\tag{8.9}$

5)   *Filtered PID$_{controller}$* $= \dfrac{0.6746 - 0.6746z^{-1} - 0.1686z^{-2}}{1 - 1.5z^{-1} + 0.5z^{-2}}$ (8.10)

6)   *Filtered PID$_{controller}$* $= \dfrac{0.1297 - 0.2064z^{-1} - 0.08456z^{-2}}{1 - 1.5z^{-1} + 0.5z^{-2}}$ (8.11)

7)   *Filtered PID$_{controller}$* $= \dfrac{0.3039 - 0.4261z^{-1} - 0.1575z^{-2}}{1 - 1.5z^{-1} + 0.5z^{-2}}$ (8.12)

8)   *Filtered PD$_{controller}$* $= \dfrac{0.11 - 0.105z^{-1}}{1 - 0.5z^{-1}}$ (8.13)

8.2.4 User-defined Parameters of Benchmarking Algorithms

The algorithms contain certain parameters that must be specified by the user. The effects of some of these parameters on the accuracy of the algorithm is tested by varying their value within a defined range. The parameters being varied are:

- **Dynamic weights in the cost function**
  The cost functions for GMV or RS-LQG benchmarks contain dynamic (transfer function) weights. Three different sets of weights have been tested, including integrator and static weights.
- **Assumed plant time delay**
  Three different plant time delays were used in the algorithms. One of them was equal to the actual plant time delay whereas the other two correspond to 2x actual plant time delay and 10x actual plant time delay
- **The size of data set**
  The norm is 1000 samples, 200 samples represents a low extreme and 2000 samples is used to test the effect of larger data sets
- **ARMA model length**
  This was varied between 1 and 90.

8.2.5 Test Criteria Summary

For the MV algorithm the optimal controller to be benchmarked was an MV controller specifically designed to minimise the error variance of the plant/noise pair under test. Consequently 32 minimum variance controllers are designed and benchmarked. While for the GMV algorithm the optimal controller to be benchmarked was a GMV controller specifically designed to minimise the error variance of the plant/noise pair under test. The designed GMV Controller was influenced by the error/control weighting, and hence three GMV Controllers resulted from the three different weightings previously defined.

Consequently 96 generalised minimum variance controllers were designed and benchmarked. For the RS-LQG algorithm the optimal controller to be benchmarked was a restricted structure controller specifically designed to minimise the error variance of the plant/noise pair under test. There are four different types of restricted structure controllers, a) P only controller , 2) PI controller 3) PD controller 4) filtered PID controller .

Furthermore the designed RS-LQG Controller was influenced by the error/control weighting, and hence 12 RS-LQG Controllers resulted from the three different weightings previously defined. Consequently 384 RS-LQG controllers are designed and benchmarked. The Benchmark test cases for the MV, GMV and RS-LQG benchmarking algorithm carried out on these data-set/system transfer functions are defined as follows:

GMV / MV Algorithm Validation

Table 8.2. Minimum variance test cases

	Number of Tests	Total No. of Simulations
Plant Model (A_0 and B_0)	8	8
Noise Model (C_1)	4 (ARMA (C0,C1), ARIMA (C2), White)	32
Controller	2 (Nominal PID, MV controller)	64
Sample rate (seconds):	1	1
		Total No. of Benchmarks Test
Delay (seconds)	3 (Actual plant delay, 2 times plant delay, 10 times plant delay)	192
Data size (samples)	3 (200, 1000, 2000)	576
ARMA model length	4(1, 5, 10, 90)	2304

Table 8.3. Generalised minimum variance test cases

	Number of Test	Total No. of Simulations
Plant Model (A_0 and B_0)	8	8
Noise Model (C_1)	4 (ARMA (C0,C1), ARIMA (C2), White)	32
Weightings	3 (IMV, Static, Integrator on error)	96
Controller	2 (Nominal PID, GMV controller)	128 Total no of Benchmark Tests
Delay (seconds)	3 (Actual plant delay, 2 times plant delay, 10 times plant delay)	384
Data size (samples)	3 (200, 1000, 2000)	1152
ARMA model length	4 (1, 5, 10, 20)	4608

For the GMV / MV algorithm validation, the following indicators were checked:

- The estimated benchmark of the GMV/MV Controller test cases should return 1 or a value within 5% of 1.
- The estimated benchmark should always be > 0 and < 1. A negative value of the benchmark should never be returned.

- The benchmark should in general be always no greater than 1, but may be deemed to be 1 if it is within 5% of 1.

Table 8.2 and Table 8.3 show a summary of the test cases for the Minimum Variance and Generalised Minimum Variance validations.

RS-LQG Algorithm Validation

For the RS-LQG algorithm validation, the indicators in Table 8.4 were checked to determine the consistency of the algorithm.

Table **8.4.** RS-LQG validation indicators

	Benchmark (against RS-LQG P-only)	Benchmark (against RS-LQG PI)	Benchmark (against RS-LQG PD)	Benchmark (against RS-LQG PID)
Nominal PID	?? (but ≤ RS-PID benchmark)	?? (but ≤ RS-PID benchmark)	?? (but ≤ RS-PID benchmark)	≤ 1
RS-LQG P-only	1	≤ 1	≤ 1	≤ 1 (and ≤ both RS-PI and RS-PD benchmarks)
RS-LQG PI	≥ 1	1	?? (but ≤ RS-PID benchmark)	≤ 1
RS-LQG PD	≥ 1	?? (but ≤ RS-PID benchmark)	1	≤ 1
RS-LQG PID	≥ 1 (and ≥ both RS-PI and RS- PD benchmarks	≥ 1 (and ≥ both RS-PI and RS- PD benchmarks	≥ 1 (and ≥ both RS-PI and RS- PD benchmarks	1

??: Value cannot be determined *a priori*

Table **8.5.** RS-LQG algorithm cases

	Number of Test	Total No. of Test Simulations
Plant Model (A₀ and B₀)	8	8
Noise Model (C₁)	4 (ARMA (C0,C1), ARIMA (C2, White)	32
Weightings	3 (plant specific, depending on approximate time constants)	96
Controller type	4 (P, PI PD, PID)	384
Controllers	2 (Nominal PID, RS-LQG controller,)	416 Total no of Benchmark Tests
Standard	2 (Nominal RS_LQG)	832
Consistency	4 (P, PI, PD, PID)	2496

Note that the relationship in the Table 8.4 are rational and would only hold for choice of controller type, RS-LQG weightings consistent with the plant/noise model. For the testing and validation of the RS-LQG algorithm, a total of 2496 validation exercises will be conducted. These exercises include the 832 standard benchmarking exercises and 1664 consistency check exercises.

8.2.6 Results : Feasible Applications of SISO Benchmarks

Extensive tests have been carried for the MV, GMV and RS-LQG benchmark modules implemented in the MATLAB® environment based on Least Square algorithms. The implemented MV algorithm was successfully applied to all but one (with non-minimum phase zeros) of the test plants. While the implemented GMV algorithm was successfully applied to all but a few cases where the plant has non-minimum phase zeros and where the combination of the disturbance and also the weightings caused the benchmark signal to have some non-stationary characteristics. For the RS-LQG algorithm there was no plant noise combination for which the algorithm could not be successfully applied.

The choice of weightings in GMV benchmarking must be consistent with the control problem and plant noise combinations. For example, when the disturbance contains an integrator, it might not be advisable to choose error weighting which also contains an integrator. This is because optimal control design when the noise includes an integrator has to include an integrator in the controller hence there is no explicit need to try to force integrating action on the controller by using weightings that contains an integrator as the optimal design in such case might be a controller with two integrators. Further the choice of data length should reflect the characteristic of the disturbance signal and design parameters. That is, if the disturbance or design parameters include integral action, it might be wise to use a larger data set.

Similarly to GMV control, the choice of weightings in RS-LQG benchmarking must be consistent with the control problem and plant noise combinations. For example, if the error weighting contains an integrator, it will be inconsistent to choose the benchmark controller to be P only controller. This is because when an integrator is chosen as the weighing one implicitly specifies the optimal control solution should include integral action and hence the choice of controller must also include an integrator thus have to be PI or PID. Another example is the case where the weighting is a static weighting, the P only controller is chosen as the benchmark controller, but the disturbance includes an integrator. The optimal controller for such a plant/noise combination must implicitly include integral action hence P only controller will not be able to stabilize the process output and is inconsistent with the control problem. Specifically the test results show that the type of noise present in a plant affects the performance of the existing controller (how well the controller performs compared with the optimal) and, in the case of MV and GMV methods, the accuracy of benchmarking result (how close to the theoretical the performance index generated by the corresponding algorithm benchmark algorithm is).

For MV and GMV algorithms:

- Correct loop delay estimation is absolutely essential to achieving accurate and valid benchmarking results.

- The ARMA model length significantly affects the accuracy of the benchmarking results. The test results show that a length of 10 is adequate for most cases to achieve the balance between benchmarking accuracy and computational load. However there is no absolute general answer as to how large this should be as it is plant-noise model and, in the case of GMV, weighting function dependent.

- In general, increasing data size improves benchmarking accuracy. However, this may not always be true due to the stochastic property of the benchmarking algorithms. The test results show that a data size between 1000 and 2000 is adequate for most cases to achieve the balance between benchmarking accuracy and computational load. Again, there is no absolute general answer as to how large this should be as it is plant-noise model and, in the case of GMV, weighting function dependent.

- It is, therefore, recommended that, in practical application, a combination of different data sizes and ARM model lengths should be tried on a plant (until the variations of the resulting indices are small) to achieve accurate benchmarking result.

-

It was found that the most common cause for the algorithms to return inaccurate or invalid benchmark result were due to:

- *Coloured process output*: The benchmark result of a process output that is more coloured could be less accurate than that which is white.

- *Using ARMA model length larger than required*: Generally, a larger value of the ARMA model length produces a more accurate benchmark result. However, when the ARMA model length used is much larger than what is required to adequately represent the associated closed loop dynamic response, there is a potential for the calculation of the estimated MV, or GMV, to be inaccurate. The reason is that the values of the extra terms might be inaccurate as the actual values might lie well below the Least Square estimation error tolerance.

- Variations in noise characteristics.

- Non-stationary characteristics of benchmark signal.

- Inappropriate choices of GMV or RS-LQG weightings.

Table 8.6 summarises the applicability of the benchmarking methods to the plants that had been tested.

For plants with non-minimum phase zeros, there is considerable amount of academic debates on whether the standard MV control theory is applicable since the MV controller will be unstable. Some argue that although the controller is unstable, the process output will have a finite variance hence a minimum value associated with the variance. When applied to processes with non-minimum phase characteristics, minimum variance and the classical generalised minimum variance theory have fundamental limitations. The GMV theory used in the benchmarking algorithm can overcome these limitations through appropriate choice of weightings. However, the weightings used in these test cases, i.e. static, integrator and IMV, are not appropriate as they do not have the dynamic properties required to make the problem solvable.

Table 8.6. SISO benchmarking algorithms applicability

Plant	Type	MV	GMV	RS-LQG
1	1st order plant, 1 sample dead time	✓	✓[3]	✓
2	1st order plant, 8 sample dead time	✓	✓	✓
3	1st order plant, 1 sample dead time	✓	✓	✓
4	1st order plant, 1 sample dead time & unstable pole	✓	✓[3]	✓
5	1st order plant, 1 sample dead time & minimum phase zero	✓	✓	✓
6	1st order plant, 1 sample dead time & non-minimum phase zero	✗	✗	✓
7	2nd order, under damped & 2 sample dead time	✓	✓[4]	✓
8	A simple double integrator, 1 sample dead-time	✓	✓	✓

8.3 Validating the MIMO Benchmarking Algorithm

In tests conducted on the feasible limits of the benchmarking algorithms using predictive control theory, the MIMO linear quadratic Gaussian predictive control (LQGPC) and generalised predictive control (GPC) benchmarking algorithms were considered together the since GPC algorithm is a special case of the LQGPC control law with the summation in the performance index limited to one step. Similarly, the Minimum Variance controller can be considered as a special case of GPC controller: with the prediction of output limited to one value and control penalty reduced to zero. The algorithms were tested by calculating the performance of four reference plants and noise models that cover different extreme conditions. For each of the plant-noise model pairs, where multi-loop controllers were implemented as the nominal controllers , the Ziegler Nichols tuning method was used to obtain the parameters for individual SISO controllers which were then empirically adjusted and used in a multi-loop control structure. For the multivariable controllers, decouplers designed to decouple inter-loop dynamics

---

[3] Except for the cases where the combination of disturbance and weightings caused the benchmark signal to have some non-stationary characteristics.

[4] Except for the case with the combination of ARIMA C2 noise model, GMV controller and static weighting. It appears that, for this plant noise combination, the algorithm has not accurately estimated the minimum cost.

were added to SISO controllers designed using the Ziegler-Nichols method to produce a full multivariable control structure.

MATLAB®/ Simulink® simulations were conducted for each case for the three controllers (nominal controller, LQGPC and GPC) and process data comprising plant output, controller output and process error will be collected and saved. Using the collected data, the actual variances and benchmark cost were calculated using the developed algorithm and then compared against those calculated theoretically. The plant models used for the validation exercise were based on approximate models of real industrial processes. Since the benchmarking algorithms require that the system be represented in discrete time state-space form, the transfer functions were discretised and converted to state-space representation. In the modelling the process it was assumed that the output signals were subjected to measurement noise with unit variance and where the noise is assumed to be uncorrelated to other disturbances. The plant and noise models are assumed to fit the following generalised description in transfer function and state space representations:

$$y_k = G(z^{-1})u_k + H(z^{-1})\xi_k \tag{8.14}$$

where $G(z^{-1})$ and $H(z^{-1})$ are the process and disturbance transfer function matrices, u_k is a vector of plant inputs and ξ_k is a vector of mutually independent zero-mean white noise sources of unit covariance.

$$x_{t+1} = A \cdot x_t + B \cdot u_t + G \cdot w_t \tag{8.15}$$

$$y_t = D \cdot x_t + v_t \tag{8.16}$$

where: x_t is a vector of system states of size n_x, u_t is a vector of control signals of size n_u, y_t is a vector of output signals of size n_y, v_t and w_t are vectors of disturbances, assumed to be Gaussian white noises with zero mean value and unit covariance matrix . A, B, G, D are constant matrices. The plant models are defined in Table 8.7.

Table 8.7. MIMO benchmarking test plants

	Plant Transfer function	Disturbance Transfer function
Plant 1	$$y(s) = \frac{0.001432s + 0.001432}{0.56s^4 + 2.12s^3 + 2.56s^2 + s}\, e^{-90s} u(s)$$	$$v(s) = \frac{1}{s+0.51}\,\xi(s)$$
Plant 2	$$\begin{bmatrix} y_1(s) \\ y_2(s) \end{bmatrix} = \begin{bmatrix} \dfrac{12.8e^{-s}}{16.7s+1} & \dfrac{-18.9e^{-3s}}{21s+1} \\[2ex] \dfrac{6.6e^{-7s}}{10.9s+1} & \dfrac{-19.4e^{-3s}}{14.4s+1} \end{bmatrix} \begin{bmatrix} u_1(s) \\ u_2(s) \end{bmatrix}$$	$$\begin{bmatrix} v_1(s) \\ v_2(s) \end{bmatrix} = \begin{bmatrix} \dfrac{1}{s+0.3} & 0 \\[2ex] 0 & \dfrac{1}{s+0.5} \end{bmatrix}\begin{bmatrix} \xi_1(s) \\ \xi_2(s) \end{bmatrix}$$
Plant 3	$$\begin{bmatrix} y_1(s) \\ y_2(s) \end{bmatrix} = \begin{bmatrix} \dfrac{0.1236s+.2472}{s-.0219}e^{-1s} & 0 \\[2ex] \dfrac{s+1}{s+.02}e^{-3s} & \dfrac{.04989s^2+.6358s+2.072}{s^2+.9731s+.218}e^{-2s} \end{bmatrix} \begin{bmatrix} u_1(s) \\ u_2(s) \end{bmatrix}$$	$$\begin{bmatrix} v_1(s) \\ v_2(s) \end{bmatrix} = \begin{bmatrix} \dfrac{1}{s+.03} & \dfrac{1}{s} \\[2ex] \dfrac{1}{s+.1} & \dfrac{1}{s+.05} \end{bmatrix}\begin{bmatrix} \xi_1(s) \\ \xi_2(s) \end{bmatrix}$$
Plant 4	$$\begin{bmatrix} y_1(s) \\ y_2(s) \end{bmatrix} = \begin{bmatrix} \dfrac{0.0407}{s+3.945\times10^{-6}} & \dfrac{-0.02931}{s^2+7.227s+2.851\times10^{-5}} & 0 \\[3ex] \dfrac{1.22\times10^{-6}}{s^3+7.227s^2+8.837\times10^{-5}s+2.361\times10^{-10}} & \dfrac{0.03083}{s^2+7.227s+5.986\times10^{-5}} & \dfrac{-0.4278s-21.47}{s^7+61.98s^6+379.2s^5+1171s^4+3107s^3+2840s^2+3019s+0.025} \end{bmatrix} \begin{bmatrix} u_1(s) \\ u_2(s) \\ u_3(s) \end{bmatrix}$$	

Plant 1, was a 1×1, 4th order plant with 90 seconds delay derived from the level control of a distillation column. A cascade control structure was utilised on the plant with the distillation level loop as the primary loop and the flow loop as the secondary/slave loop. This plant represents a simple "base" case used to test the benchmark algorithm's handling of SISO systems, and systems with long time delays. The stochastic disturbance acting on the level of the distillation column was assumed to be a Gaussian white noise driving a low pass colouring filter.

Plant 2 represents a 2×2 model of a pilot-scale binary distillation column used for methanol-water separation. The original model has been utilised in several controller performance studies [Maurath et al., 1988; Rahul and Cooper, 1997]. The plant was used to test the benchmark algorithm for MIMO systems with interaction and time delays between loops. The stochastic disturbance acting in the two loops of the binary distillation column was assumed to be uncorrelated and expressed as two mutually independent Gaussian white noise sources driving two low-pass colouring filters.

Plant 3 was a fictitious academic 2×2 model. It was chosen to test the benchmarking of MIMO systems with unstable poles and one-way interaction between loops. The disturbances were assumed to be correlated and are expressed as two mutually independent Gaussian white noise sources driving two low-pass colouring filters.

Plant 4 represents a 2×3 model of a two tank separator train used to separate gas, crude oil and water. This plant was chosen to analyse the benchmarking of MIMO systems with unequal dimensions and hence test the validity of the algorithm where there are more inputs than plant outputs.

The stochastic disturbance acting in the two loops of the separation process were assumed to be uncorrelated and expressed as a Gaussian white noise sources driving two colouring filters.

The reference was assumed to be a square wave pulse of unit magnitude as in Figure 8.1. The pulse is chosen to be long enough so that the closed loop interactions, due to set point changes, are clearly visible. This was sampled with a sampling time of 1 second. For MIMO systems, the step changes in the square waveform were arranged to occur at different intervals to highlight the interactions between loops when the outputs are observed.

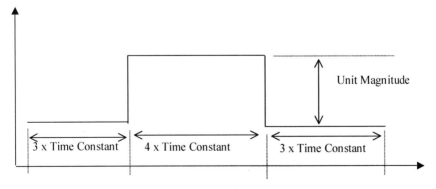

Figure 8.1. MIMO benchmarking set point input reference signal

For each of the different plant/noise model combinations, the performance of three controllers were benchmarked. The first controller is the nominal controller, which was a PID class controller, while the second and third are the corresponding optimal GPC and LQGPC controllers. The nominal controllers for each plant are defined as follows:

Plant 1: PI Controller

$$u(s) = \frac{1738.1s + 1.02}{1704s} e(s) \tag{8.17}$$

Plant 2: Multi-loop PI Controller with Decouplers

$$u(s) = \begin{bmatrix} \dfrac{(0.205s^3+0.061s^2+5.42\times10^3s+1.41\times10^4)}{(s^3+0.1293s^2+4.15\times10^3s+1.08\times10^{15})} & \dfrac{(-0.1492s^2-2.02\times10^3s+6.04\times10^4)}{(s^2+6.94\times10^3s+4.37\times10^{13})} \\[3mm] \dfrac{(4.6\times10^2s^2+9.49\times10^3s+3.47\times10^4)}{(s^2+5.98\times10^2s+1.1\times10^{13})} & \dfrac{(-0.1815s^3-0.033s^2-1.9\times10^3s-3.4\times10^5)}{(s^3+0.1293s^2+4.15\times10^3s+1.08\times10^{15})} \end{bmatrix} e(s) \tag{8.18}$$

Plant 3: Multi-loop PID Controller

$$u(s) = \begin{bmatrix} \dfrac{7.139s + 4.759}{s} & 0 \\[3mm] 0 & \dfrac{1.17s^2 + 0.149s + 0.03137}{s^2 + 2s} \end{bmatrix} e(s) \tag{8.19}$$

Plant 4: Multi-loop PI Controller

$$u(s) = \begin{bmatrix} \dfrac{0.1743s + 0.03}{s} & 0 \\[3mm] 0 & \dfrac{-0.1639s - 0.0074}{s} \end{bmatrix} e(s) \tag{8.20}$$

The optimal controllers to be benchmarked were full order LQGPC and GPC controllers designed with user supplied benchmark parameters. Each variation of the user parameters will result in new LQGPC and GPC controller designs.

8.3.1 User-defined Parameters for Benchmarking

The benchmark algorithms contain parameters that must be defined by the user. The effects of these parameters on the accuracy of the algorithm was tested by varying their value of a defined range.

Cost Function Adjustments
When the nominal controller to be benchmarked included integral action, the benchmark cost was modified to use the mean square of future process inputs deviation:

$$Benchmark\ Cost = E_{wms}(t) + \Delta U_{wms}(t) \qquad (8.21)$$

Static Weights

The effect of the process error and control weightings on the accuracy and stability of the benchmarking algorithm was tested by varying the choice of weightings. For the error weighting ($\breve{Q}_e$), two different configurations were used. For the control weighting ($\breve{Q}_u$), four different configurations were used. For the validation test, the chosen error and control variations are representative of the general characteristic of possible LQGPC error weightings and control weightings. These are normally chosen to scale the process errors and control efforts to the same magnitude and also to penalise the control, which reflect the level of emphasis in minimisation of its mean square value. For benchmarking MIMO systems, the weights can also be used to assign relative priority to each input or output. For the validation exercise, it was assumed that all inputs and outputs are normalised to unity. For the error weighting $\breve{Q}_e$ the two choices were defined as follows:

1. Nominal Error Weighting $\breve{Q}_e = \begin{bmatrix} 1 & 0 & 0 \\ \vdots & \ddots & \vdots \\ 0 & 0 & 1 \end{bmatrix}$

The above weighting assumes all process errors are of equal importance and minimises each individual process error mean square $e_i(s) \times e_i(s)$ and not the cross products $e_i(s) \times e_j(s)$.

2. Cross Error Weighting $\breve{Q}_e = \begin{bmatrix} 1 & 1 & 1 \\ \vdots & \ddots & \vdots \\ 1 & 1 & 1 \end{bmatrix}$

This weighting represents the limiting case and attaches equal importance to each individual process error mean square $e_i(s) \times e_i(s)$ as well as the cross products $e_i(s) \times e_j(s)$.

For the control weighting $\breve{Q}_u$ the choices were defined as follows:

1. MV Control Weighting $\breve{Q}_u = \begin{bmatrix} 0 & 0 & 0 \\ \vdots & \ddots & \vdots \\ 0 & 0 & 0 \end{bmatrix}$

This weighting does not penalise any control action and, for certain choices of prediction (single step prediction with $N_1 = N_2 =$ time delay) and control horizons ($N_u = 1$), results in the optimal controller design problem being equivalent to the multivariable minimum variance problem.

2. Nominal Control Weighting $\breve{Q}_u = \begin{bmatrix} 0.1 & 0 & 0 \\ \vdots & \ddots & \vdots \\ 0 & 0 & 0.1 \end{bmatrix}$

Relative to the nominal error weighting, this weighting choice imposes a 10% penalty on control action, i.e., 90% of the optimisation emphasis is on minimising the error. All controls are of equal importance and individual control mean square $u_i(s)\times u_i(s)$ not the cross products $u_i(s)\times u_j(s)$ are minimised.

3.   Extreme Control Weighting $\breve{Q}_u = \begin{bmatrix} 10 & 0 & 0 \\ \vdots & \ddots & \vdots \\ 0 & 0 & 10 \end{bmatrix}$

Relative to the nominal error weighting, this weighting choice imposes a 90% penalty on control action, i.e., 10% of the optimisation emphasis is on minimising the error. All controls are of equal importance and individual control mean square $u_i(s)\times u_i(s)$ not the cross products $u_i(s)\times u_j(s)$ are minimised.

4.   Cross Control Weighting $\breve{Q}_u = \begin{bmatrix} 1 & 1 & 1 \\ \vdots & \ddots & \vdots \\ 1 & 1 & 1 \end{bmatrix}$

The weighting represents the limiting case and attaches equal importance to minimising each individual process error mean square $u_i(s)\times u_i(s)$ as well as the cross products $u_i(s)\times u_j(s)$.

Prediction Horizon (N_2)
The prediction horizon should nominally be chosen to be two thirds of the dominant time constant plus the process time delay.

Control Horizon (N_u)
The control horizon should nominally be a value between 0 and $(N_2 - 1)$. Values close to zero result in responses that are very aggressive, producing large changes in the process outputs N_2 sample intervals before a step change. However values close to N_2 produce a smoother reference tracking. For the validation of the benchmark theory, the algorithm will be fully exercised by using the following values of user parameters:
$N_2 = 1/3$ time delay and N_2 = time delay plus two thirds of the dominant time constant. $N_u = 1$, $N_u = N_2/2$, $N_u = N_2 - 1$

8.3.2 Test Criteria Summary

The LQGPC/GPC algorithm is a theoretical method for calculating cost. It uses plant and controller models only in the cost computation. Since the method is theoretical it should provide the "true value" of the benchmark as opposed to the estimate calculated from simulation data. The case for validity of the algorithm will be strengthened if the following are seen to be true:

- When using the LQGPC benchmark, the estimated benchmark of the LQGPC controller test cases should be 1 or a value within 5% of 1.

- When using the GPC benchmark, the estimated benchmark of the GPC controller test cases should be 1 or a value within 5% of 1.

- For the nominal controller the estimated benchmark should always lie in the interval (0,1). A negative value of the benchmark should never be returned.

- Not much is known in the case of GPC controller vs. LQGPC benchmark and *vice versa*. Given the definition of the LQGPC and GPC control, it is reasonable to expect that in most cases the LQGPC should produce a lower value of both the LQGPC and GPC cost over an extended trajectory. The test results should shed more light on this.

In total 1152 simulation tests, comprising of 576 LQGPC and 576 GPC benchmark tests, were conducted on the 4 plants used in the validation exercise to determine the applicability and boundaries of the developed algorithms. The benchmark test cases for the LQGPC benchmarking algorithm carried out on these data-set/ system transfer functions are defined as follows;

Table 8.8. LQGPC / GPC MIMO benchmarking validation cases

Variable	No. of Tests	Total No. of test Simulations
Plant & Noise Model (A0 and B0)	4	4
Error Weightings	2 (Nominal and Cross)	8
Control Weightings	4 (MV, Nominal, Extreme, Cross)	32
Prediction Horizon	2 (1/3*time-delay and 2/3*time-constant plus delay)	64
Control Horizon	3	192
Reference	1 (Square Wave Pulse)	192
Controllers	3 (Nominal PID, GPC, LQGPC)	576
		Total Number of Benchmark Tests
Benchmark Type	2 (LQGPC, GPC)	**1152**

8.3.3 Results : Feasible Applications of MIMO Benchmarks

For Plant 1, two primary causes of the algorithm failure were identified as:
1. Prediction horizon shorter than the time delay: The control objective was the minimisation of the output variances. Examination of the test parameters indicates that the GPC test cases failed because the output prediction horizon $(N_2 = 30)$ was less than the real time delay in the process. Within this output horizon, because of the time delay, no controls have any influence on the output variance.

2. Control horizon longer than limit imposed by the time delay and prediction horizon when the control objective was the minimisation of the output variances. Examination of the test parameters indicates that the GPC test cases failed because there were more control sequences than required to solve the problem and hence no unique solution existed.

For Plant 2, three primary causes of test failures were identified as follows:
1. Control horizon longer than limit imposed by the time delay and the prediction horizon: Considering the user parameters for the failed test cases, the control objective was the minimisation of the output variances.

$$\text{delay matrix} = \begin{matrix} y_1 \\ y_2 \end{matrix} \Big\} \overset{\overbrace{u_1 \quad u_2}}{\begin{bmatrix} 1 & 3 \\ 7 & 3 \end{bmatrix}} \tag{8.22}$$

For the prediction horizon $N_2 = 3$, the highest output sequence was $y_1(k+3)$ and $y_2(k+3)$. Similarly for the control horizon $N_u = 1$, the optimisation involves the control sequences $u_1(k)$, $u_1(k+1)$, $u_2(k)$ and $u_2(k+1)$. Considering the time delay matrix for Plant 2 given above, notice that the control sequence $u_2(k+1)$ does not have any influence on the output variables being minimised and hence can assume any value, thus there was no unique solution for the optimisation problem. Similar observations can be made for the cases where $N_2 = 3, N_u = 2$ and $N_2 = 14, N_u = 13$.

2. Control horizon exceeds limit and trivial error weighting: The cause of this particular failure is thought to be a combination of two effects. The first was the choice of error weight which introduced an element of redundancy in the values of y_1 and y_2. Notice that with the cross error weighting as previously defined, the aim was effectively to minimise a function expressed as $x = y_1^2 + y_1 y_2 + y_2^2$. Thus if a new variable z is defined as $z = y_1 + y_2$, then observe that $z^2 = x$, the problem was thus minimised for any $y_1 = -y_2$, and the solution is trivial. A further test was run using the same control weighting, error and control horizons but an error weighting of :
$$\begin{bmatrix} 2 & 1 \\ 1 & 1 \end{bmatrix}$$
This results in a non-trivial minimisation problem ($q = z^2 + y_1^2$) and produced a satisfactory result. The additional test has also indicated the possibility that, because the control objective is the minimisation of both the output variances and covariances, the time delays in the interaction channels become factors in determining the length of the control horizon.

3. Prediction horizon - finite horizon too short for reach stabilising solution: Analysis showed that the algorithm did produce an optimal GPC controller.

However the controller was not a stabilising controller. Because the GPC solution was obtained over a finite horizon, it was possible and indeed it happened in the failed case that, for some choices of parameters (which reflected a given control objective) and depending on the dynamic characteristic of the plant, a solution that minimised the problem in the finite time was not stable in infinite time. This observation was confirmed by the fact that the LQGPC algorithm, which computes a solution over infinite time, stabilised the same system.

For Plant 3, the failed test cases, were due to the following four reasons:
1. Prediction horizon shorter than time delay: Considering the user parameters for failed test cases the control objective was the minimisation of the output variances.

$$
\text{delay matrix} = \left.\begin{matrix} y_1 \\ y_2 \end{matrix}\right\} \overbrace{\begin{bmatrix} 1 & n/a \\ 3 & 2 \end{bmatrix}}^{\displaystyle u_1 \quad u_2} \tag{8.23}
$$

For the prediction horizon $N_2 = 1$, the highest output sequence is $y_1(k+1)$ and $y_2(k+1)$. Similarly for the control horizon $N_u = 0$, the optimisation involves the control sequence $u_1(k)$ and $u_2(k)$. Given the above time delay matrix for Plant 3, notice that none of the two inputs has any influence on the output variable $y_2(k+1)$, thus the control objective cannot be achieved. Similar observations can be made for the other cases listed above.

2. Control horizon longer than limit imposed by time delay and prediction horizon: Considering the user parameters for failed test cases, the control objective was the minimisation of the output variances. For the prediction horizon $N_2 = 10$, the highest output sequence was $y_1(k+10)$ and $y_2(k+10)$. Similarly for the control horizon $N_u = 9$, the optimisation involves the control sequences $u_1(k+10)$ and $[u_2(k+9), u_2(k+10)]$. Given the above time delay matrix for Plant 3, these control sequences do not have any influence on the output variables being minimised and hence can assume any value, thus there is no unique solution for the optimisation problem.

3. Trivial control weighting and prediction horizon too short: The cause of failure was thought to be a combination of two factors. The first was the redundancy introduced because the prediction horizon $N_2 = 1$ with the effect that the control sequences $u_1(k+1)$, $u_2(k+1)$ are redundant. The second factor was introduced because the optimisation objective includes the minimisation of both the output variances and covariances of the input. The control weightings chosen introduced a double redundancy in the values of u_1 and u_2 since, as

previously stated, the effect of the weighting was the minimisation of the variance of an arbitrary variable $z = u_1 + u_2$, which has a trivial solution of $u_1 = -u_2$.

4. Prediction horizon too short for stabilising solution: Analysis of the failed GPC test cases shows that the algorithm did produce an optimal GPC controller. However the controller was not a stabilising controller. When the GPC algorithm was further tested with an increased prediction horizon, the results were satisfactory, when coupled with the fact that the LQGPC algorithm which should compute a solution over infinite time passed its own test, then this observation would appear to be validated. Also it is interesting to note that for the prediction horizons that caused the GPC test to fail, when the error weighting was changed from the initial cross weighting value of

$\begin{bmatrix} 1 & 1 \\ 1 & 1 \end{bmatrix}$ to $\begin{bmatrix} 2 & 1 \\ 1 & 1 \end{bmatrix}$, the GPC algorithm passed all the tests.

For Plant 4 two primary causes of failures were identified:
1. Control horizon longer than limit imposed by time delay and prediction horizon: Considering the user parameters for the failed test cases, the control objective was the minimisation of the output variances.

$$\text{delay matrix} = \begin{matrix} y_1 \\ y_2 \end{matrix} \Bigg\} \overbrace{\begin{bmatrix} 1 & 1 & n/a \\ 1 & 1 & 1 \end{bmatrix}}^{u_1 \quad u_2 \quad u_3} \tag{8.24}$$

For a prediction horizon $N_2 = 1$, the highest output sequence was $y_1(k+1)$ and $y_2(k+1)$. Similarly for a control horizon $N_u = 1$, the optimisation involves the control sequences, $u_1(k+1)$ and $u_2(k+1)$. Considering the above time delay matrix, notice that the control sequences do not have any influence on the output variables being minimised and hence can assume any value, consequently there is no unique solution for the optimisation problem. A similar observation can be made for the above cases when $N_2 = 8$, $N_u = 7$.

2. Trivial control weighting and prediction horizon too short : The cause of failure can be attributed to the combination of two factors, the first one of which was the redundancy introduced because the prediction horizon $N_2 = 1$ with the effect that the control sequences $u_1(k+1)$, $u_2(k+1)$ are redundant. The second factor was introduced because the optimisation objective includes the minimisation of both the output variances and covariances of the input. The chosen control weightings introduce a double redundancy in the values of u_1 and u_2 since as previously stated the effect of the weighting was the

minimisation of the variance of an arbitrary variable $z = u_1 + u_2$, which has a trivial solution of $u_1 = -u_2$.

From the analysis of the failed test cases, some general conclusions on the LQGPC/GPC benchmarks as well as guidelines for choosing benchmark parameters can be deduced:

1.   The LQGPC benchmark algorithm appears to be more robust (failed on fewer test cases) and produces more stable solutions than the GPC algorithm. However it appears that the robustness (ability to produce or find a stabilising solution) of the GPC algorithm can be improved by choosing longer output prediction horizons.

2.   Although the LQGPC benchmark algorithm was more robust than the GPC case, it was more computationally intensive and takes longer to calculate a result.

3.   For both the LQGPC and GPC algorithms, it appears that the best choice of an output prediction horizon (N_2) is one that is much larger than the biggest known or estimated time delay in the process.

4.   For both the LQGPC and GPC algorithms, it appears that the most robust and safest choice of control horizon is $N_u = 0$. This only optimises for the control sequence $u(k)$ and works in most cases, especially with very long output prediction horizons. The choice is also very practical since in reality only this sequence in any real predictive controller will be applied to the process.

8.4 Industrial Case Study: Implementation Testing Using Siemens Gas Turbine Model

The case study was carried out on the plant model of a coal fired power plant developed by SIEMENS. A coal fired power plant as shown in Figure 8.2 serves to generate electric power on the grid. In such a plant, coal is transported from ground stock, dried and milled to form a pulverised fuel which is transported through pipes to the burners by a heated air stream. It is then blown into the boiler furnace and combusted. The heat released is absorbed in the water cooled furnace walls in which the majority of the steam is generated. The steam generated in the furnace is then superheated in further stages of heat exchanger tubing before being fed to the turbine. After expansion in the High Pressure (HP) turbine stage, the steam is returned to the boiler for reheating before the final expansion in the Intermediate Pressure (IP) and Low Pressure (LP) turbine stages.

After condensation in a water or air cooled condenser, the condensate is pumped back to the boiler via a series of regenerative feed heaters which are fed by steam tappings from the main turbine.

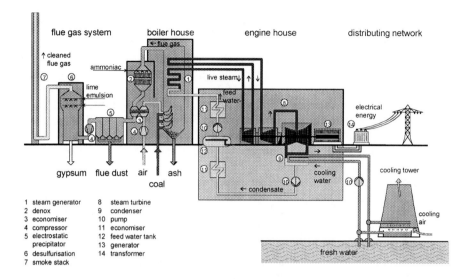

Figure 8.2. Coal fired power plant

In this case study, the focus was on the steam turbine and especially on its control system. The steam turbine control system is divided into three subsystems: speed system, load system and by-pass system (see Figure 8.3). The objectives of the speed control system are:

- to control speed during start-up and synchronisation operation,
- to take part in frequency stabilisation by parallel operation,
- to secure over-speed at load rejections

The load control system enables:

- load control during steady-state and transient operations,
- an interaction with speed controller to avoid over-speed,
- the power plant to be kept in stand-by during grid failure for fast return to electrical grid.

Turbine Steam Bypass System offers improvements in areas such as:

- starting and loading characteristics,
- independent boiler/turbine operations,
- solid particle erosion
- system stability performance

8.4.1 Siemens Power Plant Simulator

The simulator is implemented in MATLAB® version 5.3 with Simulink® toolbox and has been validated by SIEMENS with real plant data. The SIEMENS simulator includes the different models of the boiler, the turbine, the generator and a reduced model of the electrical grid. All the control systems of these components are provided. It is important to note that the version of SIEMENS simulator used for this analysis is a reduced power plant model since it only includes a hydrodynamic description of plant dynamics; the thermal dynamics are not taken into account. SIEMENS provided the simulator in three different configurations corresponding to three operating modes:

- the Speed Run-up configuration, which corresponds to the start of turbine from 0 to nominal speed (3000 rpm),
- the Load Run-up configuration, which corresponds to the loading transient from 0% to 100% of nominal power,
- the Load Rejection configuration, which corresponds to the rapid rejection of load, up to 100% of nominal power, on the electric generator.

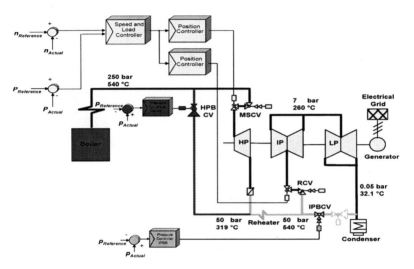

Figure 8.3. Schematic diagram of steam turbine process for coal fired power plant

8.4.2 Operating Conditions

The turbine unit is designed to work in different operating cycles, which are divided into:

- cold starts (steam turbine rotor temperature below 150°C),
- warm starts (e.g., after a weekend outage of 48 h duration),
- hot starts (e.g., after a night outage of 8 h duration),
- very hot starts (e.g., after a short outage of 2 h duration).

Since SIEMENS Power Generation is interested in the dynamic performance of the steam turbine power plant, the tests selected for this industrial study have been chosen in order to reflect the unit hydrodynamic behaviour along large transients: speed run-up, loading, and load rejection. And the controllers to be assessed are speed and load controllers. The turbine steam by-pass system is not studied.

As the SIEMENS simulator used to carry out this study does not take into account the thermal part of the plant, the SIEMENS case studies are only meaningful for "hot" and "very hot starts" cycles.

8.4.3 The Characteristics of the Load/speed Controller

For the SIEMENS power plant, a combined speed and load controller (SLC) regulates the load and speed. The controller is illustrated in Figure 8.4. It can be observed that there is only one output from the SLC which acts on the two inputs (load and speed errors) simultaneously. This output is the sum of two control signals:

$$U_{SLC}(s) = U_n(s) + U_p(s)$$

(8.25)

with the following definitions:
$U_n(s)$ = Speed Controller output
$U_p(s)$ = Load Controller output.

$$U_n(s) = K_1 \left(R_s(s) - Y_s(s) \right) + \frac{K_2}{Ts} \left(R_s(s) - Y_s(s) \right)$$

(8.26)

$$U_p(s) = K \left(R_l(s) \right) + \frac{1}{Ts} \left(R_l(s) - Y_l(s) \right)$$

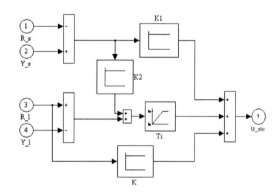

Figure 8.4. Schematic diagram of steam turbine controller

Thus the output from the speed and load controller driving the steam valve can be expressed as following:

$$U_{SLC}(s) = K \left(R_l(s) \right) + K_1 \left(R_n(s) - Y_n(s) \right) + \frac{1}{T_i s} [\left(R_l(s) - Y_l(s) \right) + K_2 \left(R_s(s) - Y_s(s) \right)]$$

where K_2 and K_1 are parameters of the speed controller and K is a parameter of the Load controller and T_i is the common integral time constant term. The meaning of the notations are:

$R_s(s)$ = Speed Reference $Y_s(s)$ = Actual Speed

$R_l(s)$ = Load Reference $Y_l(s)$ = Actual Load

This indicated that the speed loop and the load loop share a common integrator. The input of the integrator is the weighted sum of all the inputs to the controller. To be more specific, the input of the integrator can be written as:

$$e = K_2(R_s - Y_s) + (R_l - Y_l) \qquad (8.27)$$

Such an arrangement is used since the speed control and load control are used sequentially during the normal operating condition, *i.e.* during the speed run up, the reference load is set to be 0. When the turbine speed is synchronized with the electric grid, the load controller will then be used to change the turbine load. With the common integrator, the switch from speed run up to load run-up will be smooth.

Both speed controller and load controller can be considered as PI controllers. However, the speed controller has a much bigger gain than the load controller. Such controller design puts more emphasis on the speed control loop, and when a load rejection happened, the speed controller will be dominant which prevents the turbine from over-speed.

In summary, the turbine unit is controlled by speed controller only during the speed run-up. It is solely controlled by load controller during load changes. In both cases, the system can be considered as SISO systems. During the load-rejection, both speed and load controllers are put into use.

8.4.4 SISO Benchmarking Profile

Definitions of Process Variables

Table 8.9. SIEMENS process input/output variables

Outputs	Inputs
Y_1: Load	U_1: Main steam control valve
Y_2: Speed	U_2: Reheat control valve

The above variables are the main process Input/Outputs of interest. Another input is available: the grid electric demand. In speed run-up transient, it has no influence as the unit is not operated on the grid, whereas in the other cases (load rejection and load run-up) it can be used as a process disturbance. In the load rejection configuration, it is used as a parameter that indicates the level of load rejection simulated while for the load run-up case, it is fed with plant load set point.

For the purpose of SISO benchmarking analysis, was is necessary to define the input/output pairing for each SISO control loop. In **Table 8.10**, the currently applied pairings between MV-s (Manipulated Variables) and CV-s (Control Variables) on the plant are introduced.

Table 8.10. Input/output pairings

Controlled variable	Manipulated variable	
Load	Main steam and reheat control valves	$Y_1(U_1, U_2)$
Speed	Main steam and reheat control valve	$Y_2(U_1, U_2)$

Since there were no stochastic disturbances or reference signals in the SIEMENS unit, it would have been inappropriate to use the MV and GMV benchmarking indices. Although the RS-LQG benchmark algorithm is primarily designed for assessing the performance of process with stochastic disturbance and reference signals, the benchmark algorithm can still be utilised if the system does not have a stochastic disturbance, stochastic reference or neither. Therefore, the SIEMENS plant control system has only been benchmarked with the RS-LQG algorithm.

8.4.5 RS-LQG Benchmarking Test

For RS-LQG performance assessment, the existing controllers were benchmarked against the optimal RS controller in speed run up, load run up and three cases of load rejection conditions. For the load rejection tests, the performances of the speed were assessed in three different cases of load rejections: from 100% down to 10%, 50% and 80% of nominal load; this corresponds to typical abnormal operating conditions occurring in a power plant.

For the speed run up tests, the performance of the speed controller was assessed for a nominal speed trajectory provided by SIEMENS. For the load run up tests, the performance of the load controller was assessed for the trajectory provided by SIEMENS as the typical unit reference for the load run-up transient.

As discussed before, the speed controller is dominant during load rejection, so only the speed controller is assessed during the load rejection period. Of course, the original speed/load controller is coupled during load rejection, and so the benchmarking result may not be quite realistic.

8.4.6 Key Factors of RS-LQG Benchmarking

Transfer Function Model Description
As previously mentioned, to compute the performance index, the RS-LQG benchmark algorithm needs the rational transfer function describing the process, the disturbances and the reference. For the SIEMENS plant case study, SISO system identification was performed. The different simulations were run and the plant data were collected: set points, controller output and plant output.

As the turbine speed and turbine load are both controlled by the control valve, the turbine can be considered as an one input, two outputs system. So during each operating condition, a valve to speed and a valve to load model will be identified.

Corresponding to five different operating conditions, five valve/speed and five valve/load models will be identified.

As characteristic of non-linear processes, the models have different dynamics with different operating conditions. Since the RS-LQG benchmarking algorithm is model specific, then different optimal RS controllers will be obtained for these different operating conditions. However, the actual SIEMENS controller is the same for all operating conditions, normal and abnormal (i.e. it is not foreseen to have controller parameters and structure as a function of operating conditions); benchmarking the performance of the existing controller against different optimal RS controllers is unrealistic and has little physical value. To adequately benchmark such systems as the SIEMENS plant, the multiple model restricted structure design and benchmarking technique, which is more applicable to non-linear systems, would be preferable [Grimble, 2003].

However, it is still possible to use the RS-LQG benchmarking algorithm, by simply defining for each control loop, a nominal model from the set of all the identified models. The nominal model has been chosen to be the transfer function model which approximates all other and has the slower dynamic response. The two nominal models are documented in the following table.

Table 8.11. Nominal models

Control Loop	Transfer function
Speed	$z^{-1} \dfrac{3.487 \times 10^{-7} + 6.341 \times 10^{-7}\, z^{-1} - 1.162 \times 10^{-7}\, z^{-2}}{1 - 3.463\, z^{-1} + 4.425\, z^{-2} - 2.458\, z^{-3} + 0.4964\, z^{-4}}$
Load	$z^{-1} \dfrac{-3.493 \times 10^{-4} + 2.126 \times 10^{-4}\, z^{-1} + 2.062 \times 10^{-4}\, z^{-2}}{1 - 2.493\, z^{-1} - 2.001\, z^{-2} + 0.5005\, z^{-3} - 0.007613\, z^{-4}}$

Then for all the different transients studied for benchmarking, the nominal model was used as plant process model and the existing loop controller was benchmarked against the optimal restricted structure controller designed for nominal model.

Since the system does not have a stochastic disturbance, the disturbance is set to zero. Moreover, the process identification has provided models containing integral actions; therefore for numerical and computational reasons, the reference transfer function was set with its denominator equal to the denominator of the plant transfer function and its numerator equal to one. It was found that, using a reference model denominator equal to the plant denominator, it is possible to obtain a controller with good dynamic response properties.

Error and Control Weightings
The RS-LQG algorithm also requires a set of dynamic error and control weights to compute the performance index. Since SIEMENS is interested in the dynamic performance of the process, these weightings are chosen to reflect this desire to assess dynamic performance. As the system has no stochastic excitation, the error

weighting was chosen to contain an integrator in order to ensure zero steady state error, while the control weighting was chosen to be a scalar term penalising control action. The resulting transfer functions of these weightings are shown in the Table 8.12.

Table 8.12. RS-LQG weightings

Load loop weighting selection				Speed Loop weighting selection			
TEST	Qc	Rc	Noise	TEST	Qc	Rc	Noise
Load Run Up	$\dfrac{0.1}{1-z^{-1}}$	0.2	0	Load Rejection 100% to 10 %	$\dfrac{1}{1-z^{-1}}$	0.3	0
				Load Rejection 100% to 50%	$\dfrac{1}{1-z^{-1}}$	0.3	0
				Load Rejection 100% to 80%	$\dfrac{1}{1-z^{-1}}$	0.3	0
				Speed Run Up	$\dfrac{1}{1-z^{-1}}$	0.3	0

RS-LQG Benchmarking Result
As the RS-LQG benchmarking algorithm is defined for SISO systems, it may not be suitable for the SIEMENS plant. However interpreting the SLC as two independent control loops, it is possible to use the RS-LQG algorithm to benchmark the SIEMENS plant control system. The resulting RS-LQG controller is shown below

Table 8.13. RS-LQG controller

Control Loop	RS-LQG controller
Speed	89.87
Load	$\dfrac{109.5-213.2\,z^{-1}+103.9\,z^{-2}}{1-1.5\,z^{-1}+0.5\,z^{-2}}$

The simulations of load rejection, speed run-up and load-run up were run with the existing loop controllers, and all the process data were collected. The controllers were then replaced with the RS optimal controllers returned by the RS-LQG algorithm and the simulations were repeated. The simulation results are shown in the following figures. Only one load rejection test is shown, since the performance is similar in all the other cases.

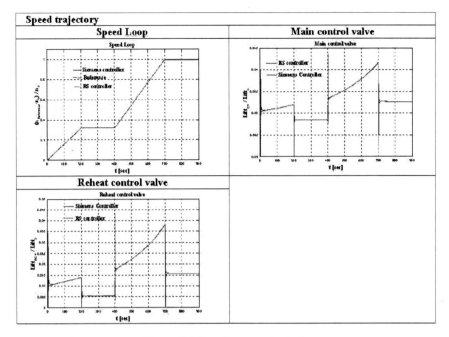

Figure 8.5. Speed run-up simulation

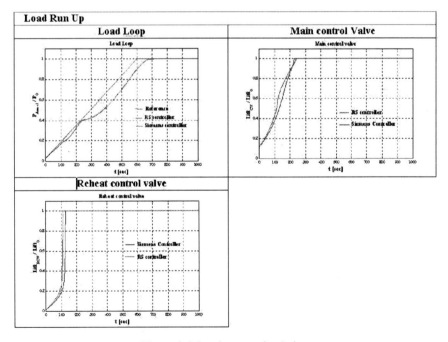

Figure 8.6. Load run-up simulation

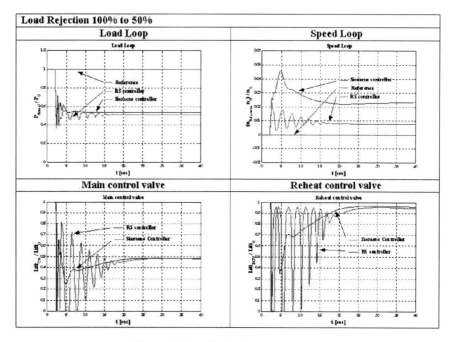

Figure 8.7. Load rejection performance

It must first be noted that the RS controllers are more aggressive in terms of control actions: for all transients, the main control valve and the reheat control valve are highly disturbed and have to work hard to achieve low benchmark cost. As far as the speed control loop is concerned, for load rejection transients, it can be observed that the RS controller can get a smaller speed steady-state error than the original controller does, and its settling time is quicker, i.e. the RS controller is really more aggressive and uses much more control action. It can also be observed that in the speed run-up transients, the original and redesigned controller has similar performance. In the case of load loop control, during loading transient, the RS controllers perform slightly better than the original controllers as the set point tracking is better for low load level. Then the performances are almost identical.

8.4.7 MIMO Benchmarking Profile

The special feature of the turbine speed/load controller has been discussed before: during the speed run-up mode, only the speed controller is in action; while during the load run-up mode, the load controller is in charge since the turbine speed is regulated by the electric network. This implies that in these two modes there is only a SISO controller in action and the controller performance has been studied using RS-LQG benchmark.

However, both speed and load controllers are involved during load rejection, and the speed/load controller can be considered as a two-input, one output system. In this case, it will be worthwhile to study the how the original controller performs when compared with a GPC MIMO controller. Thus, the benchmarking study will be focused on the controller performance during load rejection.

8.4.8 MIMO Plant Model

The original system is a highly nonlinear system, however in order to benchmark the system using the LQGPC benchmark, linearised models of both the plant and the controller are required. The MATLAB® function LINMOD is used to generate a linear model from the original nonlinear model. To guarantee the validity of the linearization, the dynamic operation range of the linearised model should be kept small. Based on the above considerations, the load rejection from 100% to 80% is selected as the scenario to investigate the performance of the existing controller. Using MATLAB® LINMOD function on each of the individual blocks, a linearized model of the simplified turbine process has been constructed. The same load rejection tests were performed on the original and on the linearized model with the original controller. The results are shown in Figure 8.8.

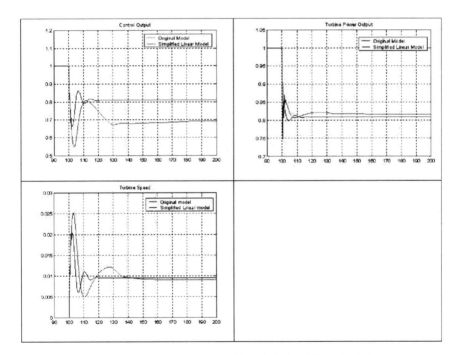

Figure 8.8. Dynamic response of the original and linearized plants

From the above figure, it can be seen that the simplified linearised model can quantitatively reflect the dynamic response of the original system. The static difference of control output between the original and the linearised model is mainly due to simplification of the pressure control loops.

8.4.9 The GPC Controller Design for the Linearised Plant

As shown in Figure 8.9, the whole turbine system can be abstracted into two parts, namely the plant and the controller. The grid requirement acted as a deterministic disturbance to the system. Since a linearised state space model is available, a state space GPC controller can be designed for the simplified linear model. It is assumed that the grid requirement is measurable. The structure of the GPC controller is depicted in Figure 8.10. A state estimator (observer) is used to estimate the system state from the plant inputs and outputs. The reference, state and disturbance gains are all static.

The key parameters of GPC designs are: error weighting matrix is $Q_e = \begin{bmatrix} 100 & 0 \\ 0 & 1 \end{bmatrix}$ and control weighting is $Q_u = 0.01$. The output prediction horizon is 50 and the control horizon is 5. By using this set of parameters, more weight is put on the difference between the reference and the actual turbine speed. This is consistent with the previous analysis.

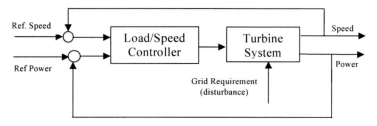

Figure 8.9. Simplified block diagram of the turbine system

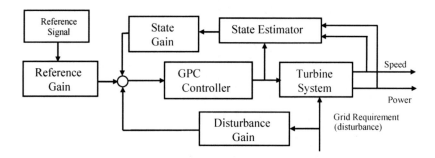

Figure 8.10. The structure of the state space GPC controller

First, a comparison between the original nonlinear controller and the GPC controller is carried out on the linearised plant. The grid requirement drops from 100% to 80% at $t=100$ sec. The results are shown in Figure 8.11.

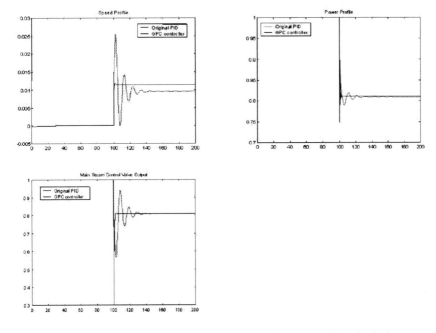

Figure 8.11. Load rejection performance comparison on the linearized plant

It can be seen that the GPC controller performs much better than the original PID controller on the simplified linear model. This result is expected, since the GPC controller is optimally tuned for this model. Then the designed GPC controller is used to control the original turbine system. The grid requirement drops from 100% to 80% at $t=100$ sec. The simulated result is shown in Figure 8.12.

It can be seen that the transient performance of the state space GPC controller is better than the original controller. However, the control action is more violent. The fast transient performance is mainly due to disturbance feed-forward compensation and the aggressive tuning of the GPC controller. It can be seen that GPC controller is a viable candidate for improving the dynamic performance of the turbine system. However, due to the non-linear nature of the plant and the specific design of the original PI controller, the LQGPC benchmark may not be very meaningful.

8.5 Conclusions : Using the Benchmark Algorithms

After conducting numerous benchmarking exercises on several models, it is possible to make the following observations about effectiveness of the algorithms.

- **Performance criteria**

The MV, GMV and RS-LQG benchmarking algorithms all use variance as the measure of controller performance assessment. These algorithms have been

designed mainly with that criterion in mind, and hence assume that the process
under test has at least one source of stochastic input either a stochastic reference or
a stochastic disturbance.

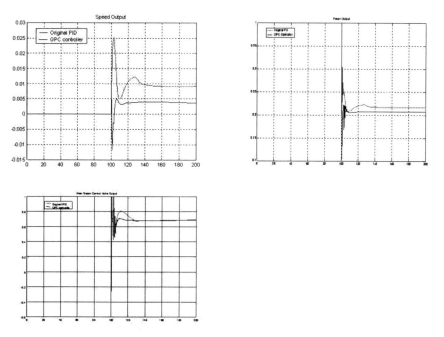

Figure 8.12. Load rejection performance comparison on the original plant

Systems in the process industry normally have at least one stochastic source
acting on the system process and normally system outputs tend to remain around a
given value for long periods of time. Also for most process systems the ability to
reduce variability in outputs often increases the possibility of pushing the process
to its safe operational limits in order to improve productivity and efficiency. For a
system with this sort of characteristic, taking variance as a measure of performance
is quite applicable and meaningful. The test conducted with the BASF model
confirmed that MV, GMV and RS-LQG benchmark algorithms could be applied
effectively for such systems.

However for systems from the power and servo industry, references and
disturbances tend to be deterministic in nature so that the systems generally do not
have any stochastic signals. Furthermore, another characteristic of such systems is
that they operate quite frequent changes to references and hence output levels. For
example, the power plants have to follow a daily load program which is fixed as
function of typical load demand and energy market requirements. So, for such
systems, taking variance as a measure of performance does not seem to be quite so
applicable. The test conducted with the SIEMENS model confirmed that the MV
and GMV benchmark algorithm could not be applied successfully to such
processes because these algorithms rely heavily on the stochastic noise present in
the input and output signals to compute the optimal performance. However if

suitable substitutions are made in the definitions of RS-LQG algorithm user parameters (especially in the definition of disturbances and references models), it is still possible to use the RS-LQG algorithm to benchmark such systems.

In the use of performance criteria, the LQGPC/GPC benchmarking algorithm is different from the other algorithms as it uses a weighted sum of variances and mean squares of process inputs and outputs as the measure of controller performance assessment. The algorithm provides an optimal benchmarking function for dynamic, steady state, deterministic or stochastic performance assessment. It uses indicators like, integral square error (ISE), integral square control action and some weighted combination of variances over a period that may includes a step change to the system inputs. It is applicable to stochastic processes as well as servo, machine, automotive and robotic systems where good reference tracking and transient behaviour are a priority.

- **Data requirement**

The MV and GMV benchmarking algorithms use only plant data to compute the benchmark index. Apart from feeding the algorithm with the plant data, for both algorithms, the user must only give two other parameters: a data length, which defines the number of data points used to compute the control system performance, and the length of the auto-regressive model used to identify the system. But testing the algorithm has shown that in most cases assuming stationary stochastic properties, the pre-defined default values of these parameters produce sufficiently accurate results. The MV algorithm can compute the benchmarking index using either system output or system error only, whereas the GMV algorithm requires both system output (or error) and system input. Moreover, the GMV algorithm requires the user to specify dynamic error and control weighting, which help determine the desired optimal controller required.

The MV and GMV algorithms both require that the system time delay be precisely known. The algorithms still return a benchmark value if only an estimate of the time delay is provided; but if the incorrect system delays are used, the index value cannot be interpreted as a MV or GMV benchmark but regarded as some sort of user defined performance index. This means that the controller against which the existing system controller is being compared might not be an MV or GMV controller. Assuming that the stochastic properties of the system are stationary, then the accuracy of the algorithms inherently depend on the accuracy to which the system time delay is known. Testing on the BASF model highlighted some of the difficulties of trying to obtain an accurate estimate of the system time-delay. Some techniques to bracket the true benchmark index by conducting a series of tests using a range of time delay values were also investigated but the tests showed the limitation of these methods.

The RS-LQG algorithm does not use plant data to compute the benchmark index. For this algorithm, the process model transfer function is required. The user has to specify the type of restricted structure (RS) controller against which the existing controller should be benchmarked. The optimal RS controllers can be one of the three following forms: P, PI or PID. The user must also define the models of the system disturbance and reference as transfer functions, and just as in the GMV algorithm, the error and control weightings must be specified. The choice of these weightings must be consistent with the choice of optimal RS controller and the objectives of the control problem. The accuracy of the results returned ultimately depends on the accuracy of the model used for benchmarking.

The LQGPC/GPC algorithm does not use plant data to compute the benchmark index. For this algorithm, a system model in state-space equation format is required. Depending on the benchmarking requirements, disturbance and measurement noise models maybe included in the system model. It also required the reference trajectory over which the benchmarking is computed. This trajectory could be a step or a ramp for dynamic assessment, or could be constant for steady state performance. It should be noted that whether the reference is a step or constant, the benchmark computes the variances obtainable by the existing controller and the optimal control as part of the benchmarking outputs.

The user also needs to specify an error and control weighting as well as the process and control prediction horizons. These are normally chosen to scale the process error and the control action to the same magnitude (using a normalisation factor for each I/O variable), but also to penalise the control action, thus reflecting the level of emphasis in minimisation of its mean square value. For benchmarking MIMO systems the weights can also be used to assign relative priority to each input or output. There are very simple guidelines to choosing the output prediction and control horizons.. The accuracy of the results returned ultimately depends on the accuracy of the model used for benchmarking. For multivariable benchmarking, issues related to the accuracy of the model and adequate interaction information can affect the usefulness of LQGPC/GPC benchmarking results.

9

A Look Back and a Look Forward to New Research Directions

Michael Johnson

Emeritus Professor, Industrial Control Centre, Department of Electronic and Electrical Engineering, University of Strathclyde, Glasgow, UK,

9.1 A Look Back

The business community has developed methods over the last twenty-five years or so, that focus on the aggressive achievement of best performance and commercial supremacy. At the centre of these approaches is the concept of the business process around which the company operates. The financial well-being of the company is directly linked to the performance of the business processes within the company. Performance is then an attribute that needs to be defined, measured, analysed and improved; from this emerges the science of performance assessment and its concomitant philosophies like business process re-engineering and business process benchmarking.

The main steps in performance assessment have generic validity across disciplines and are shown here in Figure 9.1. Within the business community, these steps have been embedded within other philosophies to enhance their wider acceptance in a commercial environment. Benchmarking is one such philosophy that is likely to survive and pass the test of time. A key activity in benchmarking is that of defining a performance standard against which to measure actual performance. In the field of business processes this leads to internal, competitive, functional, and generic benchmarking as methods for defining performance standards. Further qualification of the benchmarking activity arises when it is linked to a level in the business hierarchy, for example, performance benchmarking for a specific process, process benchmarking for a process sequence or strategic benchmarking for the managerial and corporate domains within a company. One final component of benchmarking to note is the activity wheel, typically, plan/model, search, observe and measure, analyse and propose, and then implement. It is not surprising that these are very similar in intent and purpose to the generic performance assessment steps given in Figure 9.1.

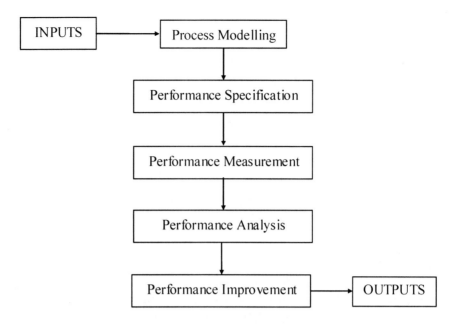

Figure 9.1. Steps in performance assessment

Business process performance assessment is a well-established paradigm and its interest for the research reported in this book is to provide inspiration for the development of performance assessment methods in the field of control engineering.

9.1.1 Economic Audits of Control Systems

Engineers and technologists involved in making large-scale industrial, utility, production and manufacturing processes work can sometimes lose sight of the fact that these technological processes are business processes too. Thus a business process view of technological processes can often give an insight into how to manage, operate, control and maintain these installations more effectively. There are not so many tools for establishing the economic links between process revenue generation and process performance and the Integrated Control and Process Revenue and Optimisation framework proposed in Chapter 2 of this book is a new method for making these links. The steps in this method are briefly described next.

Step 1 Profile and operation assessment. The financial objectives of the process are identified and the current financial operating status archived.
Step 2 Process and system assessment. A technical step; the process and its control systems are understood and documented. Current operational performance data is archived.
Step 3 Financial benefits and control strategy correlations. A key step in which company goals and financial outcomes are linked to the performance and control of

the processes. It is in this stage that a financial assessment of process and control loop performance is established.

Step 4 Optimality assessment. This is an assessment of the potential remaining in the process for enhanced performance. In many cases optimisation will provide the tool or the benchmark value to determine the under-achievement in process and control performance.

Step 5 Control system adaptation. An implementation stage concludes the process of performance improvement.

It is not difficult to make the connection between the stages in performance assessment and the steps of the economic audit of control systems as portrayed in Figure 9.2.

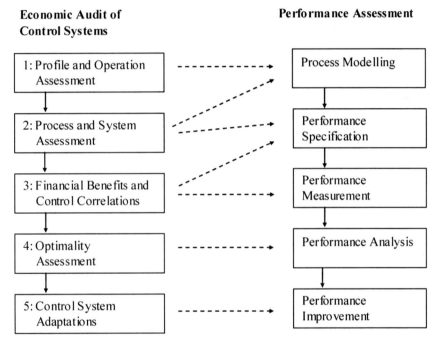

Figure 9.2. Comparison between the stages of the control system audit method and those of business process performance assessment

9.1.2 Benchmarking Control Loops

Processes in industry can be very complex and often comprise a sequence of processes where one sub-process output is the input to a subsequent sub-process. In many cases, these processes are often cross-coupled leading to highly interactive process networks. Control is therefore a demanding task that must be accomplished with sufficient top-level transparency to allow some understanding of how the process actually works. The solution in many cases is a hierarchical control structure as shown in Figure 9.3.

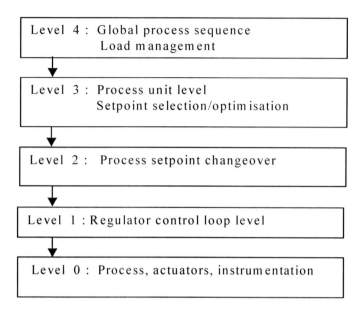

Figure 9.3. Process control hierarchy

Each level in the process control hierarchy has associated with it a performance assessment problem of which the generic steps of Figure 9.1 can be applied.

An important performance assessment problem for which significant contributions have been made in this book is that of benchmarking the control loops of Level 1 in the process control hierarchy. Although there are several algorithms to choose from, the key features of these algorithms lie in the properties of the selected cost function and the basic principle of optimality. To avoid confusion with previous notation, denote the generic cost function and its dependence on control signal as $J_c(u)$ where $u \, \varepsilon \, F$ and F is the control signal domain. The following properties are used in benchmarking studies:

   (1)   The cost function is a measure of system performance that is appropriate for optimization. For example, the system output variance would be a cost function appropriate for optimisation.

   (2)   The cost function has the mathematical property of being positive over the domain of controls, so that,

$$0 \leq J_c(u) \quad \text{for all} \quad u \, \varepsilon \, F$$

   (3)   It is assumed that there exists a control $u^\circ \, \varepsilon \, F$ that minimises the cost function, so that,

$$0 \leq J_c(u^o) \leq J_c(u) \quad \text{for } u^\circ \, \varepsilon \, F \text{ and for all } u \, \varepsilon \, F$$

Collecting these features together, if a particular control u_c ε F is selected (corresponding say to an implemented non-optimal control) then,

$$0 \leq J_c\left(u^o\right) \leq J_c\left(u_c\right)$$

or, in normalised form,

$$0 \leq \eta(u_c) = J_c(u^o) / J_c(u_c) \leq 1$$

and $\eta(u_c)$ is defined as the performance index for control u_c ε F. This normalised relationship leads directly to the interpretation that if $\eta(u_c)$ is close to "1" then the selected control is nearly optimal, but if $\eta(u_c)$ is close to "0" then the control is very non-optimal.

The development of specific methods then depends on the cost function $J_c(u)$ chosen to capture the desired features of the control design and the constraints on the information that is available to compute estimates of $\eta(u_c)$. These factors combine in different ways to lead to a variety of control loop performance assessment algorithms.

The Cost Functions
The set of cost functions for which algorithms have been devised in this book include the following:

Minimum Variance (MV):

SISO cost $\qquad$ $J(t) = E[\, y^2(t + k) \mid t]$

MIMO cost $\qquad$ $J(t) = E[\, Y^T(t)Y(t)\,]$

Generalised Minimum Variance (GMV):

SISO cost $\qquad$ $J(t) = E[\, \phi^2(t + k) \mid t]$ $\qquad$ $\phi(t) = P_c\, e(t) + F_c\, u(t)$

MIMO cost $\qquad$ $J(t) = E[\, \phi^T(t)\phi(t)\,]$ $\qquad$ $\phi(t) = P_c\, e(t) + F_c\, u(t)$

Generalised Predictive Control (GPC):

SISO cost $\quad J(t) = E[\{ \sum_{j=N_1}^{N_2} [r(t + j) - y(t + j)]^2 + \lambda^2 \sum_{j=1}^{N_u} \Delta u^2(t + j - 1)\} \mid t]$

MIMO cost $\quad J(t) = E[\sum_{j=0}^{N} \{[r(t + j +1) - y(t + j +1)]^T Q_e[r(t + j +1) - y(t + j + 1)]$

$$+\Delta u^T(t + j\,)Q_u\Delta u(t + j\,)\} \mid t]$$

Linear Quadratic Generalised Predictive Control (LQGPC):
  MIMO cost

$$J = E[\lim_{T_f \to \infty} \frac{1}{T_f+1} \sum_{t=0}^{T_f} \{\sum_{j=0}^{N} \{[r(t+j+1)-y(t+j+1)]^T Q_c [r(t+j+1)-y(t+j+1)]$$
$$+u^T(t+j)Q_u u(t+j)\}\}]$$

Linear Quadratic Gaussian (LQG)

$$\text{SISO cost } J = \lim_{T_f \to \infty} E[\frac{1}{2T_f} \int_{-T_f}^{T_f} \{(H_q e(t))^2 + (H_r u(t))^2 \}dt]$$

$$= \frac{1}{2\pi j} \oint_D \{Q_c(s)\Phi_{ee}(s) + R_c(s)\Phi_{uu}(s)\}ds$$

$$\text{MIMO cost } J = \frac{1}{2\pi j} \oint_D \{tr(Q_c(s)\Phi_{ee}(s)) + tr(R_c(s)\Phi_{uu}(s))\}ds$$

An important feature of the cost functions used is that as the function chosen to capture the control performance becomes more sophisticated then more design information has to be introduced to specify cost weightings, cost function time intervals and other parameters. This input of knowledge is generally difficult to automate, relying as it does on expert engineering input.

The Information Constraints
There are two situations for the computation of performance indices, online and off-line. In the online situation the methods devised are data-driven and avoid model identification stages. Consequently, the algorithms devised use real-time data directly to compute the performance index. Whether this can be achieved depends mainly on a mathematical re-formulation of the method. For off-line methods, computation in real-time is not a constraint and model-based techniques with sophisticated design-led cost functions and procedures are used.

In the case of model-based techniques there are some additional capabilities for benchmarking if a parameterised restricted structure controller generates the control signal. Using the cost function notation $J_c(u)$ that makes the dependence on the control signal explicit, let the control signal domain to be $u \, \varepsilon \, F$ and let the restricted structure control signal be $u_{RS} \, \varepsilon \, F_{RS} \subseteq F$, then assuming that optimal solutions exist,

$$0 \le J_c(u^\circ) \le J_c(u^\circ_{RS}) \le J_c(u_{RS})$$

where $u^\circ \, \varepsilon \, F$, u_{RS}, $u^\circ_{RS} \, \varepsilon \, F_{RS} \subseteq F$, u° is the optimal control signal for $J_c(.)$ over F and u°_{RS} is the optimal control signal for $J_c(.)$ over $F_{RS} \subseteq F$.

The relationship $0 \le J_c(u^\circ) \le J_c(u^\circ_{RS})$ holds because otherwise u°_{RS} will contradict the optimality of u° and the relationship $0 \le J_c(u^\circ_{RS}) \le J_c(u_{RS})$ defines the optimality of u^o_{RS}. The relationship between these three cost function values leads to two diagnostic situations:

Case(a) Structural Performance Index, $\eta_S = J_c(u^o) / J_c(u^o_{RS})$

where $0 \leq \eta_S = J_c(u^o) / J_c(u^o_{RS}) \leq 1$

The value of η_S indicates the closeness of the optimised restricted structure to the full optimal cost attainable.

Case(b) Controller Performance Index, $\eta_C = J_c(u^o_{RS}) / J_c(u_{RS})$

where $0 \leq \eta_C = J_c(u^o_{RS}) / J_c(u_{RS}) \leq 1$

The value of η_C indicates the closeness of the actual implemented restricted structure controller to the optimal cost obtainable if the restricted structure controller had optimised parameters.

Thus these two diagnostic indices can be used as follows: Case (a) to determine just how good the proposed controller structure is when compared to the full optimal solution and Case (b) to determine just how good the implemented controller is when compared to what could be achieved if optimal tuned values were used in the implemented controller structure. This interpretation of the use of optimal cost values and the formulation of restricted structure control leads to a combinational approach for assessing the performance of control structures of multivariable systems [Greenwood and Johnson, 2005].

A taxonomy of control loop benchmarking algorithms can be created by putting together the different types of cost functions and the type of process information available for performance index computation. The taxonomy displayed in Figure 9.4 shows the algorithms devised and illustrated in this book.

The taxonomy supports the strong conclusion that significant progress has been made in creating a body of theoretical and practical approaches to the problem of computing performance assessment indices for control loops.

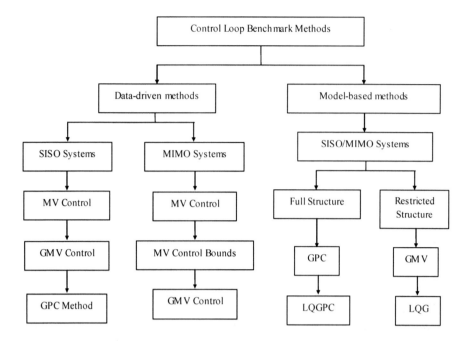

Figure 9.4. Taxonomy of control loop benchmarking methods

9.1.3 Contributions to Performance Analysis

The early chapters in this book were concerned with establishing a generic process performance assessment framework with inspiration from the business process domain. The major contributions in these chapters were the in-depth presentations of methods and algorithms for control loop benchmarking; an activity that was considered to fall into the "performance specification "and" "performance measurement" stages of the generic framework. A simple measure or index that indicates that there is a performance problem is only a beginning to the control loop benchmarking activity for there are some process problems that require more than just a simple index to identify and analyse the loss of performance. Plant-wide disturbances were introduced as an important area for performance assessment studies.

Process plants are often networks of process units which may be sequential and/or cross-coupled. Plant-wide disturbances are disturbances caused at one point in the process that ripple through the plant and upset a much larger set of process output variables. The causes of these disturbances are several, including,

(i)   Process nonlinearities (for example, stiction) can cause limit cycle oscillations in loops
(ii) Faulty measurement devices
(iii) Poorly tuned controllers can cause control loop oscillations
(iv) Control loop interactions in cross-coupled controllers

(v) External disturbances, particularly from upstream processes in a sequential process

Control-loop oscillations and oscillations in measured process variables are process disturbances for which some assessment success is claimed. Indeed it was stated that the problem of detecting an oscillation in loops is a solved problem. The real challenge is the diagnosis of the root cause. This agrees with the experience of practitioners in the business process domain where "performance analysis" and "performance improvement" are considered to be the more demanding stages of the performance assessment procedure. The techniques used for the plant-wide disturbance and oscillation problem largely depend on the operational framework selected by the researcher and the research reported in this book is summarised in Table 9.1.

Apart from the very valuable assessment of the methods available in the literature, the work reported in Chapters 6 and 7 shows just how more complicated are the stages of performance analysis and improvement. In many cases diagnostic values have to be supported by insight about the process structure or characteristics and this knowledge requirement is not so compatible with the objectives of automated procedures and the use of minimal *a priori* process information.

Table 9.1. Techniques used in the plant-wide disturbance and oscillation problem

Plant-wide disturbance methods (Chapter 6)	
Objectives	a) Automated methods used off-line and date driven b) Detect the presence of a plant-wise disturbance c) Identify the root-cause of the plant-wise disturbance
Tools	a) Detection: Power spectra, spectral principal component analysis, autocovariance functions b) Diagnosis: Harmonics in power spectra; Bicoherence tests

Oscillations in control loops (Chapter 7)	
Objectives	a) Automated methods used online that are model free, use the minimum *a priori* information about the control loops and work on output error (*Err=SP-PV*) and controller output only b) Detect the presence of loop oscillation c) Identify the root-cause of the oscillation
Tools	a) Detection: Seven methods studied based on underlying methods like the autocovariance function, spectral analysis and wavelet theory Conclusion: In almost all cases these methods give reliable detection. b) Diagnosis: Concentrated on stiction detection and studied twelve methods constructed from a wide range of underlying theory. Conclusion: No single method covers all the stiction cases reliably.

9.2 A Look Forward to New Research Directions

The paper due to Harris that initiated the area of control loop performance assessment was published in 1989. Since that time the subject has grown broader by incorporating process audits that try to link control performance to economic benefit. Further, a number of new algorithms have been developed for controller loop benchmarking by applying the Harris methodology to different controller performance indices. There is also a need for performance analysis tools and some of the new developments reported in this book have tried to establish stronger diagnostic links between indices and the underlying performance disturbance or barrier. However, the field continues to grow primarily due to industrial interest in obtaining the best performance from the installed process plant and some areas for further consideration are given next.

9.2.1 New Performance Metrics

There are two main ways that metrics are selected for performance measurement in the control field:

i)   Select a common generic mathematical metric that is readily amenable to analysis. The next step in this route is to link the metric value to a particular performance barrier thereby establishing the metric diagnostic basis.

ii)   Select a diagnostic situation (for example, the presence of oscillations in the control loop) and then devise a performance metric to purposely detect the presence of the particular barrier to performance.

Whichever route is taken to devising a performance metric, a common problem is that the chosen measure is insufficiently discriminating so that similar diagnostic values cover several different process problems. More research is needed in this basic area and the issue is revisited in the section below on Performance Analysis.

9.2.2 Algorithm Development

The taxonomy of control loop benchmarking methods shown in Figure 9.4 reveals that the theory for a wide range of methods has been devised. These procedures cover online and off-line methods and range from single index calculations to a combinational computation of index values for controller structural comparisons. The more sophisticated performance metrics need to be supported by additional design input, commonly through the use of weighting functions. Consequently, there is plenty of scope for further work and research on algorithm development. There are three particular areas that might be usefully mentioned:

i)   *Data handling.* The online tools developed use data from the process in real time and in some cases pre-processing is needed to ensure that neither poor nor false results are obtained. Algorithms to robustly pre-screen and, if necessary, condition the data are needed.

ii)   *Numerical algorithm robustness.* The routines of the benchmarking taxonomy use theory from the polynomial and state-space domains. Some prior experience [Greenwood and Johnson, 2005] has shown that many existing routines cannot always be relied upon and it is often necessary to select numerically robust routines for benchmark computations. There is scope to ensure that both online and off-line computations are achieved accurately, efficiently and robustly.

iii)   *Automated algorithm setup.* The more sophisticated cost functions such as those in the GMV control, GPC and LQG control algorithms need design input if the benchmarking procedure is to be meaningful. At the same time many of the benchmarking algorithms should be feasible without the need to call on expert design input. Similarly, in some algorithms, the actual setup of the indices and the weighting functions has to be done with care otherwise the associated computations may fail. In this book, Chapter 8 reported on some of these particular issues. A logical progression here is to devise software to automate and advise on the setup of the algorithms. Thus expert software would (a) ensure that the set-up of the algorithm is safe and that a solution is possible for the various equations being solved, and (b) automate the selection of the design input in the weighting function selection procedures.

9.2.3 Performance Analysis

Much of the research effort in the field of performance assessment has been devoted to the stages of performance specification and performance measurement. Performance analysis and the relationship between performance metrics and the diagnosis of the barrier (or problem) to performance requires more research input. There is a need for the systematic collection of know-how diagnostic knowledge in the control field. Such research will have to accommodate three groups of factors.

i)   Common diagnostic situations. There are at least four problems in process control that need diagnostics: Poor controller tuning, sluggish control loops, oscillations in control loops and their various root causes and finally other (plant-wide) process disturbances.

ii)   Common industrial control situations. Control loops in industrial process occur in different configurations and with different objectives. The measurement of performance and the diagnosis of poor performance depends on the different set ups. Common industrial process control situations include, single loops, non-interactive loop designs, cross-coupled multivariable controllers, cascade control loops, ratio control loops, feedforward control loops, and inferential control loops. In some cases, these industrial control loops are already on the process and need performance measurement and analysis, and in other cases these control loop configurations (for example, feedforward) might be part of the solution to achieve enhanced performance.

iii)   Different objectives within the process control hierarchy. The various levels of the process control hierarchy have different control performance objectives. Thus, there is a need to consider how performance measurement, analysis and improvement procedures should reflect these different hierarchical objectives. The three levels that need this type of research are the top-level supervisory control, the middle level of control changeover and the lower regulator level.

9.2.4 Industrial Implementation

An extremely important objective for the performance assessment community is industrial implementation and industrial use of the methods devised. Moving a performance assessment procedure from an academic research paper to an industrial application introduces new and interesting engineering problems. The two main areas to consider are the control system technology to be used in the implementation and how the procedure is to be used within the company.

i)   *The implications of the control system technology used.* Controller technology in the process industries tends to fall into the categories of the single-station hardware controller units and the digital computer control system (DCS; DCCS) software and hardware [Ang *et al.*, 2005; Johnson, 2005]. A decision to employ controller loop performance software in either of these two control system platforms immediately creates implementation constraints on the accuracy and the size of the computation that can be made and on the complexity of the interface to the process engineer. These constraints ensure that many of the previously listed research issues will have to be revisited, particularly under algorithm development where topics like data-handling, numerical robustness and automated setup procedures will be of critical importance in industrial versions of the procedures.

ii)   *The implications of industrial process company operations.* Two examples will be used to illustrate the implications arising from company operations for industrial implementation. Paulonis and Cox [2003] report a performance assessment application for 40 plants located at nine international sites and involving some 14,000 control loops. By way of comparison, Ericsson *et al.* [2004] describe an application involving 1,200 control loops running 24 hours a day at a single carton board mill site. Clearly these two industrial company situations will influence how the implementation should be achieved, for example, one company may construct a global solution whilst another may prefer a local plant solution.

Putting together the control system technology factors with a company's industrial structure leads to several options for the development of industrial procedures including:

i)   A top-level global solution using web-based interfaces to serve all the company's internationally located plant sites.

ii)   A top-level process plant solution using the plant-wide DCS system at a single site.

iii) A critical process unit solution using modules in the DCS system which controls individual process units

iv)   A critical process loop controller unit solution using a module to be located in selected single station controller units

v)   A single-station control loop unit solution. A manufacturer might install a controller performance assessment module in a range of single-station hardware controller units.

Each target solution is likely to require tailored algorithms for success. Hägglund [2005] recently published an example that shows how the industrial implementation of a performance assessment routine introduced new algorithmic requirements.

9.2.5 Final Conclusions

The new research directions outlined in this look forward shows that there remain many open questions and some significant areas for further development. From the academic viewpoint, the area of algorithm development and the art of performance analysis still has potential to grow, particularly in being able to diagnose the specific barriers in a process to performance attainment and then suggest constructive actions to achieve performance improvement.

The whole field of industrial implementation has considerable potential for future development. Some vendors already offer the Harris performance index procedure but the more recent performance assessment developments and the multiplicity of new indices and procedures offers exciting opportunities for further algorithm and industrial implementation studies.

References

ABB, 2005, Optimize^{IT} – Loop Performance Manager Version 2.0. User's Guide. ABB Advanced Control Solutions, www.abb.com

Ahmad M., and Benson R., 1999, *Benchmarking in the Process Industries*, IChemE, The Cromwell Press, UK

Albers, J.E., 1994, Data reconciliation with unmeasured variables, Hydrocarbon Processing (March), 73, 65-66

Albertos P and Ortega R., 1989, On generalised predictive control; two alternative formulations, Automatica, 25(5)

Andersen, B., and P-G Pettersen, 1996, *The Benchmarking Handbook : Step by Step Instructions*, Chapman and Hall, London, UK

Andersen, B., 1999, *Business Process Improvement Toolbox*, ASQ Quality Press, Milwaukee, USA

Anderson, B.D.O., and Lin Y., 1989, Controller reduction concepts and approaches, IEEE Trans. Automat. Control, 34, 802-812

Ang, K.H., Chong, G. and Li, Y., 2005, PID Control system analysis, design and technology, IEEE Trans. Control Sys. Tech., 13(4), 559-576

Aoki M, 1967, *Optimisation of Stochastic Systems*, Academic Press, London

Årzén, K-E., 1994, Grafcet for intelligent supervisory control applications, Automatica, 30, 1513-1526

Årzén, K-E., Wallén, A., and Petti, T.F., 1995, Model-based diagnosis-state transition events and constraint equations, Tzafestas and Verbruggen, Eds., *Artificial Intelligence in Industrial Decision Making, Control and Automation*, Kluwer Academic Publishers

Aspentech, 2001, Analysis of data storage technologies for the management of real-time process manufacturing data, Retrieved Jul 19th 2003, from http://www.advanced-energy.com/Upload/ symphony_wp_infoplus.pdf

Åström K. J., 1979, *Introduction to Stochastic Control Theory*, Academic Press, London

Åström, K. J., 1991, Assessment of achievable performance of simple feedback loops, International Journal of Adaptive Control and Signal Processing, 5, 3-19

Astrom K. J., and Wittermark B., 1973, On self-tuning regulators, Automatica, 9, 185- 199

Bakshi, B. R., 1998, Multiscale PCA with application to multivariate statistical process monitoring, AIChE Journal, 44, 1596-1610

Bakshi, B. R., and Stephanopoulos, G., 1994, Representation of process trends - IV. Induction of real-time patterns from operating data for diagnosis and supervisory control, Computers and Chemical Engineering, 18, 302-332

Bakshi, B. R., and Utojo, U., 1999, A common framework for the unification of neural, chemometric and statistical modeling methods, Analytica Chimica Acta, 384, 227-247

Balchen J. G., and Mumme K. I., 1988, *Process Control Structure and Application,* Van Nostrand

Basseville, M., 1988, Detecting changes in signals and systems - A survey, Automatica, 24, 309-326

Basseville, M., and Benveniste, A., 1983, Sequential detection of abrupt changes in spectral characteristics of digital signals, IEEE Transactions, IT-29, 709-724

Bauer, M., Thornhill, N.F., and Meaburn, A, 2004, Specifying the directionality of fault propagation paths using transfer entropy, DYCOPS4 conference, Boston, July 1-4, 2004

Bialkowski, W. L., 1992, Dreams vs. reality: A view from both sides of the gap, Proceedings of Control Systems 92, Whistler, BC, Canada, 283-295

Bitmead R, Gevers M and Wertz V., 1991, Adaptive optimal control and GPC; robustness analysis, European Control Conference, Grenoble, France, pp 1099

Blachuta M., and Ordys A., 1987, Optimal and asymptotically optimal linear regulators resulting from a one-stage performance index, Int. J of Systems Sci, 18, 7, 1377-1385

Bogan, C. E., and English, M. J., 1994, *Benchmarking for Best Practices: Winning Through Innovative Adaptation,* McGraw Hill Inc., New York, USA

Boullion T. L. O., 1971, *Generalized Inverse Matrices,* Wiley-Interscience

Bristol, E. H., 1990, Swinging door trending: Adaptive trend recording, ISA National Conference Proceedings, 45, 749-753

Caines P. E., 1972, Relationship between Box-Jenkins-Astrom control and Kalman linear regulator, Proceedings of IEE, 119(5), 615-20

Calligaris M., and Johnson M. A., 1999, Power system studies and new benchmarking concepts, Proc. IASTED Int. Conf. on Control and Applications, Banff, Canada, 211-216

Calligaris M., and Johnson M. A., 2000, Benchmarking for hierarchical voltage control, IFAC Symp. on Power Plants and Power Systems Control 2000, Brussels, Belgium

Camp, R. C., (Editor), 1998, *Global Cases in Benchmarking: Best Practices from Organisations Around the World,* ASQ Quality Press, Milwaukee, USA

Campo P. J., and Morari M., 1994, Achievable closed-loop properties of systems under decentralised control: conditions involving steady state gain, IEEE Trans. Automatic Control, 39(5), 932-43

Cao, S., and Rhinehart, R. R., 1995, An efficient method for on-line identification of steady state, Journal of Process Control, 5, 363-374

Chatfield, C., and Collins, A. J., 1980, *Introduction to Multivariate Analysis,* Chapman and Hall, London, UK

Chen, G., McAvoy, T. J., and Piovoso, M. J., 1998, A multivariate statistical controller for on-line quality improvement, Journal of Process Control, 8, 139-149

Chen, J., and Howell, J., 2001, A self-validating control system based approach to plant fault detection and diagnosis, Computers and Chemical Engineering, 25, 337–358

Cheung, J.T-Y., and Stephanopoulos, G., 1990, Representation of process trends - part I. A formal representation framework, Computers and Chemical Engineering, 14, 495-510

Chiang, L. H., and Braatz, R. D., 2003, Process monitoring using causal map and multivariate statistics: fault detection and identification, Chemometrics and Intelligent Laboratory Systems, 65, 159-178

Chiang, L. H., Russell, E., and Braatz, R. D., 2001, *Fault Detection and Diagnosis in Industrial Systems,* Springer-Verlag

Choudhury, M.A.A.S., 2004, Detection and Diagnosis of Control Loop Nonlinearities Using Higher Order Statistics, PhD thesis, University of Alberta

Choudhury, M.A.A.S., Thornhill, N. F., and Shah, S.L., 2005, Modelling valve stiction, Control Engineering Practice, 13, 641-658

Choudhury, M.A.A.S., Shah, S.L., and Thornhill, N.F., 2002, Detection and diagnosis of system nonlinearities using higher order statistics, 15th IFAC World Congress, Barcelona, Spain

Choudhury, M.A.A.S., Shah, S. L. and Thornhill, N. F., 2004a, Diagnosis of poor control loop performance using higher order statistics, Automatica, 40, 1719–1728

Choudhury, M.A.A.S., Shah, S. L., and Thornhill, N. F., 2004b, Detection and Quantification of Control Valve Stictio, Proceedings of DYCOPS, Boston, USA

Chow, E. Y., and Willsky, A. S., 1984, Analytical redundancy and the design of robust failure-detection systems, IEEE Transactions on Automatic Control, 29, 603-614

Clark D., Grogan W., Oates M., and Volk C., 1997, British Offshore Oil and Gas, UK Offshore Operators Association and The Natural History Museum

Clarke, D. W, 1988, Application of Generalized Predictive Control to industrial processes, IEEE Cont. Sys. Mag., 49-55

Clarke, D. W, 1991, Adaptive Generalized Predictive Control, *Chemical Process Control - CPC IV*, Arkun, Y., and Ray, W H., (Eds), American Institute of Chemical Engineers, 395-417, Padre Island, TX

Clarke D. W., and Gawthrop P. J., 1975a, Simulation of a generalised self-tuning regulator, Electronics Letters, 11(2), 41-2

Clarke D. W., and Gawthrop P. J., 1975b, Self-tuning controller, IEE Proceedings Part D: Control Theory and Applications, 122(9), 929-934

Clarke D.W., and P. J., Gawthrop, 1979, Self-tuning control, Proceedings of the IEE, 126,6, 633-640

Clarke D. W., and Hastings-James R., 1971, Design of digital controllers for randomly disturbed systems, Proceedings of the IEE, 118(10) 1503-1506

Clarke D. W., and Mohtadi C., 1987, Generalized Predictive Control, Automatica, 23, 137–148

Codling S., 1992, Best practice benchmarking, Industrial Newsletters, Dunstable, Beds., UK

Comon, P., 1994, Independent component analysis, a new concept in signal processing, Signal Processing, 36, 233-287

Crowe, C. M., 1988, Recursive identification of gross errors in linear data reconciliation, AIChE Journal, 34, 541-550

Crowe, C. M., 1996, Data reconciliation - progress and challenges, Journal of Process Control, 6, 89-98

Cutler, C. R. and Ramaker, B.L., 1980, Dynamic Matrix Control - A Computer Control Algorithm, Proc. Joint Automatic Control Conference, San Francisco, CA, Paper WP5-B

Davies, M., (Ed), 2003, Special issue: Blind signal separation, International Journal of Adaptive Control and Signal Processing, 18(3)

De Prada C., and Valentin A., 1996, Set point optimisation in multivariable constrained predictive control, Proceedings of the 13th Congress of IFAC, San Francisco, 351-356

Deibert, R., 1994, Model-based fault detection of valves in flow control loops, IFAC SAFEPROCESS '94', Espoo, Finland, 2, 445-450

Deming, W. E., 1985, *Out of the Crisis: Quality, Productivity, and Competitive Position*, Cambridge University Press, Cambridge, UK

Desborough L. D., and Harris T. J., 1992, Performance assessment measures for univariate feedback control, The Canadian Journal of Chemical Engineering, 70, 1186-1197

Desborough L. D., and Harris T. J., 1993, Performance assessment measures for univariate feedback/feedforward control, Canadian Journal of Chemical Engineering, 71, 605-615

Desborough L. D., and Harris T. J., 1994, Control performance assessment, Pulp and Paper Canada, 95(11) 441-443

Desborough, L., and Miller, R., 2002, Increasing customer value of industrial control performance monitoring – Honeywell's experience, AIChE Symposium Series No 326, 98, 153-186

Doyle J. C., 1978, Guaranteed margins for LQG regulators, IEEE Trans Automatic Control, AC-23, 756-757

Duque M, Samaan M and M'Saad, 1988, Partial state LQ and GPC adaptive control: an experimental evaluation, Bensoussan, A., and Lions, J. L., editors, *Analysis and Optimisation of Systems*, Lecture Notes in Control and Information Sciences, Springer-Verlag, Berlin

Eaton J. W., and Rawlings J. B., 1991, Model-predictive control of chemical processes: Proc. ACC, 3, 2481-5, New York, USA

Ellis R., Li X and Riggs J., 1998, Modelling and optimisation of a model IV fluidised catalytic cracking unit, AIChE Journal, 44, 9, 2068-2079

Elsayed, E. A., 2000, Perspectives and challenges for research in quality and reliability engineering, Int. Journal of Production Research, 38, 1953-1976

Ender, D. B., 1993, Process control performance: Not as good as you think, Control Engineering, 40(10) 180-190

Ender, D., (1997), Implementation of dead-band reset scheduling for the elimiation of stick-slip cycling in cotnrol valves, Process Control Electrical & Inf. Conf., 83-88

EnTech, 1998, Control Valve Dynamic Specification, Version 3.0, EnTech Control Inc., Toronto, Canada, (www.emersonprocess.com/entechcontrol/download/publications/Valvsp30_Summary.pdf)

EQI, 1993, *Business Process Management*, Ericsson Quality Institute, Gothenburg, Sweden

Ettaleb, L., Davies, M. S., Dumont, G. A., and Kwok, E., 1996, Monitoring oscillations in a multiloop system, Proc. of the 1996 IEEE Conf. Cont. App., Dearborn, MI, USA, 859-863

Famili, A., Shen, W-M., Weber, R., and Simoudis, E., 1997, Data pre-processing and intelligent data analysis, Intelligent Data Analysis, 1, http://www.elsevier.com/locate/ida

Forsman, K., 2000, On detection and classification of valve stiction, TAPPI Conf. on Process Control, Williamsburg VA, USA

Forsman, K., and Stattin, A., 1999, A new criterion for detecting oscillations in control loops, European Control Conference, Karlsruhe, Germany

Frank, P. M., 1990, Fault-diagnosis in dynamic-systems using analytical and knowledge-based redundancy - A survey and some new results, Automatica, 26, 459-474

Frank, P. M., Ding, S. X., and Marcu, T., 2000, Model-based fault diagnosis in technical processes, Transactions of the Institute of Measurement and Control, 22, 57-101

Gerry, J., and Ruel, M., 2001, How to measure and combat valve stiction online, ISA, Houston, TX, USA, http://www.expertune.com/articles/isa2001/StictionMR.htm

Gertler, J., 1998, *Fault Detection and Diagnosis in Engineering Systems*, Marcel Dekker

Gertler J, and Singer D., 1990, A new structural framework for parity equation-based failure-detection and isolation, Automatica, 26, 381-388

Gertler, J., Li, W. H., Huang, Y. B., and McAvoy, T., 1999, Isolation enhanced principal component analysis, AIChE Journal, 45, 323-334

Glassey, J., Ignova, M., Ward, A. C., Montague, G. A., and Morris, A. J., 1997, Bioprocess supervision: Neural networks and knowledge based systems, Journal of Biotechnology, 52, 201-205

Goldrat E, 1993, *The Goal: A Process of Ongoing Improvement*, Gower, Aldershot, UK

Gollmer, K., and Posten, C., 1996, Supervision of bioprocesses using a dynamic time warping algorithm, Control Engineering Practice, 4, 1287-1295

Goodwin, G., and Sin, K., 1984, *Adaptive Filtering, Prediction and Control*, Prentice-Hall

Goulding, P. R., Lennox, B., Sandoz, D. J., Smith, K. J., and Marjanovic, O., 2000, Fault detection in continuous processes using multivariate statistical methods, International Journal of Systems Science, 31, 1459-1471

Graebe, S. F., Goodwin, G. C., and Elsley, G., 1995, Control design and implementation in continuous steel casting, IEEE Control Systems Magazine, 15(4), 64-71

Greenwood, D., and Johnson, M., 2003, Multivariable controller benchmarking using LQG optimal control and a steel industry application, Proceedings of the European Control Conference

Greenwood D. and Johnson, M.A., 2005, Chapter 12, Section 12.3 et seq., In Johnson M. A. and Moradi, M. H., (Eds) PID Control, Springer Verlag London, Guildford, UK

Grimble, M. J., 1981, A control weighted minimum-variance controller for non-minimum phase systems, International J Control, 33, 4, 751-762

Grimble, M. J., Moir T. J., and Fung P. T. K., 1982, Comparison of WMV and LOG self-tuning controllers, IEEE conf. on Applications of Adaptive and Multivariable Control, Hull

Grimble, M. J., 1984, Implicit and explicit LQG self-tuning controllers, Automatica, 20(5), 661-9

Grimble, M. J., 1987, LQG multivariable controllers – minimum variance interpretation for self tuning systems, Int. J. of Control, 40(4) 83–842

Grimble, M J., 1988, Generalized minimum-variance control law revisited, Optimal Control Applications and Methods, 9, 63-77

Grimble M J., 1992, Generalised predictive optimal control – An introduction to the advantages and limitations; Int. J of Systems Sci, 23(7) 85-98

Grimble, M J, 1994, Robust Industrial Control, Optimal Design Approach for Polynomial Systems, Prentice Hall, Hemel Hempstead

Grimble M. J., 1999, Restricted structure feedforward and feedback stochastic optimal control, CDC 99, 5038-43

Grimble, M. J., 2000a, Restricted structure LQG optimal control for continuous-time systems, IEE Proc. Control Theory Applic., 147(2) 185-95

Grimble, M. J., 2000b, Controller performance benchmarking and tuning using Generalized Minimum Variance control; Report ICC-177, Industrial Control Centre, University of Strathclyde, UK

Grimble, M. J., 2001a, Industrial Control Systems Design, John Wiley, Chichester

Grimble, M. J., 2001b, Restricted structure control loop performance assessment and condition monitoring for state-space systems, Report ICC-186, Industrial Control Centre, University of Strathclyde, UK

Grimble, M. J., 2002a, Restricted structure control loop performance assessment for state-space systems, Proc. ACC, 1633-8

Grimble, M. J., 2002b Controller performance benchmarking and tuning using generalised minimum variance control, Automatica, 38 (12), 2111-2119

Grimble, M. J., 2005, Nonlinear Generalized Minimum Variance Feedback, Feedforward and Tracking Control, Automatica, 41(6) 957-969

Grimble, M. J., and Johnson, M A, 1988, Optimal multivariable control and estimation theory : theory and applications, Vols. I and II, John Wiley, London

Grimble, M. J. and Ordys A. W., 2001, Non-linear predictive control for manufacturing and robotic applications, IEEE Conference on Methods, Models in Automation, Poland

Grimble, M. J.and Uduehi, D., 2001, Process Control Loop benchmarking and revenue optimization, IEEE American Control Conference, Arlington USA

Hägglund, T., 1995, A control-loop performance monitor, Control Engineering Practice, 3 (11), 1543-1551

Hägglund, T. (2002), A Friction Compensator for Pneumatic Control Valves, J. Process Control 12, 897–904

Hale, J. C., and Sellars, H. L., 1981, Historical data recording for process computers, Chemical Engineering Progress (Nov), 38-43

Hammer, M., and Champy, J., 1993, *Re-Engineering the Corporation: A Manifesto for Business Revolution*, Harper Business, New York, USA

Harris T., 1989, Assessment of closed loop performance, Canadian Journal of Chemical Engineering 67, 856-861

Harris T. J., Boudreau, F., and MacGregor, J. F., 1996, Performance assessment of multivariable feedback controllers. Automatica, 32(11), 1505–1518

Harris T. J., and Seppala C. T., 2001, Recent developments in controller performance monitoring and assessment techniques, Chemical Process Control VI, Tuscon

Harris, T. J., Seppala, C. T., Jofreit, P. J., and Surgenor, B. W., 1996, Plant-wide feedback control performance assessment using an expert system framework, Control Engineering Practice, 9, 1297-1303

Henningsen A., Christensen A., and Ravn O., 1990, A PID Autotuner Utilizing GPC and Constraint Optimisation, Proceedings of the 29th IEEE conference on Decision and Control, 1475-80, Hawaii

Holly, W., Cook, R and Crowe, C. M., 1989, Reconciliation of mass flow rate measurements in a chemical extraction plant, Canadian Journal of Chemical Engineering, 67, 595-601

Horch, A., 1998, A simple method for detection of stiction in control valves, Control Engineering Practice, 7, 1221-1231

Horch, A., 2000, Condition Monitoring of Control Loops, PhD Thesis, Dept. of Signals, Sensors and Systems, Royal Institute of Technology, Stockholm, Sweden

Horch, A., 2002, Patents WO0239201and US2004/0078168

Horch, A., and A. J., Isaksson (1998), A method for detection of stiction in control valves, IFAC Workshop on On-Line Fault Detection and Supervision in the Chemical Process Industry, Lyon, France, Session 4B

Huang, B., Ding, S. X., and Thornhill,N., 2005, Practical solutions to multivariate feedback control performance assessment problem: reduced a priori knowledge of interactor matrix, Journal of Process Control, 15, 573-583

Huang, B., and Shah, S. L., 1998, Practical Issues in Multivariable Feedback Control Performance Assessment, Journal of Process Control, 8, pp 421-430, 1998

Huang, B and Shah S. L., 1999, *Performance assessment of control loops : Theory and Applications*, Springer Verlag, London

Huang, B., Shah, S. L., and Kwok, E. K., 1997, Good, bad or optimal? Performance assessment of multivariable processes, Automatica, 33, 1175–1183

Huang, B., Thornhill, N. F., Shah, S. L., and Shook, D., 2002, Path analysis for process troubleshooting, Proceedings of AdConIP 2002, Kumamoto, Japan, 149-154

Hyland, D C., and Bernstein D.A., 1985, The optimal projection equations for model reduction and the relationships among the methods of Wilson, Skelton and Moore, IEEE Trans. Automat. Control, 30, 1201-1211

Hyvarinen, A., and Oja, E., 2000, Independent component analysis: algorithms and applications, Neural Networks, 13, 411-430

Ignova, M., Paul, G. C., Glassey, J., Ward, A. C., Montague, G., Thomas, C. R., and Karim, M. N., 1996, Towards intelligent process supervision - industrial penicillin fermentation case study, Computers and Chemical Engineering, 20, S545-S550

Iordache, C., Mah, R.S.H., and Tamhane, A. C., 1985, Performance studies of the measurement test for detection of gross errors in process data, AIChE Journal, 31, 1187-1201

Isermann R., 1984, Process fault-detection based on modeling and estimation methods - A survey, Automatica, 20, 387-404

Jackson, J E., and Mudholkar, G. S., 1979, Control procedures for residuals associated with principal components analysis, Technometrics, 21, 341-349

Johnson, M. A., 2005, PID Control Technology, Johnson M.A. and Moradi, M.H. Editors, PID Control: New Identification and Design Methods, 3-46, Springer Verlag, London, UK

Johnson M., and Grimble M., 1999, Aspects of Performance Benchmarking for Industrial Control Systems, Proceeding Tutorial Workshop S-2, IEEE Conference on Decision and Control, Arizona

Johnson M.A. and Moradi, M. H., 2005, PID Control, Springer Verlag, London Ltd., Guildford, UK.

Johnson, R. A., and Wichern, D. W., 1992, Applied Multivariate Statistical Analysis, Prentice Hall

Jongenelen, E. M., den Heijer, C., and van Zee, G. A., 1988, Detection of gross errors in process data using Studentised residuals, Computers and Chemical Engineering, 12, 845-847

Kadali R., and Huang, B, 2002, Controller performance analysis with LQG benchmark obtained under closed loop conditions, ISA Transactions, 41, 521-537

Kano, M., Maruta, H., Kugemoto, H., and Shimizu, K., 2004, Practical Model and Detection Algorithm for Valve Stiction, IFAC Symp. on Dyn. and Control of Proc. Syst. (DYCOPS), Cambrige, USA

Kano, M., Tanaka, S., Hasebe, S., Hashimoto, I., and Ohno, H., 2003, Monitoring independent components for fault detection, AIChE Journal, 49, 969-976

Kassidas, A., Taylor, P. A., and MacGregor, J. F., 1998, Off-line diagnosis of deterministic faults in continuous dynamic multivariable processes using speech recognition methods, Journal of Process Control, 8, 381-393

Kendall, M, and Ord, J. K., 1990, Time Series (3rd Ed), Edward Arnold, Sevenoaks, Kent

Ko, B.S., and Edgar, T.F., 2001, Performance assessment of multivariable feedback control systems, Automatica, 37, 899–905

Koninckx, J., 1988, On Line Optimisation of Chemical Plants Using Steady-State Models, PhD thesis, Department of Chemical and Nuclear Engineering, University of Maryland at College Park

Kosanovich, K. A., and Piovoso, M. J., 1997, PCA of wavelet transformed process data for monitoring, Intelligent Data Analysis Journal, 1:2 on-line at http://www.elsevier.nl/locate/ida

Kourti, T., and MacGregor, J. F., 1996, Control of multivariate processes, Journal of Quality Control, 28, 409-428

Kouvaritakis B., Cannon M., Rossiter J. A., 1999, Non-linear model based predictive control, Int. J. Control, 72(10), 919-928

Kouvaritakis B., Rossiter, J. A., and Chang, A. O. T., 1992, Stable generalised predictive control: an algorithm with guaranteed stability, IEE Proceedings-D, 139(4), 349-362

Kramer, M. A., and Palowitch, B. L., 1987, A rule-based approach to fault diagnosis using the signed directed graph, AIChE Journal, 33, 1067-1078

Kresta, J. V., MacGregor, J. F., and Marlin, T. E., 1991, Multivariate statistical monitoring of process operating performance, Canadian Journal of Chemical Engineering, 69, 35-47

Ku, W. F., Storer, R. H., and Georgakis, C., 1995, Disturbance detection and isolation by dynamic principal component analysis, Chemometrics and Intelligent Laboratory Systems, 179-196

Kucera V., 1979, Discrete Linear Control, John Wiley and Sons, Chichester

Kucera V., 1980, Stochastic multivariable control a polynomial equation approach, IEEE Trans. On Auto. Contr., AC-25(5), 913-919

Lakshminarayanan, S., Shah, S. L., and Nandakumar, K., 1997, Modelling and control of multivariable processes: Dynamic PLS approach, AIChE Journal, 43, 2307-2322

Learned, R. E., and Willsky, A. S., 1995, A wavelet packet approach to transient signal classification, Applied and Computational Harmonic Analysis, 2, 265-278

Lee, D. D., and Seung, H. S., 1999, Learning the parts of objects by non-negative matrix factorization, Nature, 401,788-791

Lee, J. M., Yoo, C. K., and Lee I. B., 2004, Statistical monitoring of dynamic processes based on dynamic independent component analysis, Chemical Engineering Science, 59, 2995-3006

Lee, J. M., Yoo, C. K., and Lee, I. B., 2004, Statistical process monitoring with independent component analysis, Journal of Process Control, 14, 467-485

Levine W. S., (Ed), 1996, *The Control Handbook*, CRC Press, IEEE Press, US, Chapter 62

Li, R. F., and Wang, X. Z., 2002, Dimension reduction of process dynamic trends using independent component analysis, Computers and Chemical Engineering, 26, 467-473

Li, W. H., Raghavan, H., and Shah, S. L., 2003, Subspace identification of continuous time models for process fault detection and isolation, Journal of Process Control, 13, 407-421

Luenberger, D., 1969, *Optimisation by Vector Space Methods*, John Wiley, New York

Luo, R. F., Misra, M., and Himmelblau, D. M., 1999, Sensor fault detection via multiscale analysis and dynamic PCA, Industrial and Engineering Chemistry Research, 38, 1489-1495

Luyben W. L., 1990, *Process Modelling, Simulation and Control for Chemical Engineer*, McGraw Hill

MacGregor J. F., and Tidwell, P.W., 1977, Discrete stochastic control with input constraints, Proc. IEE, 124(8), 732-734

Maciejowski, J., 1989, *Multivariable Feedback Design*, Addison-Wesley

Mah, R. S. H., Tamhane, A. C., Tung, S. H., and Patel, A. N., 1995, Process trending with piecewise linear smoothing, Computers and Chemical Engineering, 19, 129-137

Mari, J. A., Dahlén and A. Lindequist (2000), A covariance extension approach to identification of time-series, Automatica 36(3), 379-398

Marlin T., 1990, *Designing Process and Control Systems for Dynamic Performance*, McGraw Hill

Martin, E. B., and Morris, A. J., 1996, Non-parametric confidence bounds for process performance monitoring charts, Journal of Process Control, 6, 349-358

Martin, G. D., Turpin, L. E., and Cline, R. P., 1991, Estimating control function benefits, Hydrocarbon Processing (June), 68-73

Matsuo, T., Sasaoka, H., and Yamashita, Y., 2003, Detection and Diagnosis of Oscillations in Process Plants, Lecture Notes in Computer Science, V 2773, 1258–1264

Maurath P., Laub A., Seborg D., and Mellichamp D., 1988, Predictive controller design by principal component analysis, Ind. Eng Chem Res., 27, 1204-1212

Maurya, M. R., Rengaswamy, R., and Venkatasubramanian, V., 2003a, A systematic framework for the development and analysis of signed digraphs for chemical processes: 1. Algorithms and analysis, Industrial and Engineering Chemistry Research, 42, 4789-4810

Maurya, M. R., Rengaswamy, R., and Venkatasubramanian, V., 2003b, A systematic framework for the development and analysis of signed digraphs for chemical processes: 2. Control loops and flowsheet analysis, Industrial and Engineering Chemistry Research, 42, 4811-4827

McNamee D, 1994, Reinventing the Audit: Frameworks for Change , Mc^2 Management Consulting, 15 Woodland Dr, Alamo, CA 94507 USA

Miao, T., and Seborg, D. E., 1999, Automatic detection of excessively oscillatory feedback control loops, IEEE Conference on Control Applications, Hawaii, 359-364

Miller, R., Kwok, K., Shah, S., and Wood, R., 1995, Development of a stochastic predictive PID controller, Proc. ACC , 4204-8, Evanston, USA

Moden, P. E., and Soderstrom, T., 1978, On the achievable accuracy in stochastic control, 17th IEEE Conf on Decision and Control, 490-495

Montgomery, D. C., and G.C. Runger, 2003, *Applied Statistics and Probability for Engineers*, 3rd Ed, John Wiley, New York, USA

Morari, M., and Garcia C. E, 1982, Internal model predictive control, Ind. Eng. Chem. Process Des. Dev, 21, 308

Morari, M., and Zafiriou E, 1989, *Robust Process Control*, Prentice Hall

Munoz, C and Cipriano A., 1999, An integrated system for supervision and economic optimal control of mineral processing plants, Mineral Engineering, 12(6), 627-643

Mutoh, Y., and Ortega, R., 1993, Interactor structure estimation for adaptive control of discrete- process time multivariable nondecouplable systems, Automatica, 29, 635-647

Narashiman, S., and Mah, R.S.H., 1989, Treatment of general steady-state process models in gross error identification, Computers and Chemical Engineering, 13, 851-853

Narashiman, S., Mah, R.S.H., Tamhane, A.C., Woodward, J.W., and Hale, J.C., 1988, A composite statistical test for detecting changes of steady-states, AIChE Journal, 1409-1418

Olsson, H., 1996, Control Systems with Friction, PhD Thesis, Department of Automatic Control, Lund Institute of Technology, Lund, Sweden

Omar, A. A., 1997, Multivariable Predictive Control Design with Application to a Gas Turbine Power Plant, MPhil Thesis, University of Strathclyde, Glasgow

Ordys, A., and Clarke D., 1993, A state space description of GPC controller, Int J. of Systems Sci, 24(9), 1727-44

Ordys, A., 1993, Model-system parameter mismatch, GPC Control, Int. J of Adaptive Control and Signal Processing, 7, 239-253

Ordys, A. W., and Grimble M. J, 2001, Predictive control design for systems with state dependent non-linearities, Presented at 5th SIAM Conference on Control and its Applications, San Diego, California, pp 223

Ordys A., Hangstrup M. E., and Grimble M., 2000, Dynamic algorithm for linear quadratic Gaussian predictive control, Int. J. of Applied Mathematics and Computer Science, 10(2), 227-44

Ordys A., Tomizuka, M., and Grimble, M. J., 2006, State-space Dynamic Performance Preview/Predictive Controller, to appear in: ASME Journal of Dynamic Systems, Measurement and Control

Owen, J., 1997, Automatic Control Loop Monitoring and Diagnostics, Patent. Int. Publ. No. WO 97/41494. Int. App. No. PCT/CA97/00266

Owen, J.G., Read, D., Blekkenhorst, H., and Roche, A.A., 1996, A mill prototype for automatic monitoring of control loop performance, Proceedings of Control Systems 96, Halifax, Novia Scotia, 171-178

Patton, R., Frank, P., and Clark, R., 2000, *Issues of Fault Diagnosis for Dynamic Systems*, Springer Verlag

Paulonis, M.A., and Cox, J.W., 2003, A practical approach for large-scale controller performance assessment, diagnosis, and improvement, Journal of Process Control, 13, 155-168

Pearson, R., 2001, Exploring process data, Journal of Process Control, 11, 179-194

Peng, Y., and M. Kinnaert, 1992, Explicit solution to the singular LQ regulation problem, IEEE Transactions on Automatic Control, 37, 633–636

Perkins, J. D., Barton G. W., and Chan W. K., 1991a, Interaction between process design and process control: the role of open-loop indicators, Journal of Process Control, 1(3), 161-70

Perkins, J. D., Narraway L. T. and Barton G. W., 1991b, Interaction between process design and process control: economic analysis of process dynamics, Journal of Process Control, 1(5), 243-50

Persson, P., and Astrom K. J, 1993, PID control revisited, IFAC 12th Triennial World Congress, Sydney Australia, 951-60

Petersen, J., 2000, Causal reasoning based on MFM, Proceedings of the Conference on Cognitive Systems Engineering in Process Control (CSEPC 2000), Taejon, Korea, 36-43

Petti, T.F., Klein, J., and Dhurjatu, P.S., 1990, Diagnostic model processor - using deep knowledge for process fault-diagnosis, AICHE Journal, 36, 565-575

Piiponen, J., 1998, Nonideal valves in control: tuning of loops and selection of valves, Pulp and Paper Canada, 99(10), 64-69

Plumbley, M., 2002, Conditions for non-negative independent component analysis, IEEE Signal Processing Letters, 9, 177-180

Poggio, T., and Girosi, F., 1990, Networks for approximation and learning, Proceedings of the IEEE, 78, 1481-1497

Pota, H. R., 1996, MIMO Systems-Transfer Function to State-Space, IEEE Transactions on Education, 39(1), 97-99

Pryor, C., 1982, Autocovariance and power spectrum analysis, Control Engineering (Oct), 103-106

Qin, S.J., 1998, Control performance monitoring - a review and assessment, Computers and Chemical Engineering, 23, 173-186

Qin, S.J., and McAvoy, T.J., 1992, Nonlinear PLS modeling using neural networks, Computers and Chemical Engineering, 16, 379-391

Rahul, S. and Cooper D., 1997, A tuning strategy for unconstrained SISO model predictive control, Ind. Eng. Chem. Res., 36, 729-746

Rao, C. R. M., 1971, *Generalized Inverse of Matrices and its Applications*, Wiley, Canada

Rengaswamy, R., and Venkatasubramanian, V., 1995, A syntactic pattern-recognition approach for process monitoring and fault-diagnosis, Engineering Applications of Artificial Intelligence, 8, 35-51

Rengaswamy, R., T. Hägglund and V. Venkatasubramanian 2001, A qualitative shape analysis formalism for monitoring control loop performance, Eng. App. of Artificial Intell, 14, 23-33

Richalet, J., 1993, Industrial applications of the model based predictive control, Automatica, 29(5), 1251-74

Richalet, J., Rault A., Testud J.L. and Papon J., 1978, Model predictive heuristic control-application to industrial processes, Automatica, 14 , 413 – 428

Rivera, D.E., and Morari M., 1990, Low order SISO controller tuning methods for the H_2, H_∞ and μ objective functions, Automatica, 26(2), 361-369

Rogozinski, M.W., A.P. Paplinski, and M.J. Gibbard, 1987, An algorithm for the calculation of a nilpotent interactor matrix for linear multivariable systems, IEEE Transactions on Automatic Control, 32, 234–237

Rollins, D.K., and Davis, J.F., 1993, Gross error detection when variance-covariance matrices are unknown, AIChE Journal, 39, 1335-1341

Rolstadas, A. (Editor), 1995, *Performance Management: A Business Process Benchmarking Approach,* Chapman and Hall, London, U.K

Rossi, M., and C. Scali, 2004, Automatic detection of stiction in actuators: A technique to reduce the number of uncertain cases, Proceedings of DYCOPS, Boston, USA

Ruel, M., 2000, Stiction – The hidden menace, Control Engineering, 11

Ruel, M., 2001, *TOP Control Optimization List*, http://www.topcontrol.com/en/ papers.htm

Ruel, M., and Gerry, J., 1998, Quebec quandary solved by Fourier transform, Intech (August), 53-55

Ryan, T.P., 2000, *Statistical Methods for Quality Improvement*, 2nd Ed., John Wiley, NY, USA

Saeki, M., and Aimoto, K., 2000, PID controller optimisation for H_∞ control by linear programming, International Journal of Robust Nonlinear Control, 10, 83-99

Saeki, M., and Kimura J., 1997, Design method of robust PID controller and CAD system, 11th IFAC Symposium on System Identification, 3, 1587-1593

Saez, D., Cipriano A, Ordys A., 2002, *Optimisation of Industrial Processes at the Supervisory Level*, Advances in Industrial Control, Springer-Verlag London

Salsbury, T.I. and A. Singhal, 2005, A new approach for arma pole estimation using higher-order crossings, American Control Conference 2005, Portland, Oregon, USA, Session FrB14.1

Scali, C., and F. Ulivari, 2005, A comparison of techniques for automatic detection of stiction: simulation and application to industrial data, J. Proc. Cont., 15(5) 505-514

Schley, M., V. Prasad, L.P. Russo and B.W. Bequette, 2000, Nonlinear Model Predictive Control of a Styrene Polymerization Reactor, Nonlinear Model Predictive Control, F. Allgower and A. Zheng (eds.), Progress in Systems and Control

Schreiber, T., 2000, Measuring information transfer, Physical Review Letters, 85, 461-464

Shao, R., Jia, F., Martin, E.B., and Morris, A.J., 1999, Wavelets and non-linear principal components analysis for process monitoring, Control Engineering Practice, 7, 865-879

Sharif, M. A., and R.I. Grosvenor, 2000, Condition monitoring of pneumatic value actuators: multiple fault detection and analysis of effects on process variables, Comadem 2000 13th International Congress on Condition Monitoring and Diagnostic Engineering Management, Houston USA, 93-117 ISSN 0963545027

Shi, R.J., and MacGregor, J.F., 2000, Modeling of dynamic systems using latent variable and subspace methods, Journal of Chemometrics, 14, 423-439

Shinskey, F.G., 2000, The three faces of control valves, Control Engineering (July), 83

Shunta, J.P., 1995, *Achieving World Class Manufacturing Through Process Control*, Prentice-Hall PTR, Englewood Cliffs, NJ, USA

Singhal, A. and T.I. Salsbury, 2005, A simple method for detecting valve stiction in oscillating control loops, J. Proc. Control, 15, 4, 371-382

Skogestad, S., and Postlethwaite I., 1996, *Multivariable Feedback Control*, John Wiley and Sons

Soeterboek, R., 1992, *Predictive Control-A Unified Approach*, Prentice Hall

Srinivasan, R., R. Rengaswamy and R. Miller 2005, Control performance assessment, I. Qualitative approach for stiction diagnosis, Submitted to 'Industrial and Engineering Chemistry Research'

Srinivasan, R., R. Rengaswamy, S. Narasimhan and R. Miller, (2005, Control loop performance assessment: a hammerstein model approach for stiction diagnosis, Industrial and Engineering Chemistry Research 44, 6719-6728

Stanfelj, N., Marlin T.E. and MacGregor J.F., 1993, Monitoring and diagnosing process control performance : The single loop case, Ind. Eng. Chem, Res., 32, 301-314

Stanfelj, N., Marlin, T.E., and MacGregor, J.F., 1993, Monitoring and diagnosing process control performance: The single loop case, Industrial and Engineering Chemistry Research, 32, 301-314

Stenman, A., F. Gustafsson and K. Forsman, 2003, A segmentation-based method for detection of stiction in control valves, Int. J. of Adapt. Cont. and Sig. Proc, 17, 7-9, 625-634

Stephanopoulos, G., Locher, G., Duff, M.J., Kamimura, R., and Stephanopoulos, G., 1997, Fermentation data base mining by pattern recognition, Biotchnoogy and Bioengineering, 53, 443-452

Taha, O., G.A. Dumont and M.S. Davies, 1996, Detection and diagnosis of oscillations in control loops, 35th IEEE Conf. On Decision and Control, Kobe, Japan, 2432-2437

Tan, C.C., Thornhill, N.F., and Belchamber, R.M., 2002, Principal components analysis of spectra, with application to acoustic emissions from mechanical equipment, Transactions of the Institute of Measurement and Control, 24, 333-353

Tangirala, A.K., and Shah, S.L., 2005, Non-negative matrix factorization for detection of plant-wide oscillations, submitted to IEEE Transactions on Knowledge and Data Mining

Tham, M.T., and Parr, A., 1994, Succeed at on-line validation and reconstruction of data, Chemical Engineering Progress (May), 46-56

Thompson, M.L., and Kramer, M.A., 1994, Modelling chemical processes using prior knowledge and neural networks, AIChE Journal, 40, 1328-1340

Thornhill, N.F., 2005, Finding the source of nonlinearity in a process with plant-wide oscillation, IEEE Transactions on Control System, Technology, 13(3), 434-443

Thornhill, N.F., Choudhury, M.A.A.S., and Shah, S.L., 2004, The impact of compression on data-driven process analysis, Journal of Process Control, 14, 389-398

Thornhill, N.F., and Hägglund, T., 1997, Detection and diagnosis of oscillation in control loops, Control Engineering Practice, 5, 1343-1354

Thornhill, N.F., Huang, B., and Zhang, H., 2003, Detection of multiple oscillations in control loops, Journal of Process Control, 13, 91-100

Thornhill, N.F., Oettinger, M., and Fedenczuk, P., 1998, Performance assessment and diagnosis of refinery control loops, AIChE Symposium Series 320, 94, 373-379

Thornhill, N.F., Oettinger M. and Fedenczuk P., 1999, Refinery-wide control loop performance assessment, Journal of Process Control, 9, 109-124

Thornhill, N.F., Shah, S.L., Huang, B., and Vishnubhotla, A., 2002, Spectral principal component analysis of dynamic process data, Control Engineering Practice, 10, 833-846

Tjoa, I.B., and Biegler, L.T., 1991, Simultaneous strategies for data reconciliation and gross error detection of non-linear systems, Computers and Chemical Engineering, 15, 679-690

Tong, H.W., and Crowe, C.M., 1996, Detecting persistent gross errors by sequential analysis of principal components, Computers and Chemical Engineering, 20, S733-S738

Tugnait, J.K., 1982, Detection and estimation for abruptly changing systems, Automatica, 18, 607-615

Turner, P., Montague, G., and Morris, J., 1996, Dynamic neural networks in non-linear predictive control (an industrial application), Computers and Chemical Engineering, 20, S937-S942

Tyler, M.L., and Morari, M., 1996, Performance monitoring of control systems using likelihood methods, Automatica, 32, 1145-1162

Vedam, H., and Venkatasubramanian, V., 1999, PCA-SDG based process monitoring and fault diagnosis, Control Engineering Practice, 7, 903-917

Venkatasubramanian, V., 2001, Process fault detection and diagnosis: Past, present and future, Proceedings of CHEMFAS4, Korea, 3-15

Venkatasubramanian, V., Rengaswamy, R., Yin, K., and Kavuri, S.N., 2003a, A review of process fault detection and diagnosis Part I: Quantitative model-based methods, Computers and Chemical Engineering, 27, 293-311

Venkatasubramanian, V., Rengaswamy, R., Kavuri, S.N., 2003b, A review of process fault detection and diagnosis Part II: Quantitative model and search strategies, Computers and Chemical Engineering, 27, 313-326

Venkatasubramanian, V., Rengaswamy, R., Kavuri, S.N., Yin, K., 2003c, A review of process fault detection and diagnosis Part III: Process history based methods, Computers and Chemical Engineering, 27, 327-34

Verneuil Jr, V.S., and Madron, F., 1992, Banish bad plant data, Chemical Engineering Progress (Oct), 45-51

Wallén, A., 1997, Valve Diagnostics and Automatic Tuning, Proceedings of the American Control Conference, Albuquerque, New Mexico, USA, 2930-2934

Watson, M.J., Liakopoulos, A., Brzakovic, D., and Georgakis, C., 1998, A practical assessment of process data compression techniques, Industrial and Engineering Chemistry Research, 37, 276-274

Wise, B.M., and Gallagher, N.B., 1996, The process chemometrics approach to process monitoring and fault detection, Journal of Process Control, 6, 329-348

Wise, B.M., and Ricker, N.L., 1992, Identification of finite impulse response models by principal components regression: Frequency response properties, Process Control and Quality, 4, 77-86

Wold, S., Esbensen, K., and Geladi, P., 1987, Principal component analysis, Chemometrics and Intelligent Laboratory Systems, 2, 37-52

Wolovich W.A. and P.L. Falb, 1976, Invariants and canonical forms under dynamic compensation, SIAM Journal of Control and Optimization,14, 996–1008

Xia, C., 2003, Control Loop Measurement Based Isolation of Faults and Disturbances in Process Plants, PhD Thesis, University of Glasgow

Xia, C. and Howell, 2005, Isolating multiple sources of plant-wide oscillations via spectral independent component analysis, Control Engineering Practice, 13, 1027-1035

Xia, C., and Howell, J., 2003, Loop status monitoring and fault localization, Journal of Process Control, 13, 679-691

Xia, C., Howell, J., and Thornhill, N.F., 2005, Detecting and isolating multiple plant-wide oscillations via spectral independent component analysis, Automatica, 41, 2067-2075

Xia, H., P. Majecki, A. Ordys and M. Grimble, 2005, Performance assessment of MIMO systems based on I/O delay information, Journal of Process Control, to appear

Yamashita, Y., 2004, Qualitative Analysis for Detection of Stiction in Control Valves, Lecture Notes in Computer Science, 3214, Part II, 391-397

Yoon, S., and MacGregor, J.F., 2000, Statistical and causal model-based approaches to fault detection and isolation, AIChE Journal, 46, 1813-1824

Youla, D.C., Jabr H.A. and Bongiorno, J. J., 1976, Modern Wiener-Hopf design of optimal controllers - part II : the multivariable case, IEEE Trans on Automatic Control, AC-12, 3, 319-338

Yukimoto, M., Y. Lino, S. Hino and K. Takahashi, 1998, A new PID controller tuning system and its application to a flue gas temperature control in a gas turbine power plant, IEEE Conference on Control Applications, Trieste, Italy

Zhang, J., 2001, Developing robust neural network models by using both dynamic and static process operating data, Industrial and Engineering Chemistry Research, 40, 234-241

Zheng L, Skurikhin V, Procenko N. and Sapunova N., 1995, Design of an Adaptive Robust and Careful Controller for a Thermal Power Plant. Proceedings of the IFAC conference on Control of Power Plants and Power Systems, SIPOWER'95, Cancun, pp 221-226

Index

Other titles published in this Series (continued):

Adaptive Voltage Control in Power Systems
Giuseppe Fusco and Mario Russo

Advanced Control of Industrial Processes
Piotr Tatjewski

Modelling and Analysis of Hybrid Supervisory Systems
Emilia Villani, Paulo E. Miyagi and Robert Valette

Magnetic Control of Tokamak Plasmas
Marco Ariola and Alfredo Pironti
Publication due May 2007

Continuous-time Model Identification from Sampled Data
Hugues Garnier and Liuping Wang (Eds.)
Publication due May 2007

Model-based Process Supervision
Belkacem Ould Bouamama and
Arun K. Samantaray
Publication due June 2007

Process Control
Jie Bao, and Peter L. Lee
Publication due June 2007

Distributed Embedded Control Systems
Matjaž Colnarič, Domen Verber and
Wolfgang A. Halang
Publication due October 2007

Optimal Control of Wind Energy Systems
Iulian Munteanu, Antoneta Iuliana
Bratcu, Nicolas-Antonio Cutululis and
Emil Ceanga
Publication due November 2007

Model Predictive Control Design and Implementation Using MATLAB®
Liuping Wang
Publication due November 2007

Printed in the United States
70673LV00001B/22-27

9 781846 286230